T0176760

Introduction to Porous Materials

Inorganic Chemistry

A Wiley Series of Advanced Textbooks
ISSN: 1939-5175

Editorial Board

David Atwood, University of Kentucky, USA
Bob Crabtree, Yale University, USA
Gerd Meyer, Iowa State University, USA
Derek Woollins, University of St. Andrews, UK

Previously Published Books in this Series

Metals in Medicine, 2nd edition
James C. Dabrowiak; ISBN 978-1-119-19130-8

The Organometallic Chemistry of N-heterocyclic Carbenes
Han Vinh Huynh; ISBN: 978-1-118-59377-6

Bioinorganic Chemistry: Inorganic Elements in the Chemistry of Life, An Introduction and Guide, 2nd Edition
Wolfgang Kaim, Brigitte Schwederski, Axel Klein; ISBN: 978-0-470-97523-7

Structural Methods in Molecular Inorganic Chemistry
David Rankin, Norbert Mitzel and Carole Morrison; ISBN: 978-0-470-97278-6

Introduction to Coordination Chemistry
Geoffrey Alan Lawrance; ISBN: 978-0-470-51931-8

Chirality in Transition Metal Chemistry
Hani Amouri & Michel Gruselle; ISBN: 978-0-470-06054-4

Bioinorganic Vanadium Chemistry
Dieter Rehder; ISBN: 978-0-470-06516-7

Inorganic Structural Chemistry 2nd Edition
Ulrich Muller; ISBN: 978-0-470-01865-1

Lanthanide and Actinide Chemistry
Simon Cotton; ISBN: 978-0-470-01006-8

Mass Spectrometry of Inorganic and Organometallic Compounds: Tools-Techniques-Tips
William Henderson & J. Scott McIndoe; ISBN: 978-0-470-85016-9

Main Group Chemistry, Second Edition
A.G. Massey; ISBN: 978-0-471-19039-5

Synthesis of Organometallic Compounds: A Practical Guide
Sanshiro Komiya; ISBN: 978-0-471-97195-5

Chemical Bonds: A Dialog
Jeremy Burdett; ISBN: 978-0-471-97130-6

The Molecular Chemistry of the Transition Elements: An Introductory Course
Francois Mathey & Alain Sevin; ISBN: 978-0-471-95687-7

Stereochemistry of Coordination Compounds
Alexander von Zelewsky; ISBN: 978-0-471-95599-3

For more information on this series see: www.wiley.com/go/inorganic

Introduction to Porous Materials

Pascal Van Der Voort
Ghent University
Ghent, Belgium

Karen Leus
Ghent University
Ghent, Belgium

Els De Canck
Recticel NV Insulation
Belgium

WILEY

This edition first published 2019
© 2019 John Wiley & Sons Ltd

All rights reserved. No part of this publication may be reproduced, stored in a retrieval system, or transmitted, in any form or by any means, electronic, mechanical, photocopying, recording or otherwise, except as permitted by law. Advice on how to obtain permission to reuse material from this title is available at http://www.wiley.com/go/permissions.

The right of Pascal Van Der Voort, Karen Leus and Els De Canck to be identified as the authors of this work has been asserted in accordance with law.

Registered Offices
John Wiley & Sons, Inc., 111 River Street, Hoboken, NJ 07030, USA
John Wiley & Sons Ltd, The Atrium, Southern Gate, Chichester, West Sussex, PO19 8SQ, UK

Editorial Office
John Wiley & Sons Ltd, The Atrium, Southern Gate, Chichester, West Sussex, PO19 8SQ, UK

For details of our global editorial offices, customer services, and more information about Wiley products visit us at www.wiley.com.

Wiley also publishes its books in a variety of electronic formats and by print-on-demand. Some content that appears in standard print versions of this book may not be available in other formats.

Limit of Liability/Disclaimer of Warranty
In view of ongoing research, equipment modifications, changes in governmental regulations, and the constant flow of information relating to the use of experimental reagents, equipment, and devices, the reader is urged to review and evaluate the information provided in the package insert or instructions for each chemical, piece of equipment, reagent, or device for, among other things, any changes in the instructions or indication of usage and for added warnings and precautions. While the publisher and authors have used their best efforts in preparing this work, they make no representations or warranties with respect to the accuracy or completeness of the contents of this work and specifically disclaim all warranties, including without limitation any implied warranties of merchantability or fitness for a particular purpose. No warranty may be created or extended by sales representatives, written sales materials or promotional statements for this work. The fact that an organization, website, or product is referred to in this work as a citation and/or potential source of further information does not mean that the publisher and authors endorse the information or services the organization, website, or product may provide or recommendations it may make. This work is sold with the understanding that the publisher is not engaged in rendering professional services. The advice and strategies contained herein may not be suitable for your situation. You should consult with a specialist where appropriate. Further, readers should be aware that websites listed in this work may have changed or disappeared between when this work was written and when it is read. Neither the publisher nor authors shall be liable for any loss of profit or any other commercial damages, including but not limited to special, incidental, consequential, or other damages.

Library of Congress Cataloging-in-Publication Data

Names: Voort, P. van der (Pascal), author. | Leus, Karen, author. | Canck,
    Els de, author.
Title: Introduction to Porous Materials / Pascal Van Der Voort (Ghent
    University, Ghent, Belgium), Karen Leus (Ghent University, Ghent,
    Belgium), Els De Canck (Recticel NV Insulation, Belgium).
Other titles: Porous materials
Description: First edition. | Hoboken, NJ : Wiley, [2019] | Includes
    bibliographical references and index. |
Identifiers: LCCN 2019008194 (print) | LCCN 2019009891 (ebook) | ISBN
    9781119426585 (Adobe PDF) | ISBN 9781119426707 (ePub) | ISBN 9781119426608
    (hardcover)
Subjects: LCSH: Porous materials. | Mesoporous materials. | Silicates.
Classification: LCC TA418.9.P6 (ebook) | LCC TA418.9.P6 V66 2019 (print) |
    DDC 620.1/16–dc23
LC record available at https://lccn.loc.gov/2019008194

Cover Design: Wiley
Cover Images: Courtesy of Pascal Van Der Voort

Set in 10/12pt TimesLTStd by SPi Global, Chennai, India
Printed and bound in Singapore by Markono Print Media Pte Ltd

10 9 8 7 6 5 4 3 2 1

Contents

Preface

This book is the first of its kind to discuss the development and applications of porous materials since their early start in the 1950s until now, the end of 2018. It is intended for students interested in materials science at the Masters or undergraduate level, but it is also intended for Ph.D. students who start their research in the field of porous materials.

Surprisingly, we noted that a comprehensive overview of the most important inorganic porous materials, combined with the necessary theory and characterization methods that can be used as a complete course or introduction to the field, does not exist. There are many books and monographs, but they tend to be either a collection of individual papers with less consistency, or they are high level monographs focusing on one type of material only.

The chapters in this book are almost chronological, starting with nature's materials, and then followed by the zeolites, silicas, aluminas, and carbons. After that, we discuss the materials of the twenty-first century; advanced carbons, PMOs (Periodic Mesoporous Organosilicas), MOFs (Metal-Organic Frameworks), and COFs (Covalent Organic Frameworks).

We have included one "theoretical" chapter, providing the reader with a solid introduction to the models of adsorption, heterogeneous catalysis, and surface area and pore volume measurements. We believe that this chapter is necessary to be able to understand the rest of the book.

We have made the choice to integrate other theoretical sections in the materials chapters at the point where we need these techniques first. So, solid state infrared spectroscopy is discussed in the silica chapter, XRD (X-Ray Diffraction) in the zeolite chapter, TEM (transmission electron microscopy) in the MOF chapter, and so on.

The year 1999 was a magical one; in that year, many synthetic new classes of porous materials were reported, including the very famous MOFs and the PMOs. These materials are referred to as hybrid materials, as they contain both inorganic and organic functionalities. The book ends with the COFs. These are basically organic materials, but are often researched in inorganic groups, illustrating that the divisions "organic," "inorganic," and "analytical" become more and more obsolete in modern science.

We hope you will enjoy this book and that you will learn a lot from it. We are open to any comments and suggestions you may have regarding the content of this book.

Pascal Van Der Voort
Karen Leus
Els De Canck

April 2019

About the Authors

Pascal Van Der Voort (1967) started his professional career in 1989 as an assistant at the University of Antwerp, preparing a Ph.D. thesis on the surface decoration of silica, using a technique called "Chemical Surface Coating." This technique is in fact very similar to the Molecular Layering technique used in the former USSR (Malygin – Saint Petersburg) and the Atomic Layer Deposition method described by Suntola.

Throughout my research career, I have worked on almost every material that is described in this book except for the zeolites, although I was surrounded by zeolite researchers in the group in the early stages and was almost literally walled by large models of zeolites.

So, during my post-doctoral career, lasting an astonishing 10 years, I continued to work on the beloved silica supports and went to Georgia Tech to use the surface decoration method "Molecular Designed Dispersion," a name created by Mark White at the time. During that period the MCM-materials (and later the SBA-materials) broke through, and I remember looking for the original Mobil patent at Georgia Tech, using these microfilm machines that are guaranteed to give you an incredible headache. In collaboration with Galen Stucky's group, I wrote on my first paper on the MCM-48 material for catalysis, as I already understood then the benefits of a 3D open pore structure. During a short stay in Montpellier with François Fajula and Anne Galarneau, we continued to work on this topic.

A few years later, I discovered the so-called "Plugged Hexagonal Mesoporous Silica" (PHTS), a name too difficult to be remembered by anyone. It was a partially blocked and a partially open SBA-15 variant and I was so proud when it got its own hysteresis loop in the latest IUPAC report on porous materials (hysteresis loop H5).

After a break, I returned to the academic world in 2006, when I was appointed Assistant Professor at Ghent University, where I founded the COMOC (Center for Ordered Materials, Organometallics, and Catalysis) in 2007. We decided to work on hybrid materials, starting with PMOs and porous phenolic resins. The group needed to be built up from scratch, but soon became sizeable with the necessary work force and instruments. So, the research on MOFs started a few years later and the most recent addition to the research are covalent organic materials. Currently, still at Ghent University as a full professor, I have published 265 papers indexed in Web of Science on silica and alumina, clays and layered double hydroxides, mesoporous ordered silicas, PMOs, MOFs, carbons and phenolic resins, COFs, and CTFs.

Els De Canck (1985) started her career at Ghent University in the COMOC group. Her Ph.D. (2013) was on the development of novel Periodic Mesoporous Organosilicas (PMOs) for heterogeneous catalysis and adsorption. She worked closely with Abdel Sayari (Ottawa) on the development of PMOs for CO_2 capture and with Dolores Esquivel (Córdoba) for the catalytic applications.

The PMO group in COMOC made remarkable discoveries. We were the first to create a diastereoisomeric pure ethene bridged PMO, and were the first to report on the thiol PMO, the allyl-ring PMO and on the easy thiol-ene click reaction in PMO modification. We work closely with Shinji Inagaki (one of the inventors of PMOs) and with several European groups on PMO materials. Els is now (since 2017) International Lab Officer at Recticel NV Insulation, a Belgian-founded international company researching and producing polyurethane-based insulation materials.

Karen Leus (1985) also started her career at Ghent University in the COMOC group. Her Ph.D. (2012) was on the development of Metal–Organic Frameworks for heterogeneous catalysis. During her post-doctorate at COMOC, she went to the labs of Herme Garcia and Avelino Corma at ITQ-Valencia to learn more about nanoparticles in MOFs and oxidation reactions. She has focused a lot on V-based MOFs, nanoparticles, and ALD, and on mixed-metal MOFs to tune the breathing behavior of flexible MOFs.

In recent years, she has been focusing on Covalent Triazine Frameworks (CTFs). She did an extensive stay at the labs of Professor Markus Antonietti (MPI Potsdam) and is now initiating a research line on electrocatalysis. In her young career, she has published 50 papers so far, all indexed in the Web of Science and all on MOFs and COFs. She won the Belgian Incentive Award for young researchers.

1 Nature's Porous Materials: From Beautiful to Practical

Porous materials are materials that contain voids, channels, holes, or basically pores. This type of material has always attracted a lot of attention as the presence of pores means that the material possesses an internal surface area of interest for all type of applications (see Chapter 2). Nowadays, many porous materials are made in the laboratory and can even be produced on a large industrial scale (see Chapters 3 and 4). However, many porous materials are naturally occurring and were first produced in "Nature's laboratory" without any human influence. In fact, mankind has often based the preparation procedures of synthetically porous materials on processes that occur in nature.

Nature has found a way to produce beautiful and practical porous materials and they can be very diverse: tissue or bones in the human body and animals, rocks, fruit, and so on. A general overview with some examples is presented in Figure 1.1. Besides that, mankind has found its own way to introduce porosity in many materials as some examples clearly demonstrate (Figure 1.2). Ceramics, bricks, and clothing are a few items that were developed very early.

This chapter describes a few carefully selected naturally occurring porous materials. It aims to give the reader a taste of what is available in nature. These materials are also the foundation for development of synthetic porous materials that are more elaborately described in Chapters 3–9 of this book. Silicas and zeolites are also materials that were originally found in nature before a synthetic procedure was discovered to produce them. They will not be covered in this chapter, as they are described in depth in Chapters 3 and 4.

1.1 Living Porosity

1.1.1 Butterflies

Porous materials can be found in animal and human bodies. The bones and lungs of humans are famous examples of ingenious porous structures. In particular, the bones of a human skeleton are very robust, despite their high porosity, as they must support and protect our body and vital organs, respectively. Animals can also create porous structures of very diverse and beautiful shapes. For example, sponges are multicellular organisms that have an entire body containing pores. The wings of butterflies are not only colorful and useful to fly, but they are also porous (Figure 1.3). The cuticle on the scales of these butterflies' wings is composed of nano- and microscale,

Introduction to Porous Materials, First Edition.
Pascal Van Der Voort, Karen Leus and Els De Canck.
© 2019 John Wiley & Sons Ltd. Published 2019 by John Wiley & Sons Ltd.

Figure 1.1
Examples of naturally occurring porous materials: lemons, snowflakes, sea sponges, coral reef, egg shells, butterfly wings (European peacock butterfly), soil, and sandstones. Source: All photographs are public domain.

Figure 1.2
Synthetic porous materials, all made by mankind: Concrete road, paper, fabric of clothes, chalk, ceramics, cake, bread, pottery, bricks, and artificial sponges for cleaning. Source: All photographs are public domain.

Figure 1.3

(a) Optical image of *M. menelaus*; (b,c) Scanning Electron Microscope (SEM) image of the nanostructure of the wing under different magnification. (d) Optical image of *P. u. telegonus*; (e) SEM image of the nanostructure of the blue region; the insert in (e) is the high magnification of SEM image; (f) SEM image of the nanostructure of the fiber region; and, the insert in (f) is the high magnification of SEM image. (g) Optical image of *O. c. lydius*; (h,i) the SEM image of the nanostructure of the wing according to different magnification. Source: Reproduced with permission. Taken from Ref. [1], open access: https://creativecommons.org/licenses/by/4.0/.

transparent, chitin-and-air layered structures. Rather than absorb and reflect certain light wavelengths as pigments and dyes do, these multiscale structures cause light that hits the surface of the wing to diffract and interfere. Cross ribs that protrude from the sides of ridges on the wing scale diffract incoming light waves, causing the waves to spread as they travel through spaces between the structures. The diffracted light waves then interfere with each other so that certain color wavelengths cancel out (destructive interference) while others are intensified and reflected (constructive interference). The varying heights of the wing scale ridges appear to affect the interference such that the reflected colors are uniform when viewed from a wide range of angles.

1.1.2   Algae

Single-celled diatoms can also produce porous structures, however, on a very different scale. Diatoms are microalgae that can be abundantly found in, for example, oceans all around the world. They are part of the phytoplankton family and contribute a staggering 20% of total oxygen produced on our planet every year. They are very unique and useful small creatures and, moreover, they produce a porous cell wall or protective shell called a frustule [2]. The frustule consists of two overlapping structures with identical shapes but slightly different in size. They are called the thecae or valve, and a girdle band or expansion joint holds the two thecae together.

The frustule is entirely made from silica, with a very well-defined structure and unique for every diatom species. It is estimated that approximately 200 000 separate species exist with very different frustules [3]. The dimensions of the frustules can be very different depending on the species. Pore sizes range from 3 nm up to a few hundred nm [4].

A few examples of different species are presented in Figures 1.4–1.6. These figures clearly show the different morphologies, but also diverse types of porosity. These frustules do not only have beautiful porous structures, they can also be used practically.

These algae can be produced on a large industrial scale as they possess a very fast growth rate and only need a limited amount of space. Moreover, they use carbon for

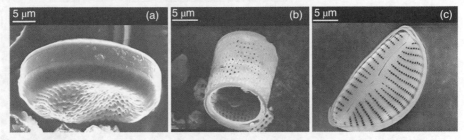

Figure 1.4
SEM images of purified diatom frustules of *Coscinodiscus* sp. (a), *Melosira* sp. (b) and *Navicula* sp. (c). Scale bar = 5 μm. Source: Reproduced with permission of John Wiley & Sons, Ltd. Taken from Ref. [4c]

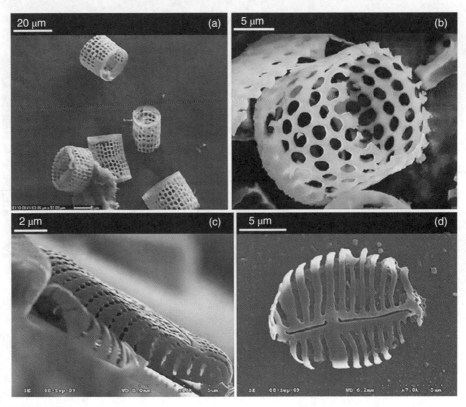

Figure 1.5
SEM images of diatom frustules after 1% HF treatment: (a) and (b) *Melosira* after 2 and 3 h, respectively; (c) and (d) *Navicula* after 1 and 2 h, respectively. Source: Reproduced with permission of Springer Nature. Taken from Ref. [4c].

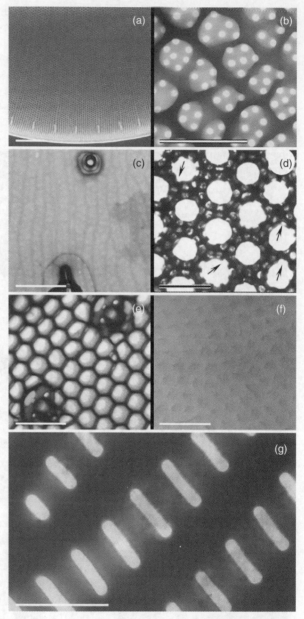

Figure 1.6

Electron micrographs of the pore structures of different diatom species: (a) *Lauderia borealis*; (b) *Odontella sinensis*; (c) *Thalassiosira weissflogii*; (d) *Coscinodiscus granii*; (e) *Navicula salinarum*; (f) *Nitzschia sigma*; (g) *Stauroneis constricta*. Scale bar = 5 μm (a) and 0.1 μm (b–g). Source: Reproduced with permission of the RSC. Taken from Ref. [4b].

photosynthesis, which also makes them very interesting. It is believed that diatoms for these reasons are a very promising alternative biomass resource to produce biofuels. Additionally, they present a new source of porous silica with very defined pore sizes and distinct morphologies. The silica source can be further used as support for all kinds of applications (Chapter 5).

As an example, here we show how we extracted the silica from algae and used it as a photocatalyst for air purification [5]. Diatom frustules were extracted from a sample containing a cultivation of *Thalassiosira pseudonana* in its salt water medium. After an initial washing procedure to remove the majority of the salts, an acid treatment was used to remove any remaining carbonates and partially digest the organic matter. After washing away the acid, calcination in air at 550 °C was used to completely free the frustules of organic components. The resulting pure silica sheets are shown in Figure 1.7.

It can be clearly seen that these silica sheets contain very uniform pores. We then deposited titania nanoparticles onto these frustule sheets. The results are shown in Figure 1.8.

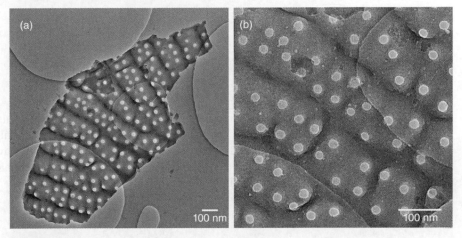

Figure 1.7
Silica extracted as diatom frustules from the algae species *Thalassiosira pseudonana*. Source: Reproduced with permission of Elsevier. Taken from Ref. [5].

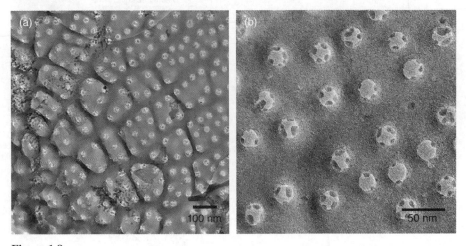

Figure 1.8
TEM images of the optimized titanium functionalized frustules, showing an overview of the nanoparticles (a) and a detail of the nanoparticles contained inside the pores. Source: Reproduced with permission of Elsevier. Taken from Ref. [5] with permission.

It is remarkable how all the titania nanoparticles are situated in the pores of the silica nanosheets. These materials were shown to be very active photocatalysts for ambient air purification, outperforming the current commercial benchmarks.

1.1.3 Bamboo

Another example of an organic source that has a high silica content are bamboo leaves. An amount of 1 g of bamboo leaves contains 0.03 g of silica. A careful extraction is again key to extract the beautiful and fluffy silica flakes as presented in Figure 1.9.

As we zoom in closer on the silica that is extracted from the bamboo leaves, we can see nicely in Figure 1.10 how the silica is the exact negative replicate of the bamboo leaf.

1.2 Clay Minerals

1.2.1 Natural Clays

Clay minerals are yet a completely different type of porous materials [6]. They naturally occur in mud, soils, rocks, sediments, and so on. The materials are formed in the presence of water and are most of the time fine-grained. They have been used by mankind for ages, especially in early civilization, to produce ceramics, but now are also used in cosmetic and pharmaceutical applications.

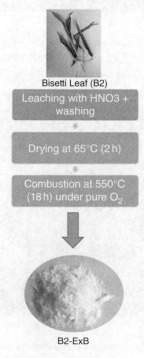

Bisetti Leaf (B2)

Leaching with HNO3 + washing

Drying at 65°C (2 h)

Combustion at 550°C (18 h) under pure O_2

B2-ExB

Figure 1.9
Extraction of silica out of bamboo.

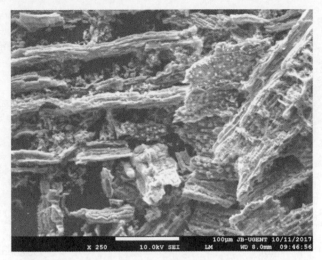

Figure 1.10
SEM picture (×250) of pure silica extracted from a bamboo leaf.

Most clay minerals are phyllosilicates or sheet silicates (Figure 1.11). They represent an entire family of silicate materials that mainly contain aluminum, silicon, and oxygen atoms. They possess a sheet-like structure with layers of corner sharing SiO_4 tetrahedra and AlO_4 octahedra (Figures 1.12 and 1.13) where the oxygen atoms

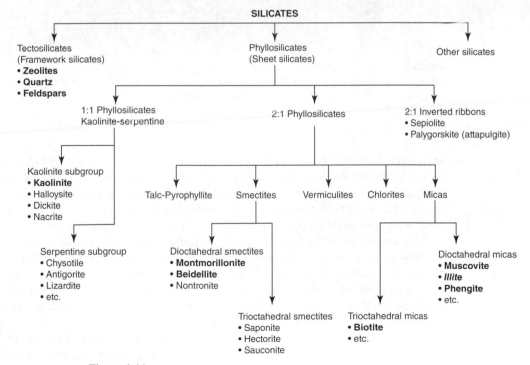

Figure 1.11
Classification of silicates. Source: Adapted from Z. Adamis et al. [7].

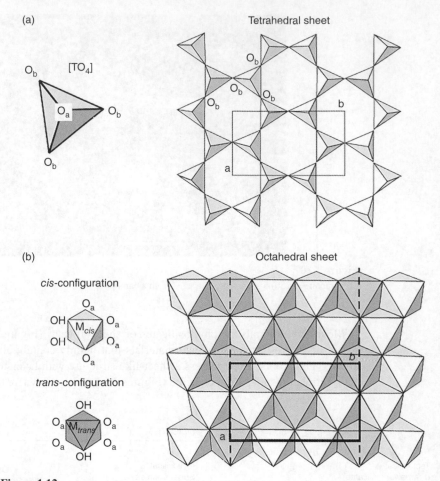

Figure 1.12
Schematic representation (a) of a tetrahedron unit and tetrahedral sheet; and (b) of an octahedron unit and octahedral sheet. Source: Adapted from G. E. Christidis et al. [6a].

are commonly used. The clay minerals are categorized according to the type of connection of the tetrahedral and octahedral layers. A 1 : 1 clay mineral consists of one tetrahedral and one octahedral group in each layer. The smectite structure, shown in Figure 1.13, is a typical example of a phyllosilicate and a 2 : 1 type of clay mineral. Here, two tetrahedral sheets are connected to each side of an octahedral sheet.

The composition of the tetrahedral and octahedral sheets will determine whether the entire layer is negatively charged or not. There are also negatively balanced clays, the Layered Doubled Hydroxides (LDH); hydrotalcites are an example. In cationic clays, a charged layer will be balanced out by small cationic species, such as Na^+ or K^+ atoms, that are present in the so-called interlayer of the clay mineral. This interlayer can also contain water (Figure 1.13).

The stacking of the different layers (containing tetrahedral and octahedral sheets), vary with the interlayer will form a crystalline phyllosilicate material. Examples of 1 : 1 clay minerals are the kaolinite and serpentine group with kaolinite (Figure 1.14a) being the most famous mineral of that group. The clay minerals of this group have a

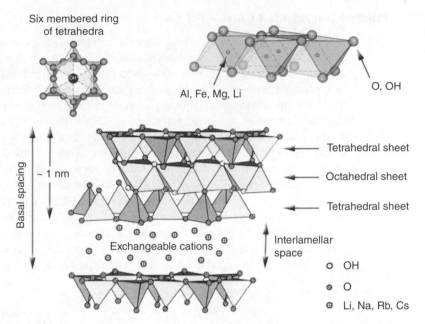

Six membered ring
of tetrahedra

Al, Fe, Mg, Li

O, OH

Basal spacing

~ 1 nm

Tetrahedral sheet

Octahedral sheet

Tetrahedral sheet

Exchangeable cations

Interlamellar
space

o OH

o O

⊕ Li, Na, Rb, Cs

Figure 1.13
An example of a clay mineral: the Smectite structure (2 : 1 clay) with two tetrahedral sheets that sandwich
one octahedral sheet within the clay-stacking pattern. Source: Reproduced with permission of the RSC.
Redrawn from Ref. [8].

general formula of $Al_4Si_4O_{10}(OH)_8$ and normally do not have any layer charge. Halloysite is the hydrated form of kaolinite and can exhibit a wide range of morphologies, such as the tubular crystals as shown in Figure 1.14b.

Due to the possibility of containing cations in the interlayer, clay minerals are highly suited to being ion exchange materials in many environmental but also industrial applications [6a]. They play an important role in, for example, catalytic cracking processes. They can act as an acid catalyst as they contain both Lewis and Brønsted acid sites on the clay mineral surface [9]. Also, biological and biomedical applications have been explored with clay minerals [8]. These materials are closely related to zeolites and zeotypes (Chapter 3).

Figure 1.14
SEM images of the clay minerals (a) Kaolinite (pseudo hexagonal crystals) and (b) halloysite (tubular
crystals). Source: Taken from Ref. [6a].

1.2.2   Pillared Interlayered Clays – PILCs

As ever, humans have tried to tune such clays to their specific demands. One famous example was the research into the synthesis of PILCs. This was based on the pioneering work of Richard Maling Barrer in 1955 [10]. He obtained microporous materials by replacing the interlayer exchangeable cations in the smectite montmorillonite with tetraalkylammonium ions. However, organic pillared clay minerals of this type are thermally unstable. At temperatures above 250 °C the interlayers collapse. When used below the decomposition temperature, these clays can, however, still be applied as catalysts [11] or adsorbents [12]. In the late 1970s, scientists, still in search for an answer to the need for large-pore nanoporous materials (see Chapter 5), started investigating the possibility of creating porosity in the interlayer space of layered clay hosts by inserting inorganic polymers [13]. These PILCs are prepared by exchanging the charge-compensating cations (e.g. Na^+, K^+, and Ca^{2+}) between the swelling phylosilicate clay layers with larger polymeric or oligomeric hydroxy metal cations. Upon heating, these metal hydroxy cations undergo dehydration and dehydroxylation, whereby stable metal oxide clusters are formed (e.g. Al_2O_3, TiO_2, Fe_2O_3, …).

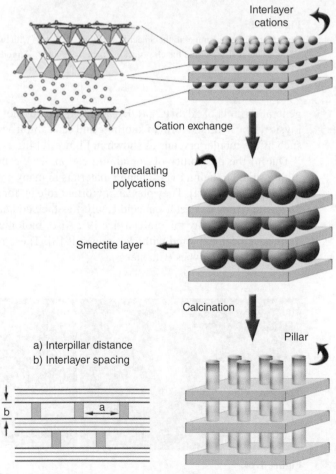

Figure 1.15
Concept of pillaring.

These metal oxide clusters act as *pillars*, keeping the silicate layers permanently separated and creating an interlayer space. In Figure 1.15 a schematic overview is given, illustrating the concept of pillaring. With these novel two-dimensional aluminosilicates, an easy method to develop materials with pores ranging from 0.5 to 2.0 nm and surface areas up to 500 m^2 g^{-1} was introduced. Moreover, clay minerals intercalated with inorganic species retain their porosity above 300 °C.

Ever since the first announcement of the commercial availability of PILC in 1979 [13c] their use in petroleum cracking alone has exceeded that of other catalysts. With an acidity as strong as zeolite Y (Chapter 3), PILCs demonstrate a high activity in cracking, while displaying a selectivity for larger product molecules. However, due to their low hydrothermal stability, the original objective of using these materials as catalysts for Fluid Catalytic Cracking (Chapter 3) was not achieved [14]. Nevertheless, PILCs have proved to be very interesting acid catalysts in various organic reactions [15].

However, with the discovery of a new class of ordered mesoporous aluminosilicates in 1992, namely M41S (Chapter 5) [16], a new era in ordered porous materials commenced and much of the interest in PILCs faded away.

References

1. Elbaz, A., Lu, J., Gao, B. et al. (2017). *Polymers* 9: 386.
2. Hamm, C.E., Merkel, R., Springer, O. et al. (2003). *Nature* 421: 841–843.
3. Mann, D.G. and Droop, S.J.M. (1996). *Hydrobiologia* 336: 19–32.
4. (a) Vrieling, E.G., Beelen, T.P.M., van Santen, R.A., and Gieskes, W.W.C. (2000). *J. Phycol.* 36: 146–159; (b) Vrieling, E.G., Beelen, T.P.M., Sun, Q.Y. et al. (2004). *J. Mater. Chem.* 14: 1970–1975; (c) Zhang, D.Y., Wang, Y., Zhang, W.Q. et al. (2011). *J. Mater. Science* 46: 5665–5671; (d) Gelabert, A., Pokrovsky, O.S., Schott, J. et al. (2004). *Geochim. Cosmochim. Acta* 68: 4039–4058.
5. Ouwehand, J., Van Eynde, E., De Canck, E. et al. (2018). *Appl. Catal. B-Environ.* 226: 303–310.
6. (a) Christidis, G.E. (2010). *Advances in the Characterization of Industrial Minerals* (ed. G.E. Christidis). European Mineralogical Union; (b) Kerr, P.F. (1952). *Clays Clay Miner.* 1: 19–32; (c) Theng, B.K.G. (ed.) (1979). Clay minerals, Ch. 1. In: *Developments in Soil Science - Formation and Properties of Clay-Polymer Complexes*, vol. 9, 3–36. Elsevier; (d) Velde, B. (1995). *Origin and Mineralogy of Clays: Clays and the Environment* (ed. B. Velde), 8–42. Berlin, Heidelberg: Springer.
7. Adamis, Z., Williams, R.B., and Fodor, J. I. L. Organisation, U. N. E. Programme, W. H. Organization, I. P. o. C. Safety, I.-O. P. f. t. S. M. o. Chemicals (2005). *Bentonite, Kaolin, and Selected Clay Minerals*. World Health Organization.
8. Ghadiri, M., Chrzanowski, W., and Rohanizadeh, R. (2015). *RSC Adv.* 5: 29467–29481.
9. Wu, L.M., Zhou, C.H., Keeling, J. et al. (2012). *Earth-Sci. Rev.* 115: 373–386.
10. Barrer, R.M. and Macleod, D.M. (1955). *Trans. Faraday Soc.* 51: 1290.
11. Vaccari, A. (1999). *Appl. Clay Sci.* 14: 161–198.
12. (a) Zhu, L.Z., Ren, X.G., and Yu, S.B. (1998). *Environ. Sci. Technol.* 32: 3374–3378; (b) Meier, L.P., Nueesch, R., and Madsen, F.T. (2001). *J Colloid Interf. Sci.* 238: 24–32.
13. (a) Brindley, G.W. and Sempels, R.E. (1977). *Clay Miner.* 12: 229–237; (b) Lahav, N., Shani, U., and Shabtai, J. (1978). *Clays Clay Miner.* 26: 107–115; (c) D. Vaughan, R. Lussier, J. Magee, US Patent: Pillared interlayered clay materials useful as catalysts and sorbents. Number: US4176090, 1979; (d) D. Vaughan, R. Lussier, J. Magee, US Patent: Stabilized pillared interlayered clays. Number: US4248739, 1981; (e) D. Vaughan, R. Lussier, J. Magee, US Patent: Pillared interlayered clay products. Number: US 4271043, 1981.
14. Magee, J.S. and Mitchell, M.M. (1993). *Fluid Catalytic Cracking: Science and Technology*, vol. 76. Elsevier.
15. Ding, Z., Kloprogge, J.T., Frost, R.L. et al. (2001). *J. Porous Mat.* 8: 273–293.
16. Beck, J.S., Vartuli, J.C., Roth, W.J. et al. (1992). *J. Am. Chem. Soc.* 114: 10834–10843.

2 Theory of Adsorption and Catalysis: Surface Area and Porosity

2.1 Determination of Surface Area and Porosity by Gas Sorption

2.1.1 Introduction

In fields like heterogeneous catalysis and sorption (sorbents for gases, metals, organic matter), we are interested in the surface of the adsorbent or catalyst, and less in the bulk of the material, because the bulk is inaccessible to liquids or gases. Often, it is found that the surface properties of a material are different from the bulk properties. For instance, silica as bulk material is known as SiO_2, whereas the surface of silica is usually described as $\equiv SiOH$ (see Chapter 5). For applications in heterogeneous catalysis or adsorption, it is important that the material has a high *specific surface area* (expressed as $m^2\,g^{-1}$), and preferably pores of uniform size with a high *pore volume* (expressed as $ml\,g^{-1}$). The word *specific* in the surface area means the value is normalized to 1 g of material.

Most materials are irregular in shape, and indirect methods are required to assess the surface area. The most used technique is the adsorption of a well-known gas. If the *mean cross-sectional area* (i.e. the amount of space that the adsorbed molecule occupies on the surface of the solid material, expressed in nm^2) of the gas molecule is known, and if the amount of gas required to form exactly one adsorbed monolayer on the surface can be determined, the surface area can be easily calculated. However, the tricky part is the determination of the amount of gas, required to form *exactly one monolayer*.

2.1.2 Chemisorption and Physisorption

Depending on the strength of the interaction, all adsorption processes can be divided into two categories of chemical and physical adsorption, *chemisorption* and *physisorption*. Chemisorption is characterized by large interaction energies, which lead to high heats of adsorption often approaching the value of chemical bonds. Examples of chemisorption include the modification of the surface of supports by a surface reaction with other chemicals, we will see examples of this in the next chapters.

Introduction to Porous Materials, First Edition.
Pascal Van Der Voort, Karen Leus and Els De Canck.
© 2019 John Wiley & Sons Ltd. Published 2019 by John Wiley & Sons Ltd.

Physisorption has the following typical indicators:

1. Physical adsorption is accompanied by low heats of adsorption.
2. Physical adsorption may lead to surface coverage by more than one layer of adsorbate.
3. Physical adsorption equilibrium is achieved rapidly since no activation energy is required as in chemisorption.
4. Physical adsorption is fully reversible.
5. Physically adsorbed molecules are not restrained to specific sites and are free to cover the entire surface.

An important interaction at the gas-solid interface during physical adsorption is due to dispersion forces. These forces often account for the major part of the adsorbate-adsorbent potential. The most important dispersion forces include:

1. ion–dipole interactions: an ionic solid and electrically neutral but polar adsorbate;
2. ion–induced dipole interactions: a polar solid and a polarizable adsorbate;
3. dipole–dipole interactions: a polar solid and a polar adsorbate;
4. quadrupole interactions: symmetrical molecules with atoms of different electronegativities, such as CO_2, possess no dipole moment, but do have a quadrupole ($^-O - {}^+C^+ - O^-$); which can lead to interactions with polar surfaces.

Adsorption forces are similar in nature and origin to the forces that lead to liquefaction of vapors. Thus, those vapors with a high boiling point and, therefore strong intramolecular interactions, will also tend to be strongly adsorbed.

2.1.3 Reversible Monolayer Adsorption – The Langmuir Isotherm

Irvin Langmuir attended his early education at various schools and institutes in America and Paris (1892–1895). He graduated with a Bachelor of Science degree in metallurgical engineering (Met.E.) from the Columbia University School of Mines

in 1903. He earned his Ph.D. degree in 1906 under Nobel Laureate Walther Nernst in Göttingen. He later did postgraduate work in chemistry. Langmuir then taught at Stevens Institute of Technology in Hoboken, New Jersey, until 1909, when he began working at the General Electric research laboratory (Schenectady, New York). He joined Katharine B. Blodgett to study thin films and surface adsorption. They introduced the concept of a monolayer (a layer of material one molecule thick) and the two-dimensional physics that describes such a surface. In 1932 he received the Nobel Prize in Chemistry "for his discoveries and investigations in surface chemistry."

The Langmuir model [1] describes the adsorption (either chemisorption or physisorption) of *exactly one monolayer* of molecules on the surface of a material. Knowing the amount of inert gas (nitrogen or argon) that is required to form exactly one monolayer combined with the knowledge of the space taken by each molecule gives then simply the surface area of a material.

Using the kinetic approach, Langmuir was able to describe the type I isotherm with the assumption that adsorption was limited to a monolayer. According to the kinetic theory of gases, the number of molecules N, striking each square centimeter of surface per second is given by (see Figure 2.1):

$$N = \frac{N_A P_A}{\sqrt{2\pi MRT}} \tag{2.1}$$

with N_A: being Avogadro's number and M the adsorbate molecular weight.

If θ^* (or θ_0) is the fraction of the surface unoccupied, that is, with no adsorbed molecules, then the number of collisions with bare or uncovered surface per square centimeter of surface each second is written as

$$N = \frac{N_A P_A}{\sqrt{2\pi MRT}}(1 - \theta) = \frac{N_A P_A}{\sqrt{2\pi MRT}}\theta^* = kP_A\theta^* \tag{2.2}$$

The constant k is $N_A/\sqrt{2\pi MRT}$. The number of molecules striking and adhering to each square centimeter of surface is

$$N_{ads} = kP_A\theta^* A_1 \tag{2.3}$$

The term A_1 is the *sticking coefficient* and represents the probability of a molecule being adsorbed upon collision with the surface.

The rate at which adsorbed molecules leave each square centimeter of surface is given by

$$N_{des} = N_m\theta_1 v_1 e^{-E/RT} \tag{2.4}$$

where N_m is the number of adsorbate molecules is a complete monolayer (of one square centimeter) and θ_1 is the fraction of the surface occupied by the adsorbed

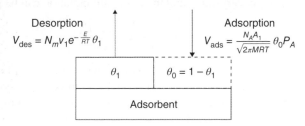

Figure 2.1
The Langmuir model.

molecules, E the energy of adsorption and v_1 is the adsorbate's vibrational energy normal to the surface when adsorbed. Actually, the product $N_m\theta_1$ is the number of molecules adsorbed per square centimeter. Multiplication by v_1 converts this number of molecules to the maximum rate at which they can leave the surface.

At equilibrium the rates of adsorption and desorption are equal. In other words,

$$N_{ads} = N_{des},$$

or

$$N_m\theta_1 v_1 e^{-E/RT} = kP_A\theta^* A_1 \tag{2.5}$$

Recognizing that

$$\theta_0 = \theta^* = 1 - \theta_1 \tag{2.6}$$

yields

$$N_m\theta_1 v_1 e^{-E/RT} = kP_A A_1 - \theta_1 kP_A A_1 \tag{2.7}$$

then

$$\theta_1 = \frac{kP_A A_1}{N_m v_1 e^{-E/RT} + kP_A A_1} \tag{2.8}$$

If we define K_A, the *adsorption coefficient* as

$$K_A = \frac{k_{ads}}{k_{des}} = \frac{kA_1}{N_m v_1 e^{-E/RT}} \tag{2.9}$$

and substituting (2.8) in (3.9), then we find

$$\theta_1 = \frac{K_A P_A}{1 + K_A P_A} \tag{2.10}$$

This equitation is the famous *law of Langmuir*. It is an asymptotic approach toward monolayer capacity as P/P_0 becomes 1 (see Figure 2.2).

So, the Langmuir model describes the monomolecular adsorption of a molecule onto the surface or the chemisorption of the molecule.

The Langmuir isotherm can also be written in terms of the uncovered surface:

$$\theta^* = 1 - \theta_A = \frac{1}{1 + K_A P_A} \tag{2.11}$$

and then follows

$$\theta_A = K_A P_A \theta^* \tag{2.12}$$

Looking at the isotherm, we can see specific cases:

At *low loadings* or *low pressure*, when either P_A or K_A is very small, the product $K_A P_A$ is very small and the Langmuir equation can be simplified to

$$\theta_A = \frac{K_A P_A}{1 + K_A P_A} \rightarrow \theta_A = K_A P_A \tag{2.13}$$

In this area of the Langmuir isotherm, there is a linear relationship between the amount of substrate adsorbed and the pressure (or concentration) of the substrate. This is often referred to as the *Henry regime*, alluding to Henry's Law ($P = k_H C$). Working in the Henry regime allows an easy estimation of the adsorption coefficient.

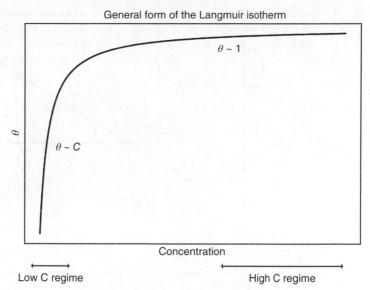

Figure 2.2
The Langmuir isotherm.

At *high loadings* or *high pressure*, the Langmuir isotherm behaves asymptotically. In this case:

When either P_A or K_A are large and when θ_A approaches 1:

$$\theta^* = 1 - \theta_A = \frac{1}{1 + K_A P_A} \approx \frac{1}{K_A P_A} \tag{2.14}$$

In order to determine the surface area, the Langmuir equation is usually linearized.

As the exact value of the monolayer capacity in an asymptotic function is difficult to establish exactly, one often uses the linearized form of the Langmuir equation. θ_A can be rewritten as V_A/V_m, where V_A is het volume of gas that is adsorbed at the pressure P_A and V_m is the volume of gas that is adsorbed at saturation pressure, in other words the volume of one monolayer of gas. We obtain

$$\frac{V_A}{V_m} = \frac{KP_A}{1 + KP_A} \tag{2.15}$$

Which gives after rearrangement the linear form of the Langmuir equation

$$\frac{P_A}{V_A} = \frac{1}{KV_m} + \frac{P_A}{V_m} \tag{2.16}$$

A plot of P_A/V_A versus P_A yields a straight line with intercept $1/(KV_m)$ and a slope $1/V_m$. From this, the exact value of V_m can be determined. This is visualized in Figure 2.3. Once the monolayer capacity is known, the calculation of the specific surface area is done by

$$S[m^2 g^{-1}] = m_m N_A A \cdot 10^{-18} \tag{2.17}$$

in which m_m is the number of moles adsorbate that is adsorbed per gram of adsorbent; it is obtained by dividing V_m by 22 400 (*why?*); and A is the "mean cross-sectional area" of the adsorbate. For the dinitrogen molecule, the mean cross-sectional area is

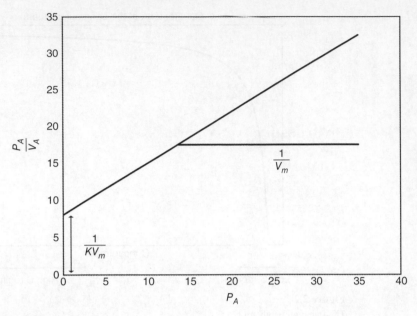

Figure 2.3
The linear Langmuir equation.

Table 2.1 Cross-sectional areas for common absorbents.

Adsorbent	Mean cross-sectional area (nm^2)
Argon	0.142
Ammonia	0.146
Carbon dioxide	0.195
Krypton	0.195
Nitrogen	0.162
Oxygen	0.141
Water	0.108
Xenon	0.025

0.162 nm^2. Values for other adsorbates can be found in the literature. Some of them are listed in Table 2.1.

Note 1
In practical calculation, the pressure is usually not expressed in absolute units, but in relative pressures, expressed at P/P_0, in which P_0 is the saturation pressure. In the case of nitrogen, adsorbing at 77 K, the temperature of liquid nitrogen, the saturation pressure therefore equals the atmospheric pressure. In the remainder of this chapter, we will use the term *relative pressure*.

Note 2
A very good estimation of the mean cross-sectional area of molecules is obtained by the formula of Emmett and Brunauer:

$$A = 1.092 \cdot \frac{M^{\frac{2}{3}}}{N_A \rho} \cdot 10^{14} \tag{2.18}$$

In this equation, A is expressed as nm^2, and the density, ρ, is expressed as $g \cdot ml^{-1}$. Emmett and Brunauer made the assumption that the adsorbate molecules are spherical. M/ρ stands thus for V_{mol}, the liquid molar volume of the adsorbate. The factor 1.092 is a packing correction factor, that accounts for the molecular packing of these spherical molecules in a hexagonally closed packing.

2.2 The BET (Brunauer, Emmet, Teller) Model

2.2.1 The BET Equation

During the process of physical adsorption, at very low relative pressures, the first sites to be covered are the more energetic ones. But this does not mean that no adsorption occurs on sites of less potential. As the adsorbate pressure is allowed to increase, the surface becomes progressively occupied and the probability increases that a gas molecule with strike on an already adsorbed molecule. So, *prior to complete surface coverage, the formation of second and higher adsorbed layers is already happening. In reality, there is no specific pressure at which the surface is covered with exactly one completed physically adsorbed layer.*

The *Brunauer, Emmett, and Teller theory* (1938) extended the Langmuir model to multilayer adsorption. The BET theory assumes that the uppermost molecules in adsorbed stacks are in dynamic equilibrium with the vapor. This means that, where the surface is covered with only one layer of adsorbate, an equilibrium exists between that layer and the vapor, and where two layers are adsorbed, the upper layer is in equilibrium with the vapor, and so forth.

Edward Teller, Paul Emmett and Stephen Brunauer, circa 1960s.

Paul Emmett (1900–1985), a friend and colleague of Linus Pauling, graduated in 1922 from Oregon Agricultural College and completed his doctoral work on hetero-geneous catalysis at Caltech in 1925. Emmett was elected to the National Academy of Sciences in 1955 and worked at a handful of institutions, including The Johns Hopkins University, where he chaired the Chemical Engineering Department until his retirement in 1971.

Edward Teller (1908–2003) was a Hungarian-American theoretical physicist who is known as "the father of the hydrogen bomb." He made numerous contributions to nuclear and molecular physics, spectroscopy (in particular the Jahn–Teller and Renner–Teller effects). His extension of Enrico Fermi's theory of beta decay, in the form of Gamow–Teller transitions is also very well known.

Stephen Brunauer (1903–1986) was born in a Jewish family in Budapest, Hungary who emigrated to the U.S.A. in 1921. He was an American research chemist, government scientist, and university teacher. He resigned from his position with the U.S. Navy during the McCarthy era.

Photograph: http://scarc.library.oregonstate.edu/coll/emmett/emmett-lifetime/page2.html

Using the Langmuir theory and Eq. (2.6) as a starting point to describe the equilibrium between the vapor and the adsorbate in the first layer

$$N_m \theta_1 v_1 e^{-\frac{E_1}{RT}} = kP\theta_0 A_1 \tag{2.19}$$

By analogy, for the fraction of surface covered by two layers, one may write

$$N_m \theta_2 v_2 e^{-\frac{E_2}{RT}} = kP\theta_1 A_2 \tag{2.20}$$

In general, for layer n, one obtains

$$N_m \theta_n v_n e^{-\frac{E_n}{RT}} = kP\theta_{n-1} A_n \tag{2.21}$$

The BET theory assumes that the term v, E, and A remain constant for the second and higher layers. This assumption is justifiable on the grounds that the second and higher layers are all equivalent to the liquid state. This undoubtedly approaches the truth as the layers proceed away from the surface but is somewhat questionable for the layers nearer the surface because of the polarizing forces. Nevertheless, using this assumption one can write a series of equations, using L as the heat of liquefaction

$$N_m \theta_1 v_1 e^{-\frac{E_1}{RT}} = kP\theta_0 A_1 \tag{2.22}$$

$$N_m \theta_2 v_2 e^{-\frac{L}{RT}} = kP\theta_1 A \tag{2.23}$$

$$N_m \theta_3 v_3 e^{-\frac{L}{RT}} = kP\theta_2 A \tag{2.24}$$

$$N_m \theta_n v_n e^{-\frac{L}{RT}} = kP\theta_{n-1} A \tag{2.25}$$

A tedious mathematical manipulation of this set of equations yields [2]

$$\frac{N}{N_m} = \frac{C\left(\frac{P}{P_0}\right)}{\left(1 - \left(\frac{P}{P_0}\right)\right) \cdot \left[1 - \left(\frac{P}{P_0}\right) + C\left(\frac{P}{P_0}\right)\right]} \tag{2.26}$$

Realizing that $N/N_m = W/W_m = V/V_m$ or V_a/V_m at pressure a and rearranging Eq. (2.26), one obtains

$$\frac{1}{V_a\left[\left(\frac{P}{P_0}\right)-1\right]} = \frac{1}{V_mC} + \frac{C-1}{V_mC}\frac{P}{P_0} \qquad (2.27)$$

Equation (2.27) is the famous BET equation. In this equation, C, the BET constant, is defined as

$$C = \frac{A_1 v_2}{A_2 v_1}e^{\frac{(E_1-L)}{RT}} \qquad (2.28)$$

C is therefore a value for the strength of interaction between the adsorbate and the adsorbent. One can show that [2]

$$\theta_0 = \frac{\sqrt{C}-1}{C-1} \qquad (2.29)$$

From this equation, it is evident that when sufficient adsorption has occurred to form a monolayer, there is still always some fraction of surface unoccupied. Only for C values approaching infinity will θ_0 will approach zero and in such cases the high adsorbate- surface interaction can only result from chemisorption. For nominal C values, say near 100, the fraction of unoccupied surface, when exactly sufficient adsorption has occurred to form a monolayer, is 0.091. It also means that for C values that are much lower than 100, the physical meaning of the BET surface area becomes more and more questionable.

2.2.2 Multipoint BET Analysis

The determination of surface areas from the BET theory is a straightforward application of Eq. (2.27). A plot of $1/[V_a[(P_0/P)-1]]$ versus P/P_0 will yield a straight line, usually in the range $0.05 \leq (P/P_0) \leq 0.30$. Caution, however, as this is not automatically the case, and the operator should always inspect the BET plot and manually select the points that best produce a linear BET plot! The slopes and intercepts of the BET plot are

$$s = \frac{C-1}{V_mC} \qquad (2.30)$$

$$i = \frac{1}{V_mC} \qquad (2.31)$$

Solving the preceding equations yields

$$V_m = \frac{1}{s+i} \qquad (2.32)$$

$$C = \frac{s}{i} + 1 \qquad (2.33)$$

V_m is then used, similar as for the Langmuir model, to calculate the surface area, by means of Eq. (2.17).

In Figure 2.4 we give an example. On the left side, we show the raw isotherm of the nitrogen adsorption on a silica material, as a function of the relative pressure.

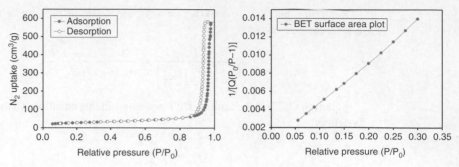

Figure 2.4

Isotherm (left) and BET plot (right) of a silica material.

On the right side, we show the BET transformation, again as a function of the relative pressure. We see that all points of the BET transformation lie almost perfectly on a straight line, which is an important indication that the BET model will yield very accurate data and that no unwanted side effects other than the multilayer formation are occurring.

Also, typically, automatic adsorption devices will produce the other parameters of the BET analysis as well:

BET Report

BET Surface Area:	95.4051 ± 0.8745 m^2/g
Slope:	0.045476 ± 0.000411 g/cm^3 STP
Y-Intercept:	0.000146 ± 0.000077 g/cm^3 STP
C:	312.664468
Qm:	$21{,}9192$ cm^3/g STP
Correlation Coefficient:	0.9995917
Molecular Cross-Sectional Area:	0.1620 nm^2

These data show that we indeed have a very low error on the BET value (based on the fit, weighing errors and other errors are not included here), we get the s and i values. The value of C is well above the magical number of 100 (312.6), referring to a good interaction between adsorbate and adsorbent and a valid BET model.

2.2.2.1 Single Point BET Analysis

By reducing the experimental requirement to only one data point, the single point BET method offers the advantages of simplicity and speed with often little loss in accuracy. From Eqs. (2.30) and (2.31), it follows that

$$\frac{s}{i} = C - 1 \tag{2.34}$$

For reasonably high values of C, the intercept is small compared to the slope and may in many instances be taken as zero. With this approximation, the BET equation (2.27) becomes

$$\frac{1}{V_a\left[\left(\frac{P_0}{P}\right) - 1\right]} = \frac{C - 1}{V_m C}\frac{P}{P_0} \tag{2.35}$$

Table 2.2 Relatives errors for different values of C.

C	Relative error
1	0.70
10	0.19
50	0.04
100	0.02
1000	0.002

and since $1/(V_m C)$, the intercept, is assumed to vanish, Eq. (2.35) reduces to

$$V_m = V_a \left(1 - \left(\frac{P}{P_0} \right) \right) \tag{2.36}$$

As a rule of thumb, the pressure $(P/P_0) = 0.3$ is selected for a single point BET measurement. For most samples, the relative error created by the use of the one-point BET method is less than 5%.

When the value C is known, following formula can be used to correct the single point BET method. This is especially useful for large batches of similar samples (production control).

$$\frac{(V_m)_{mp} - (V_m)_{sp}}{(V_m)_{mp}} = \frac{1 - \left(\frac{P}{P_0} \right)}{1 + (C - 1) \left(\frac{P}{P_0} \right)} \tag{2.37}$$

Table 2.2 shows the relative errors at $(P/P_0) = 0.3$ for different values of C.

It is very important not to use the software of the analysis device as a black box! The isotherm, converted to the BET plot, has to show a linear area somewhere between $P/P_0 = 0.05$ to 0.3. However, this is very variable and dependent on the sample. So, the operator should always check to choose the correct point for the calculation of the BET area. Also, the presence of micropores can bias the analysis. It may be impossible to separate the processes of monolayer-multilayer adsorption and micropore filling. With microporous adsorbents, the linear range of the BET plot may be very difficult to locate. A useful procedure [3] allows one to overcome this difficulty and avoid any subjectivity in evaluating the BET monolayer capacity. This procedure is based on the following main criteria:

1. the quantity C should be positive (i.e. a negative intercept on the ordinate of the BET plot is the first indication that one is outside the appropriate range);
2. application of the BET equation should be restricted to the range where the term $n(1 - P/P_0)$ continuously increases with P/P_0;
3. the P/P_0 value corresponding to n_m should be within the selected BET range.

2.3 Capillary Condensation and Pore Size, the Type IV Isotherm

2.3.1 The Kelvin and the Halsey Equation

Adsorption studies leading the measurements of mesopore size and pore size distribution (PSD) generally make use of the *Kelvin equation*, which relates the equilibrium vapor pressure of a curved surface, such as that of a liquid in a capillary or pore, to the equilibrium pressure of the same liquid on a plane surface. Equation (2.38) is a convenient form of the Kelvin equation:

$$\ln\left(\frac{P}{P_0}\right) = -\frac{(2\gamma\overline{V})}{rRT}\cos\theta \tag{2.38}$$

where P is the equilibrium vapor pressure of the liquid contained in a narrow pore of radius r and P_0 is the equilibrium pressure of the same liquid exhibiting a plane surface. The terms γ and $\overline{V}$ are the liquid surface tension and molar volume, respectively, and θ is the contact angle at which the liquid meets the pore wall.

In a smaller pore, the overlapping potentials of the walls more readily overcome the energy of an adsorbate molecule so that condensation will occur at a lower pressure. Thus, as the relative pressure is increased, condensation will occur first in pores of smaller radii and will progress into the larger pores. Conversely, as the relative pressure decreases, evaporation will occur progressively out of pores with decreasing radii.

For nitrogen adsorption at 77 K, $\gamma = 8.85 \cdot 10^{-3}$ Nm^{-1}, $\theta = 0$ and $V = 34.64 \cdot 10^{-3}$ dm^3 mol^{-1}. Therefore, Eq. (2.38) reduces to

$$r_k = -\frac{0.415}{\log\left(\dfrac{P}{P_0}\right)} \tag{2.39}$$

The term r_k (*Kelvin radius*) indicates the radius into which the condensation occurs at the required relative pressure. *This radius is not the actual pore radius* because some adsorption has occurred on the pore wall, prior to condensation, leaving a center core of radius r_k. Alternatively, during desorption, an adsorbed film remains on the pore wall when evaporation of the center core takes place. If the depth of the adsorbed film when condensation or evaporation occurs is t, then the actual pore radius r_p is given by

$$r_p = r_k + t \tag{2.40}$$

This is further visualized in Figure 2.5. At lower pressures, first the monolayer and then several multilayers of nitrogen are adsorbed on the pore walls. At the moment of

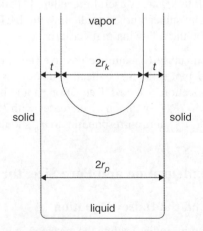

Figure 2.5
Relation between pore radius (r_p), Kelvin radius (r_k) and the statistical thickness of the adsorbed film (t).

capillary condensation, the nitrogen molecules do "see" the pore radius r_k, however, this radius is smaller than the actual pore radius r_p due the layers of nitrogen that are already adsorbed on the walls.

So, how do we determine the statistical thickness t? Using the assumption that the adsorbed film depth in a pore is the same as on a plane surface for any value of relative pressure, one can write

$$t = \left(\frac{V_a}{V_m} \right) \tau \tag{2.41}$$

where V_a and V_m are the quantity adsorbed at a particular relative pressure and the volume adsorbed corresponding to a single monolayer. Essentially, *this equation indicates that the thickness of the adsorbed film is simply the number of layers times the thickness τ of one layer*, regardless of whether the film is in a pore or on a plane surface.

The value of τ can be calculated by considering the area and the volume occupied by one mole of liquid nitrogen if it was spread over a surface as one monolayer.

$$\tau = \frac{\overline{V}}{S} = \frac{3.46 \cdot 10^{20} \text{ nm}^3}{(6.02 \cdot 10^{23})(0.162 \text{ nm}^2)} = 0.354 \text{ nm} \tag{2.42}$$

Equation (2.41) can now be written as

$$t = \left(\frac{V_a}{V_m} \right) 0.354 \text{ nm} \tag{2.43}$$

The value of t, the statistical thickness is approximated by the *empirical Halsey equation*, which for nitrogen can be written as

$$t = 0.354 \left[-\frac{5}{\ln \frac{P}{P_0}} \right]^{\frac{1}{3}} \tag{2.44}$$

The Halsey equation and similar equations are *empirical* formulas, they have no physical meaning. In the case of Halsey, he adsorbed nitrogen on a completely non-porous silica (to avoid any capillary condensation) and fitted this isotherm as a function of the t-value. The adsorbed nitrogen could only be the result of multilayer formation, as no pores or other artifacts are present. To account for differences in surface energy, formulas exist of oxide surface (the equation shown here), but similar equations exist for carbon surfaces and so on. The Eq. (2.44) is the best fit of these data. The *t-value* is often referred to as the *statistical thickness*.

2.3.2 Barrett, Joyner, Halenda (BJH) Pore Size Distributions

(This is an adaption of the methodology introduced by Lowell in 1979 in ref [2])

Let us look at a real isotherm in Figure 2.6. The adsorbed volumes are from a real porous material, called SBA-15 (see Chapter 6). We see different parts. In the first area ($P/P_0 = 0 - 0.1$), we see a Langmuir shaped isotherm, meaning that in this range two things occur: *a (sub)monolayer is formed* and the *micropores are being filled*. Both phenomena are not separable, we will use the t-plot method (see later) to extract the exact amount of micropores.

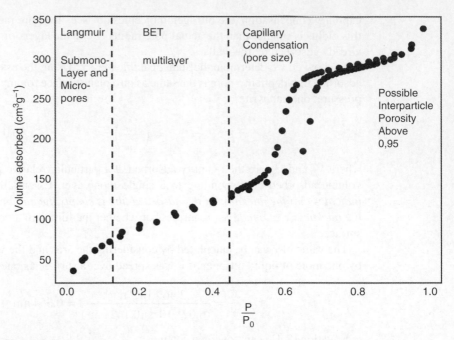

Figure 2.6

The "raw isotherm," adsorbed volume as a function of the normalized pressure, showing the three regions, $P/P_0 < 0.1$: Langmuir region (submonolayer and micropore filling); $0.1 < P/P_0 < 0.4$: BET region (multilayer formation); $P/P_0 > 0.5$ capillary condensation region (liquefaction of nitrogen in the pores).

In a second part, we see a very linear increase of the volume of adsorbed nitrogen as a function of the pressure. For this material, this part expands to $P/P_0 = 0.4$ or a bit more. This is *the region of the multilayer formation*, or the *BET region*.

The last part of the isotherm shows a very steep increase in the adsorbed volume of nitrogen. This is *the region of the capillary condensation* where the nitrogen liquefies in the pores, according to the Kelvin equation.

Let us now turn to the calculation. To calculate the PSD consider the first lines of the worksheet, shown in Table 2.3, and the corresponding explanation of each column. The data are the same as graphically shown in Figure 2.6. This procedure used is the numerical integration method of *Barret, Joyner, and Halenda*. The BJH model assumes that all pores are open cylindrical pores.

The method uses either data form the adsorption or desorption isotherm. However, as will be argued later on, the desorption curve is usually employed except is those cases where the adsorption curve corresponds to the thermodynamically more stable condition. In either case, for the case of presentation, the data are always evaluated downward, from high to low relative pressures.

Columns 1 and 2 contain the data obtained directly from the isotherm. The adsorbed volumes are normalized for 1 g of adsorbent. Relative pressures are chosen using small decrements at high values, where r_k is very sensitive to small changes in relative pressure and where the slope of the isotherm is large, such that small changes in relative pressure produce a large change in volume.

Column 3, the Kelvin radius, is calculated from Eq. (2.38), the Kelvin equation. If nitrogen is the adsorbate, Eq. (2.39) can be used.

Table 2.3 BJH worksheet for pore size distribution.

$\frac{P}{P_o}$ (1)	$V_{gas}\ STP$ (cm³g⁻¹) (2)	r_k (Å) (3)	t (Å) (4)	r_p (Å) (5)	$\bar{r}_k$ (Å) (6)	$\bar{r}_p$ (Å) (7)	Δt (Å) (8)	$\Delta V_{gas}\ STP$ (cm³g⁻¹) (9)	$\Delta V_{liq} \times 10^3$ (cm³g⁻¹) (10)	$\Delta t \Sigma S$ (cm³g⁻¹) (11)	$V_p \times 10^3$ (cm³g⁻¹) (12)	S (m²) (13)	ΣS (m²) (14)
0.99	161.7	950	28.00	978	711	737	5.80	0.20	0.31	0	0.33	0.01	0.01
0.98	161.5	473	22.20	495	394	414	2.80	0.50	0.77	0.00	0.85	0.04	0.05
0.97	161.0	314	19.40	333	250	268	3.20	0.80	1.23	0.02	1.39	0.10	0.15
0.95	160.2	186	16.20	202	138	153	3.40	1.40	2.16	0.05	2.59	0.34	0.49
0.90	158.8	90.7	12.80	104	74.8	87.0	1.70	1.60	2.46	0.08	3.22	0.74	1.23
0.85	157.2	58.8	11.10	69.9	50.8	61.4	1.10	2.00	3.08	0.14	4.29	1.40	2.63
0.80	155.2	42.8	10.00	52.8	39.7	49.5	0.50	2.30	3.54	0.13	5.30	2.14	4.77
0.77	152.9	36.6	9.50	46.1	34.9	44.3	0.30	4.00	6.16	0.14	9.70	4.38	9.15
0.75	148.9	33.2	9.20	42.4	31.8	40.9	0.30	3.80	5.85	0.27	9.23	4.51	13.66
0.73	145.1	30.4	8.90	39.3	29.2	38.0	0.20	4.20	6.47	0.27	10.50	5.53	19.19
0.71	140.9	27.9	8.70	36.6	26.9	35.4	0.30	5.00	7.70	0.58	12.33	6.97	26.16
0.69	135.9	25.8	8.40	34.2	24.9	33.2	0.20	5.90	9.09	0.52	15.24	9.18	35.34
0.67	130.0	23.9	8.20	32.1	–	–	–	–	–	–	–	–	–

Column 4, the film depth, t, is calculated using Eq. (2.44), the Halsey equation.

Column 5, the pore radius is obtained from Eq. (2.40).

Column 6 and 7, $\overline{r}_p$ and $\overline{r}_k$ are calculated using the mean value in each decrement from successive entries.

Column 8, the change in the film depth, is calculated by taking the difference between the successive values of t.

Column 9, ΔV_{gas}, is the change in adsorbed volume between successive P/P_0 values and is determined by subtracting successive values from column 2.

Column 10, ΔV_{liq}, is the volume of liquid corresponding to ΔV_{gas}. The most direct way to convert ΔV_{gas} to ΔV_{liq} is to calculate the moles of gas, and multiply by the liquid molar volume. For nitrogen at standard temperature and pressure, this is given by

$$\Delta V_{liq} = \frac{\Delta V_{gas}}{22\,400} \cdot 34.6 \, \text{cm}^3 \tag{2.45}$$

Column 11 represents the volume change of the adsorbed film remaining on the walls of the pores from which the center core has previously evaporated. This volume is the product of the film area ΣS and the decrease in the film depth Δt. By assuming no pores are present, larger than 95 nm (at $P/P_0 = 0.99$), the first entry in column 11 is zero since there exist no film area from previously emptied pores. The error this assumption introduces in negligible because the area produces by pores larger than 95 nm will be small compared to their volume. Subsequent entries in column 11 are calculated as the product of Δt for a decrement and ΣS from the row above corresponding to the adsorbed film area exposed by evaporation of the center cores during all previous decrements.

Column 12, the actual pore volume, is evaluated by recalling that the volume of liquid, column 10, is composed of the volume evaporated out of the center cores plus the volume desorbed from the film left on the pore walls. As the BJH model *assumes cylindrical pores*, we obtain:

$$\Delta V_{liq} = \pi \overline{r}_k^2 L + \Delta t \Sigma S \tag{2.46}$$

and since

$$V_p = \pi \overline{r}_p^2 L \tag{2.47}$$

then, by combining the preceding equations,

$$V_p = \left(\frac{\overline{r}_p}{\overline{r}_k}\right)^2 \cdot [\Delta V_{liq} - \Delta t \Sigma S] \tag{2.48}$$

Column 13 is the surface area of the pore walls calculated form the pore volume by

$$S = \frac{2V_p}{\overline{r}_p} \tag{2.49}$$

It is this value of S that is summed in column 14. The summation is multiplied by Δt from the following decrement to calculate the film volume decrease in column 11. The worksheet discloses that the volume of all pores greater than 1.36 nm is 0.39 cm^3 g^{-1}. This does not mean that micropores with smaller radii are absent, but rather that the validity of the Kelvin equation becomes questionable because of

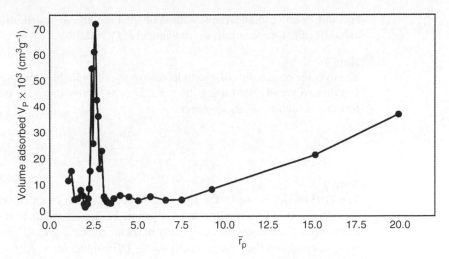

Figure 2.7
Pore size distribution.

the uncertainty regarding molar volumes and surface tension when only one or two molecular diameters are involved. Termination of the analysis is necessary, therefore, when the relative pressure approaches 0.3. The volume of micropores, if present, can be evaluated by the difference between the total pore volume and the sum of column 12. The total pore volume is usually evaluated at $P/P_0 = 0.99$, by the following equation

$$V_{tot} = \frac{V_{gas}}{22\,400} \cdot 34.6 \, \text{cm}^3 \qquad (2.50)$$

Equation (2.50) is commonly referred to as the *Gurvitch rule*. A PSD plot is shown in Figure 2.7: PSD. It is the plot of the data, of which a portion was presented in the worksheet. We note a very nice narrow PSD around 2.5 nm. The pores are very narrowly distributed. This could already be deducted from the original isotherm (Figure 2.6), where the capillary condensation at $P/P_0 = 0.65$ (adsorption) is also very steep.

The isotherm has seemingly a lot of pores in the range of 15–20 nm. One has to be very careful with the interpretation of these data. The Kelvin equation (2.38) gives a logarithmic relation between gas pressure and pore size. At larger pressures, this law becomes very inaccurate. If we look at Table 2.3, we see that from $P/P_0 = 0.98$ to $P/P_0 = 0.99$, the r_k jumps from 47,3 to 95,0 nm; in other words, the PSD data have steps of 50 nm here. At the same time, the amount of gas adsorbed only increases with 0.2 cm^3g^{-1} *at STP*.

Looking at the original isotherm in Figure 2.6, we see an increase in adsorbed nitrogen above $P/P_0 = 0.9$. This is most often due to *interparticle porosity*. These pressures correspond to capillary condensation in the size of 25–100 nm with a very low resolution. In many cases, this condensation occurs when the particles are a few times bigger, as the voids in between the particles will create pores in which the nitrogen condenses. This has therefore nothing to do with the real porosity inside the particles and should be omitted.

Care is needed in those cases, as many automatic routines calculate the total pore volume at $P/P_0 = 0.99$. In that case, the interparticle porosity will also be taken into account, leading to an overestimation of the porosity. *As a general rule, we recommend to calculate the total pore volume at* $P/P_0 = 0.95$.

Note 1

The average pore radius of a sample, assuming a sample has purely cylindrical pores, can be easily calculated using the *formula of Wheeler*. It is very easy to derive this formula, which is left as an exercise.

$$\hat{r}_w = \frac{2V_p}{S_{BET}} \tag{2.51}$$

Note 2

The BJH model is one of the simplest models that can be calculated by hand. It is still mostly used for every routine analysis of PSD with excellent results. Obviously, more sophisticated models exist; many of them are "model-less," meaning they make no assumption on the pore geometry as the BJH model does. Also, (non-linear) density functional theory models are now often available in the software packages of instruments.

Several more specialized books and reviews exist on the topic. Matthias Thommes, Alex V. Neimark, James P. Olivier, Francisco Rodriguez-Reinoso, Katsumi Kaneko, Jean and Françoise Rouquerol, Kenneth Sing, and many others are the current trendsetters in this field.

2.3.3   Types of Adsorption Isotherms

In the 1985, the IUPAC published its recommendations on the interpretation of physisorption isotherms [4]. However, over the past 30 years, various new characteristic types of isotherms have been identified. Therefore, in 2015 the original IUPAC classifications of physisorption isotherms and associated hysteresis loops were refined [5]. The proposed updated classification of physisorption isotherms is shown in Figure 2.8. In the context of physisorption, it is necessary to classify pores according to their size (IUPAC recommendation, 1985):

  (i) pores with widths exceeding about 50 nm are called *macropores*;
 (ii) pores of widths between 2 nm and 50 nm are called *mesopores*;
(iii) pores with widths not exceeding about 2 nm are called *micropores*.

In the following, we present a shortened version of the IUPAC 2015 classification and recommendations [5].

Reversible *Type I isotherms* ("Langmuir isotherms") are given by microporous solids having relatively small external surfaces (e.g. some activated carbons, molecular sieve zeolites and certain porous oxides). A Type I isotherm is concave to the P/P_0 axis and the amount adsorbed approaches a limiting value. This limiting uptake is governed by the accessible micropore volume rather than by the internal surface area. A steep uptake at very low P/P_0 is due to enhanced adsorbent-adsorptive interactions in narrow micropores (micropores of molecular dimensions), resulting in micropore filling at very low P/P_0. For nitrogen and argon adsorption at 77 and 87 K,

Type I(a) isotherms are given by microporous materials having mainly narrow micropores (of width $< \sim 1$ nm);

Type I(b) isotherms are found with materials having PSDs over a broader range including wider micropores and possibly narrow mesopores ($< \sim 2.5$ nm).

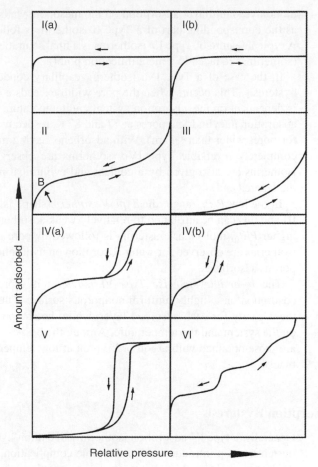

Figure 2.8
New 2015 IUPAC classification of isotherms.

Reversible *Type II isotherms* are given by the physisorption of most gases on nonporous or microporous adsorbents. The shape is the result of unrestricted monolayer-multilayer adsorption up to high P/P_0. If the knee is sharp, Point B – the beginning of the middle almost linear section – usually corresponds to the completion of monolayer coverage. A more gradual curvature (i.e. a less distinctive Point B) is an indication of a significant amount of overlap of monolayer coverage and the onset of multilayer adsorption. The thickness of the adsorbed multilayer generally appears to increase without limit when $P/P_0 = 1$.

In the case of a *Type III isotherm*, there is no Point B and therefore no identifiable monolayer formation; the adsorbent–adsorbate interactions are now relatively weak and the adsorbed molecules are clustered around the most favorable sites on the surface of a nonporous or macroporous solid. In contrast to a Type II isotherm, the amount adsorbed remains finite at the saturation pressure (i.e. at $P/P_0 = 1$).

Type IV isotherms are given by mesoporous adsorbents (e.g. many oxide gels, industrial adsorbents and mesoporous molecular sieves). The adsorption behavior in mesopores is determined by the adsorbent-adsorptive interactions and also by the interactions between the molecules in the condensed state. In this case, the initial monolayer-multilayer adsorption on the mesopore walls, which takes the same path as the corresponding part of a Type II isotherm, is followed by pore condensation. A typical feature of Type IV isotherms is a final saturation plateau, of variable length (sometimes reduced to a mere inflection point).

In the case of a Type IVa isotherm, capillary condensation is accompanied by hysteresis. This occurs when the pore width exceeds a certain critical width, which is dependent on the adsorption system and temperature (e.g. for nitrogen and argon adsorption in cylindrical pores at 77 and 87 K, respectively, hysteresis starts to occur for pores wider than $\sim$4 nm). With adsorbents having mesopores of smaller widths, completely reversible Type IVb isotherms are observed. In principle, Type IVb isotherms are also given by a conical and cylindrical mesopores that are closed at the tapered end.

In the low P/P_0 range, the *Type V isotherm* shape is very similar to that of Type III and this can be attributed to relatively weak adsorbent–adsorbate interactions. At higher P/P_0, molecular clustering is followed by pore filling. For instance, Type V isotherms are observed for water adsorption on hydrophobic microporous and mesoporous adsorbents.

The *reversible stepwise Type VI isotherm* is representative of layer-by-layer adsorption on a highly uniform nonporous surface. The step-height now represents the capacity for each adsorbed layer, while the sharpness of the step is dependent on the system and the temperature. Among the best examples of Type VI isotherms are those obtained with argon or krypton at low temperature on graphitized carbon blacks.

2.3.4   Adsorption Hysteresis

Generally, a type IV adsorption isotherm will exhibit hysteresis. The presence of the hysteresis loop introduces a considerable complication, in that within the region of the hysteresis loop, there are two relative pressure values corresponding to a given quantity adsorbed with the lower value always residing on the desorption isotherm. The pore radius, however, must be single valued and some criteria have to be established as to which value of relative pressure is to be employed in the Kelvin equation. The molar free energy change accompanying the condensation of vapor into a pore during adsorption is given by

$$\Delta G_{ads} = RT(\ln P_{ads} - \ln P_0) \tag{2.52}$$

For the same quantity on the desorption branch, the corresponding free energy chance is

$$\Delta G_{des} = RT (\ln P_{des} - \ln P_0) \tag{2.53}$$

Since $P_{des} \leq P_{ads}$, it follows that $\Delta G_{des} \leq \Delta G_{ads}$. Therefore, the desorption value of relative pressure corresponds to the more stable adsorbate condition and the desorption isotherm should, with certain exceptions, be used for pore size analysis.

Reproducible, permanent hysteresis loops, which are located in the multilayer range of physisorption isotherms, are generally associated with capillary condensation. In an open-ended pore (e.g. of cylindrical geometry), delayed condensation is the result of metastability of the adsorbed multilayer. It follows that in an assembly of such pores the adsorption branch of the hysteresis loop is not in thermodynamic equilibrium. Since evaporation does not involve nucleation, the desorption stage is equivalent to a reversible liquid–vapor transition.

Therefore, if the pores are filled with liquid-like condensate, thermodynamic equilibration is established on the desorption branch.

In more complex pore structures, the desorption path is often dependent on network effects and various forms of *pore blocking*. These phenomena occur if wide pores only have access to the external surface through narrow necks (e.g. *ink-bottle pore shape*). The wide pores are filled as before and remain filled during desorption until the narrow necks empty at lower vapor pressures. In a pore network, the desorption vapor pressures are dependent on the size and spatial distribution of the necks. If the neck diameters are not too small, the network may empty at a relative pressure corresponding to a characteristic percolation threshold. Then, useful information concerning the neck size can be obtained from the desorption branch of the isotherm.

Theoretical and experimental studies have revealed that if the neck diameter is smaller than a critical size (estimated to be around 5–6 nm for nitrogen at 77 K), the mechanism of desorption from the larger pores involves *cavitation* (i.e. the spontaneous nucleation and growth of gas bubbles in the metastable condensed fluid). Cavitation controlled evaporation has been found, for instance, with certain micro-mesoporous silicas, mesoporous zeolites, clays, and also some activated carbons. Contrary to the situation of pore blocking/percolation-controlled evaporation, no quantitative information about the neck size and neck size distribution can be obtained in the case of cavitation.

So, in practice, *every nitrogen hysteresis loop closes* at $P/P_0 = 0.42$ due to cavitation when the liquid meniscus is no longer stable. This leads to a sharp peak in the PSD based upon desorption. Much too many papers have been published giving this peak a physical meaning. So be careful, the cavitation peak in the desorption based PSD is an artifact and meaningless.

Many different shapes of hysteresis loops have been reported, but the main types are shown in Figure 2.9. Types H1, H2(a), H3, and H4 were identified in the original IUPAC classification of 1985, which is now extended in the light of more recent findings. Each of these six characteristic types is fairly closely related to particular features of the pore structure and underlying adsorption mechanism.

The *Type H1 loop* is found in materials that exhibit a narrow range of uniform mesopores, as, for instance, in templated silicas (e.g. MCM-41, MCM-48, SBA-15) (see Chapter 6), some controlled pore glasses and ordered, mesoporous carbons. Usually, network effects are minimal and the PSD can be calculated using the desorption isotherm.

Hysteresis loops of *Type H2* are given by more complex pore structures in which network effects are important. The very steep desorption branch, which is a characteristic feature of H2(a) loops, can be attributed either to pore blocking/percolation in a narrow range of pore necks or to cavitation-induced evaporation.

H2(a) loops are for instance given by many silica gels (Chapter 5), some porous glasses (e.g. vycor) as well as some ordered mesoporous materials (e.g. SBA-16 and KIT-5 silicas) (Chapter 6). The *Type H2(b)* loop is also associated with pore blocking, but the size distribution of neck widths is now much larger.

There are two distinctive features of the *Type H3 loop*: (i) the adsorption branch resembles a Type II isotherm, and (ii) the lower limit of the desorption branch is normally located at the cavitation-induced (= 0.42 for nitrogen). Loops of this type are given by non-rigid aggregates of plate-like particles (e.g. certain clays) and also

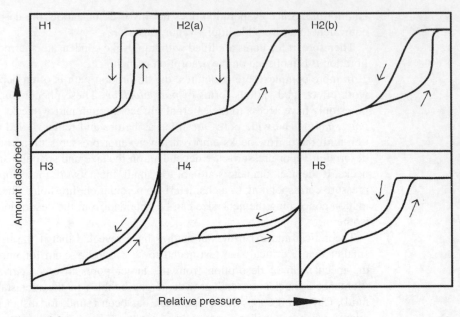

Figure 2.9
New 2015 IUPAC classification of hysteresis loops.

if the pore network consists of macropores that are not completely filled with pore condensate.

The *H4 loop* is somewhat similar, but the adsorption branch is now a composite of Types I and II, the more pronounced uptake at low P/P_0 being associated with the filling of micropores. H4 loops are often found with aggregated crystals of zeolites, some mesoporous zeolites, and micro-mesoporous carbons.

Although the *Type H5* loop is unusual, it has a distinctive form associated with certain pore structures containing both open and partially blocked mesopores (e.g. plugged hexagonal templated silicas). This type of materials was discovered in 2002 by Van Der Voort [6].

As already indicated, the common feature of H3, H4, and H5 loops is the sharp step-down of the desorption branch. Generally, this is located in a narrow range of P/P_0 for the particular adsorptive and temperature (e.g. at $P/P_0 \sim 0.42$ for nitrogen at temperatures of 77 K).

2.3.5 Evaluation of Micropores

The analysis of micropore distributions is more complex than the analysis of mesopore distributions. Although several sophisticated models exist, and are even available commercially, this text will be restricted to one method that yields the total micropore volume, but not the micropore distribution. Micropores are pores with a diameter < 2 nm.

The *t-plot method* of De Boer is the most used. It consists of plotting the volume of gas adsorbed versus t, the statistical thickness of the adsorbed film. t is a function of P/P_0, as measured in the standard isotherm according to the Halsey equation. In any normal case of multimolecular adsorption, the experimental points (t vs. P/P_0) should fall on a straight line through the origin. If the adsorbent contains

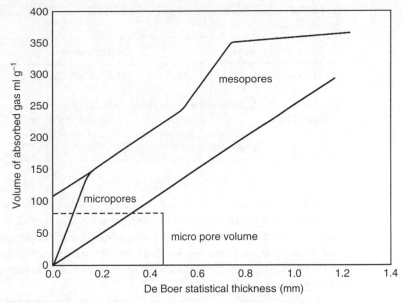

Figure 2.10
t-plot method of De Boer.

micropores, the uptake is enhanced in the low-pressure region and the isotherm is correspondingly distorted. In the *t*-plot, extrapolation of the curve gives a positive intercept, which is equivalent to the micropore volume by means of the Gurvitch rule (Eq. (2.50)).

When mesopores are present, capillary condensation will occur according to the Kelvin equation (Eq. (2.39)). A *t*-plot will therefore show an upward deviation, starting at the relative pressure at which the finest pores are just being filled. Figure 2.10 schematically illustrates the effect of micro- and mesoporosity on a *t*-plot.

2.4 Liquid Phase Adsorption – Langmuir and Freundlich Isotherms

Many adsorption processes occur in the liquid phase, the adsorbent then being a solid. The purpose is usually the recovery of the adsorbate from the liquid [7]. Water purification filters work on this principle and many other processes to selectively or non-selectively separate salts, heavy metals, toxic metal, organic components, and many more from the water (or the air).

The performance of an adsorbent is assessed by its adsorption capacity for a specific adsorbate, which depends on the efficiency of the interactions [8]. The key factors that affect the efficiency of these interactions and, therefore, the adsorption capacity are: (i) adsorbent characteristics, which include porosity, particle size, specific surface area, presence of functional groups, stability, and selectivity; (ii) adsorbate chemistry, such as size charge and speciation; and (iii) the operating conditions of the system, including pH, temperature, contact time, agitation, occurrence of competing species, and coexistence of other pollutants [7a, 9].

As an example, for the adsorption of metal ions from a liquid matrix, the interactions between the adsorbate and the adsorbent surface can be described by Pearson's Hard–Soft Acid–Base (HSAB) principle. This states that hard acids easily bind to

Table 2.4 Classification of hard and soft Lewis acids and bases according to the HSAB principle.

	Hard	Intermediate	Soft
Lewis-acid	Li^+, Na^+, K^+, Mg^{2+}, Ca^{2+}, Al^{3+}	Ni^{2+}, Cu^{2+}, Zn^{2+}	Cu^+, Ag^+, Au^{3+}, In^{3+}, Rh^{3+}
Lewis-base Adsorbent has	Binding groups: hydroxyl, carbonyl, carboxyl, sulfonate, phosphonate Ligand atom: O	Binding groups: amine, amide, imine Ligand atom: N	Binding groups: thiol, thioether, imidazole Ligand atom: N, S

hard bases and soft acids to soft bases [3]. Table 2.4 shows the classification of some metals as hard, soft, or intermediate acids and ligand atoms as hard and soft bases. According to this principle, hard acid–base bonds are of ionic character and soft acid–base bonds are covalent [7a]. Intermediate metal ions can bind ligands with different preferences, as influenced by other factors such as metal concentration, competing ions, and pH [7a].

A well-designed solid adsorbent should have the required particle size, shape, morphology, and proper selective ligands attached. In addition, in case of dynamic high-pressure separation systems, the solid sorbent should be characterized by a well-developed pore structure with tunable pore size, pore connectivity, and surface properties.

We will show examples of selective metal sequestration from water and gas purification in the next chapters.

A variety of models is proposed for the fitting of adsorption isotherms and the adsorption kinetics. The most commonly used models will be described briefly in the following section.

2.4.1 Adsorption Kinetics

In general, the amount of adsorbate adsorbed at equilibrium is often calculated as follows:

$$q_e = \frac{(C_0 - C_e).V}{m} \qquad (2.54)$$

C_0 ($mg\,l^{-1}$) is the initial concentration of the adsorbate and C_e ($mg\,l^{-1}$) is the adsorbate concentration at equilibrium, m (g) is mass of the used adsorbent, and V (l) is the volume of the adsorbate solution. The adsorption performance is commonly expressed in mg of adsorbate adsorbed per gram of adsorbent.

The initial adsorption rate can be very fast (within a couple of min) if there is a high affinity between the adsorbent and adsorbate. This has been for example observed in the case for the adsorption of heavy metals on biochars in which 80.6–96.9% was removed after only one minute contact time [10]. For the adsorption of inorganic and organic contaminants onto porous adsorbents it can take longer to reach equilibrium (from a couple of days to several weeks) in comparison to nonporous sorbents. This is due to the fact that other adsorption mechanisms occur in porous materials of which pore filling is the most common one together with several types of interactions, for example π-π interaction, electrostatic interactions, and hydrogen bonding [11]. For this reason, kinetic studies play a crucial role in identifying the required time necessary to reach equilibration and to elucidate the optimal contact time between the adsorbent and adsorbate.

Two types of equations have been commonly used to represent the kinetics [11]. The first one, which corresponds to a diffusion-controlled process, is the *intra-particle diffusion equation*. In this model, the diffusion of the adsorbent in the pores of the adsorbate is the rate-limiting step.

The second model assumes that the process is controlled by the adsorption reaction at the liquid–solid interface in the adsorbent. So, the actual adsorption step is considered as the rate-limiting step. Two types of kinetics are generally used and compared, namely the *pseudo-first-order* and *pseudo-second-order rate laws*.

The pseudo-first-order kinetics were first proposed at the end of the nineteenth century by Lagergren [12]. The assume that the reaction rate is first-order, and thus the adsorbent plays no role in the kinetics.

Pseudo-second-order kinetics was introduced in the middle of the 1980s [13]. However, it was not very popular until 1999 when Ho and McKay [14] analyzed a number of experimental results taken from the literature, and arrived at the conclusion that, "for all of the systems studied, [...] the pseudo-second-order reaction kinetics provide the best correlation of the experimental data" This model assumes that the adsorbent plays also a role in the kinetics, making the kinetics of a second-order.

The *pseudo-first-order rate equation*, presented by Lagergren for the first time in 1898 for the adsorption of oxalic acid and malonic acid onto charcoal, is expressed as follows [12]:

$$\ln(q_e - q_t) = -k_1 t + \ln(q_e) \tag{2.55}$$

in which q_e and q_t are the amounts of adsorbate adsorbed per mass of adsorbent at equilibrium and at any time t *(min)* respectively and k_1 *(1 min^{-1})* is the rate constant.

This equation is widely applied in adsorption kinetics; however, it is important to note that the latter equation can only be used during the initial 20–30 minutes of contact time and not for the whole range [15].

Blanchard et al. introduced in 1984 a *pseudo-second-order rate equation* for the removal of heavy metals from water using zeolites [13], here we present the notation of Ho and McKay [14]

$$\frac{t}{q(t)} = \frac{t}{q_e} + \frac{t}{k_2 q_e^2} \tag{2.56}$$

Or rewritten as

$$q(t) = q_e \frac{k_2 q_e t}{1 + k_2 q_e t} \tag{2.57}$$

in which q_e and q_t are the amount of adsorbate adsorbed at equilibrium and at any time respectively expressed in mg g^{-1} and k_2 expressed in g (mg min)$^{-1}$ is the rate constant of the pseudo-second-order equation.

Usually, the data are fitted (preferably using the linearized form of the equations) to establish which model best represents the kinetics of adsorption.

Although the pseudo-second-order equation can be very useful in describing the adsorption kinetic data, it cannot be used to reveal the occurring adsorption mechanisms.

Roginsky and Zeldovich introduced in 1934 an empirical equation for the adsorption of carbon monoxide onto manganese dioxide, which is known nowadays as the *Elovich equation* [16]:

$$\frac{dq_t}{dt} = \alpha \, e^{-\beta q_t} \qquad (2.58)$$

In which q_t is the amount of adsorbate adsorbed per mass of adsorbent at time t (min); α is the initial rate constant expressed in mg g^{-1} min^{-1} and β (mg g^{-1}) is the desorption constant.

The *Elovich equation* is often used for *activated adsorption*, meaning that the rate of adsorption decreases exponentially with the increase of the amount of adsorbate that is adsorbed. A very typical case of such activated adsorption is clearly *chemisorption*, as the rate quickly decreases with the disappearance of the reacting groups on the solid.

The linear form of the Elovich equation, introduced by Chien and Clayton in 1980 [17], can be written as follows:

$$q_t = \frac{1}{\beta} \ln(t) + \frac{1}{\beta} \ln(\alpha\beta) \qquad (2.59)$$

The Elovich equation is typically used for the analysis of chemisorption data. Moreover, it can be used to describe several reaction mechanisms such as bulk and surface diffusion [11].

In 1963 Weber and Morris presented the linearized form of the *intra-particle diffusion model* [18]:

$$q_t = k_p \sqrt{t} + C \qquad (2.60)$$

in which k_p (mg g^{-1} min$^{1/2}$) is the rate constant of the intra-particle diffusion model and C (mg g^{-1}) is the constant which correspond to the thickness of the boundary layer.

This equation is used to describe the adsorption mechanisms and to determine the rate-determining step. *More specifically, if a plot of q_t versus $t^{0.5}$ is linear and passes through the origin, the adsorption is governed by intra-particle diffusion.* However, if this not the case and the plot gives multiple linear regions, then the adsorption process is characterized by a multistep process. Walter proposed that four steps of transport occur during the adsorption process conducted by porous materials (Figure 2.11) [20]. The first step, which occurs quickly, is the transport in the solution phase, also known as the bulk transport. This process occurs immediately after the addition of the adsorbent into the adsorbate solution. The second step is the so-called "film diffusion," which is rather slow. In this step, the adsorbate molecules are transferred from the bulk liquid to the adsorbent's external surface via a hydrodynamic boundary layer or film. The third step, known as the intra-particle diffusion, also occurs rather slowly. This process holds the diffusion of the adsorbate from the external surface into the pores of the adsorbent. The last stage, the "adsorptive attachment," takes place very quickly and is for this reason not significant in engineering design.

2.4.2 Adsorption Isotherms

A large variety of models of adsorption isotherms are described in literature. These models can be classified as follows: one parameter isotherms (i.e. Henry isotherm), two-parameter isotherms (e.g. Langmuir, Freundlich, Dubinin-Radushkevich, Temkin, Flory-Huggins, and Hill), three-parameter isotherms (e.g. Redlich-Peterson,

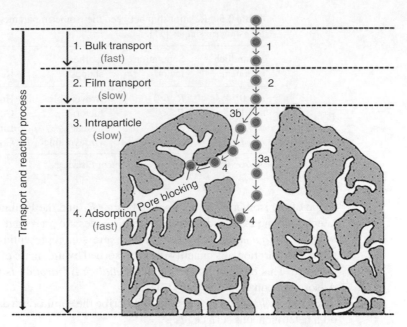

Figure 2.11
Four steps of transport during the adsorption process using a porous adsorbent, taken from Ref. [19].
Source: Reproduced with permission of ACS.

Sips, Toth, Koble-Corrigan, Khan, Fritz-Schluender, Vieth-Sladek, and Radke-Prausnitz) and more than three parameter isotherms (Weber-van Vliet, Fritz-Schlunder, and Baudu). [11] Of all these models, the *Langmuir* and *Freundlich* models are the most commonly used because of their simplicity and ease of interpretation.

In Eq. (2.10) we described the *Langmuir adsorption* isotherm for a gas A adsorbing on a surface. We can adapt same Langmuir isotherm now for liquid phase adsorption: Langmuir described in 1918 the nonlinear form of the Langmuir equation [1]:

$$q_e = q_{max}^0 \frac{K_L C_e}{1 + K_L C_e} \tag{2.61}$$

in which $q°_{max}$ (mg g^{-1}) is the maximum saturated monolayer adsorption capacity, C_e (mg l^{-1}) is the adsorbate concentration at equilibrium, q_e (mg g^{-1}) is the amount of adsorbate uptake at equilibrium and K_L (l mg^{-1}) is the constant related to the affinity between the adsorbate and the adsorbent. When an adsorbent is good, one expects a high adsorption capacity $q°_{max}$ and a steep initial sorption isotherm slope (or in other words a high K_L).

If the Langmuir model can adequately describe the experimental adsorption data, it is important to calculate in a following step the separation factor or equilibrium parameter R_L, which was introduced by Hall et al. [21] in 1966:

$$R_L = \frac{1}{1 + K_L C_0} \tag{2.62}$$

R_L is the dimensionless separation factor, K_L is the Langmuir equilibrium constant, and C_0 (mg l^{-1}) is the initial adsorbate concentration. It is important to note that the

Table 2.5 Relationship between the isotherm parameters and the isotherm shapes [11].

Freundlich exponent	Separation factor	Isotherm shapes	Remarks
$n = 0$	$R_L = 0$	Irreversible	Horizontal
$n < 1$	$R_L < 1$	Favorable	Concave
$n = 1$	$R_L = 1$	Linear	Linear
$n > 1$	$R_L > 1$	Unfavorable	Convex

Source: Reproduced with permission of Elsevier.

shape of the isotherm can be predicted using R_L and the Freundlich exponent n (see later). In Table 2.5, the different isotherm shapes are presented.

In 1909, *Freundlich* gave an empirical expression representing the isothermal variation of adsorption of a quantity of gas adsorbed by unit mass of solid adsorbent with pressure. This equation is known as *Freundlich Adsorption Isotherm* or Freundlich Adsorption equation.

The *Freundlich equation* is used to describe the equilibrium data for the adsorption on a heterogeneous surface [22].

The nonlinear form of the Freundlich equation is expressed as follows:

$$q_e = K_F C_e^n \qquad (2.63)$$

in which q_e (mg g^{-1}) is the amount of adsorbate uptake at equilibrium, C_e (mg l^{-1}) is the adsorbate concentration at equilibrium, K_F (mg g^{-1})/(mg l^{-1})^n is the Freundlich constant and n (dimensionless) is the Freundlich intensity parameter indicating the magnitude of the adsorption driving force or the surface heterogeneity.

The reasoning behind this very old and simple model is simple. At very low pressures, the extent of the adsorption is directly proportional to the pressure, raised to the power one. Remember that this was also (later) described by Henry (*Henry regime*). At very high pressure, again similar to the Langmuir model, the extent of adsorption is independent of the pressure, so is a function of the pressure, raised to the power zero.

In intermediate pressures, adsorption is proportional to a pressure raised to a power $(1/m)$, with m being a number larger than one. $(1/m)$ is expressed as n in Eq. (2.63).

2.5 Heterogeneous Catalysis

2.5.1 Introduction

A catalyst accelerates a chemical reaction. It does so by forming bonds with the reacting molecules and by allowing these to react to a product that detaches from the catalyst and leaves it unaltered so it is available for the next reaction. In fact, we can describe the catalytic reaction as a cyclic event in which the catalyst participates and is recovered in its original form at the end of the cycle.

Consider the catalytic reaction between two molecules A and B to give a product P, see Figure 2.12. The cycle starts with the bonding of molecules A and B to the catalyst. A and B then react within this complex to give a product P, which is also bound to the catalyst. In the final step, P separates from the catalyst, thus leaving the reaction cycle in its original state.

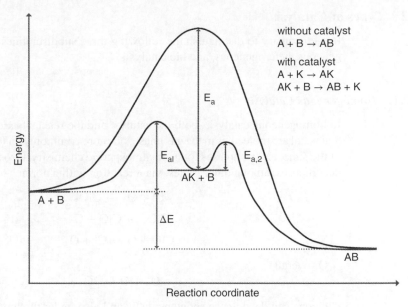

Figure 2.12

Potential energy diagram of a heterogeneous catalytic reaction, with gaseous reactants and products and a solid catalyst. Note that the uncatalyzed reaction has to overcome a substantial energy barrier, whereas the barriers in the catalytic route are much lower.

To see how the catalyst accelerates the reaction, we need to look at the free energy diagram in Figure 2.12 that compares the non-catalytic and the catalytic reaction. For the non-catalytic reaction, the figure is simply the familiar way to visualize the Arrhenius equation: the reaction proceeds when A and B collide with sufficient energy to overcome the activation barrier. The change in Gibbs free energy between the reactants, A + B, and the product P is ΔG.

The catalytic reaction starts by bonding of the reactants A and B to the catalyst, in a spontaneous reaction. Hence, the formation of this complex is exothermic, and the free energy is lowered. There then follows the reaction between A and B while they are bound to the catalyst. This step is associated with an activation energy; however, it is significantly lower than that for the uncatalyzed reaction. Finally, the product P separates from the catalyst in an endothermic step.

The energy diagram of Figure 2.12 illustrates several important points:

- The catalyst offers an alternative path for the reaction, which is obviously more complex, but energetically much more favorable.
- The activation energy of the catalytic reaction is significantly smaller than that of the uncatalyzed reaction; hence, the rate of the catalytic reaction is much larger.
- The overall change in free energy for the catalytic reaction equals that of the uncatalyzed reaction. Hence, the catalyst does not affect the equilibrium constant for the overall reaction of A + B to P. Thus, if a reaction is thermodynamically unfavorable, a catalyst cannot change this situation. *A catalyst changes the kinetics but not the thermodynamics.*
- The catalyst accelerates both the forward and the reverse reaction to the same extent. In other words, if a catalyst accelerates the formation of the product P from A and B, it will do the same for the decomposition of P into A and B.

2.5.2   Types of Catalysis

It is customary to distinguish the following three subdisciplines in catalysis: homogeneous, heterogeneous, and bio catalysis.

2.5.2.1   *Homogeneous Catalysis*

In homogeneous catalysis, both the catalyst and the reactants are in the same phase; that is, all are molecules in the gas phase, or, more commonly, in the liquid phase. One of the simplest examples is found in atmospheric chemistry. Ozone in the atmosphere decomposes, among other routes, via a reaction with chlorine atoms:

$$Cl + O_3 \rightarrow ClO_3$$
$$ClO_3 \rightarrow ClO + O_2$$
$$ClO + O \rightarrow Cl + O_2$$

Or overall

$$O_3 + O \rightarrow 2O_2$$

Ozone can decompose spontaneously, and also under the influence of light, but a Cl atom accelerates the reaction tremendously. As it leaves the reaction cycle unaltered, the Cl atom is a catalyst. Because both reactant and catalyst are both in the same phase, namely the gas phase, the reaction cycle is an example of homogeneous catalysis (this reaction was historically important in the prediction of the hole in the ozone layer).

Industry uses a multitude of homogeneous catalysts in all kinds of reactions to produce chemicals. The catalytic carbonylation of methanol to acetic acid:

$$CH_3OH + CO \rightarrow CH_3COOH$$

by $[Rh(CO)_2I_2]-$ complexes in solution is one of many examples. In homogeneous catalysis, often aimed at the production of delicate pharmaceuticals, organometallic complexes are synthesized in procedures employing molecular control, so that the judicious choice of ligands directs the reacting molecules to the desired products.

2.5.2.2   *Biocatalysis*

Enzymes are nature's catalysts. For the moment, it is sufficient to consider an enzyme as a large protein, the structure of which results in a very shape-specific active site. Having shapes that are optimally suited to guide reactant molecules (usually referred to as substrates) in the optimum configuration for reaction, enzymes are highly specific and efficient catalysts. For example, the enzyme catalase catalyzes the decomposition of hydrogen peroxide into water and oxygen at an incredibly high rate of up to 107 hydrogen peroxide molecules per second. Enzymes allow biological reactions to occur at the rates necessary to maintain life, such as the build-up of proteins and DNA, or the breakdown of molecules and the storage of energy in sugars.

2.5.2.3   *Heterogeneous Catalysis*

In heterogeneous catalysis, solids catalyze reactions of molecules in gas or solution. As solids – unless they are porous – are commonly impenetrable, catalytic reactions

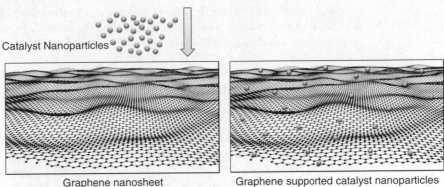

Catalyst Nanoparticles

Graphene nanosheet

Graphene supported catalyst nanoparticles

Figure 2.13

Catalysts are nanomaterials and catalysis is nanotechnology. Here we show an example of gold nanoparticles on a graphene support (artistic representation). Source: Taken from Ref. [23]. Reproduced with permission of Elsevier.

occur at the surface. To use often-expensive materials (e.g. platinum) in an economical way, catalysts are usually nanometer-sized particles, supported on an inert, porous structure (see Figure 2.13). Heterogeneous catalysts are the workhorses of the chemical and petrochemical industry.

As an introductory example we take one of the key reactions in cleaning automotive exhaust, the catalytic oxidation of CO on the surface of noble metals such as platinum, palladium, and rhodium. To describe the process, we will assume that the metal surface consists of active sites, denoted as "*." The catalytic reaction cycle begins with the adsorption of CO and O_2 on the surface of platinum, whereby the O_2 molecule dissociates into two O atoms (X* indicates that the atom or molecule is adsorbed on the surface, that is, bound to the site *):

$$O_2 + 2(*) \rightleftharpoons 2O^*$$
$$CO + (*) \rightleftharpoons CO^*$$

The adsorbed O atom and the adsorbed CO molecule then react on the surface to form CO_2, which, being very stable and relatively unreactive, interacts only weakly with the platinum surface and desorbs almost instantaneously:

$$CO + O^* \rightleftharpoons CO_2 + 2(*)$$

Note that in the latter step the adsorption sites on the catalyst are liberated, so these become available for further reaction cycles.

The activation energy of the gas phase reaction will be roughly equal to the energy required to split the strong O—O bond in O_2, that is, about $500 \, \text{kJ mol}^{-1}$. In the catalytic reaction, however, the O_2 molecule dissociates easily – in fact, without an activation energy – on the surface of the catalyst. The activation energy is associated with the reaction between adsorbed CO and O atoms, which is of the order of 50–$100 \, \text{kJ mol}^{-1}$. Desorption of the product molecule CO_2 costs only about 15–$30 \, \text{kJ mol}^{-1}$ (depending on the metal and its surface structure). Hence, if we compare the catalytic and the uncatalyzed reaction, we see that the most difficult step of the homogeneous gas phase reaction, namely the breaking of the O—O bond, is easily performed by the catalyst. Consequently, the ease with which the CO_2

molecule forms determines the rate at which the overall reaction from CO and O_2 to CO_2 proceeds. This is a very general situation for catalyzed reactions, hence the expression: *a catalyst breaks bonds, and allows other bonds to be formed.*

2.5.3   Toward Green and Sustainable Industrial Chemistry

2.5.3.1   *Concept of Green Chemistry*

The chemical industry of the twentieth century could not have developed to its present status on the basis of non-catalytic, stoichiometric reactions alone. Reactions can in general be controlled on the basis of temperature, concentration, pressure, and contact time. Raising the temperature and pressure will enable stoichiometric reactions to proceed at a higher rate of production, but the reactors in which such conditions can be safely maintained become progressively more expensive and difficult to make. In addition, there are thermodynamic limitations to the conditions under which products can be formed; for example, the conversion of N_2 and H_2 into ammonia is practically impossible above 600 °C. Nevertheless, higher temperatures are needed to break the very strong N≡N bond in N_2. Without catalysts, many reactions that are common in the chemical industry would not be possible, and many other processes would not be economical.

James Clark is a Professor at the University of York (UK) and is the founding Editorial Board Chair of the highly reputed journal Green Chemistry. *He was one of the pioneers that put forward the principles of Green Chemistry in 1999.*

Photograph: https://blogs.fco.gov.uk/wp-content/uploads/Pic_James-Clark.jpg

Catalysts accelerate reactions by orders of magnitude, enabling them to be carried out under the most favorable thermodynamic regime, and at much lower temperatures and pressures. In this way, efficient catalysts, in combination with optimized reactor and total plant design, are the key factor in reducing both the investment and operation costs of a chemical processes. But that is not all.

Technology is called "green" if it uses raw materials efficiently, so that the use of toxic and hazardous reagents and solvents can be avoided while formation of waste or undesirable byproducts is minimized. Catalytic routes often satisfy these criteria. A good example is provided by the selective oxidation of ethylene to ethylene epoxide, an important intermediate toward ethylene glycol (antifreeze), and various polyethers and polyurethanes.

The 12 principles of Green Chemistry were first published by Anastas and Warner in *Green Chemistry: Theory and Practice* (Oxford University Press, New York, 1998, p. 30). They are summarized as follows:

1. Prevent Waste
2. Atom Economy (every atom of the starting molecules should end up in the product)
3. Less Hazardous Synthesis
4. Design Benign Chemicals
5. Benign Solvents and Auxiliaries ("the best solvent is no solvent")
6. Design for Energy Efficiency
7. Use of Renewable Feedstocks
8. Reduce Derivatives ("selectivity")
9. Use Catalysts (instead of stoichiometric reactions)
10. Design for Degradation (biodegradable products)
11. Real-Time Analysis for Pollution Prevention
12. Inherently Benign Chemistry for Accident Prevention.

The ACS (American Chemical Society) Green Chemistry Institute[®] rephrased this as follows:

1. Green Chemistry uses renewable, biodegradable materials that do not persist in the environment.
2. Green Chemistry is using catalysis and biocatalysis to improve efficiency and conduct reactions at low or ambient temperatures.
3. Green Chemistry is a proven systems approach.
4. Green Chemistry reduces the use and generation of hazardous substances.
5. Green Chemistry offers a strategic pathway to build a sustainable future.

The old, non-catalytic route (called the epichlorohydrine process) follows a three-step synthesis:

$$Cl_2 + NaOH \rightarrow HOCl + NaCl$$

$$C_2H_4 + HOCl \rightarrow CH_2 - Cl - CH_2OH$$

$$CH_2Cl - CH_2OH + \frac{1}{2}Ca(OH)_2 \rightarrow \frac{1}{2}CaCl_2 + C_2H_4O + H_2O$$

or in total:

$$Cl_2 + NaOH + \frac{1}{2}Ca(OH)_2 + C_2H_4 \rightarrow C_2H_4O + \frac{1}{2}CaCl_2 + NaCl + H_2O$$

Hence, for every molecule of ethylene oxide, one molecule of salt is formed, creating a waste problem that was traditionally solved by dumping it in a river. Such practice is of course now totally unacceptable. The catalytic route, however, is simple and clean, although it does produce a small amount of CO_2. Using silver promoted by small amounts of chlorine as the catalyst, ethylene oxide is formed directly from C_2H_4 and O_2 at a selectivity of around 90%, with about 10% of the ethylene ending up as CO_2. Nowadays, all production facilities for ethylene oxide use catalysts.

2.5.3.2   *Atom Efficiency, E Factors and Environmental Friendliness*

Roger Sheldon received a Ph.D. in organic chemistry from the University of Leicester (UK) in 1967. This was followed by a postdoctoral fellowship with Jay Kochi at Case Western Reserve University and Indiana University (USA). From 1969 to 1980 he was with Shell Research Laboratories in Amsterdam and from 1980 to 1990 he was Vice President for R&D with DSM-Andeno.

In 1991 he was appointed Professor of Biocatalysis & Organic Chemistry at Delft University of Technology. He is a globally recognized expert on catalysis and Green Chemistry and the (co)author of several books on the subject of catalysis as well as more than 400 professional papers and around 50 granted patents.

His books Metal Catalyzed Oxidations *(together with Jay Kochi) and* Chirotechnology *have become desk references for professionals in the field. More recently, he has written, together with Isabel Arends and Ulf Hanefeld, the book* Green Chemistry & Catalysis. *He was the first Chairman of the Editorial Board of the highly successful RSC journal Green Chemistry. He is widely known for developing the concept of E factors for assessing the environmental footprint of chemical manufacturing processes.*

Photograph: https://royalsociety.org/people/roger-sheldon-12270

Numerous organic syntheses are based on stoichiometric oxidations of hydrocarbons with sodium dichromate and potassium permanganate, or on hydrogenations with alkali metals, borohydrides, or metallic zinc. In addition, there are reactions such as aromatic nitrations with H_2SO_4 and HNO_3, or acylations with $AlCl_3$ that generate significant amounts of inorganic salts as byproducts.

Fine chemicals are predominantly (but not exclusively!) the domain of homogeneous catalysis, where solvents present another issue of environmental concern. It is generally accepted that the best solvent is no solvent, but if a solvent is unavoidable, then water is a good candidate. Several indicators have been introduced to measure the efficiency and environmental impact of a reaction. The *atom efficiency* is the

Table 2.6 Environmental acceptability of products in different segments of the chemical industry.

Industry segment	Product tonnage	E factor (kg waste/kg product)
Oil refining	10^6–10^8	<0.1
Bulk chemicals	10^4–10^6	<1–5
Fine chemicals	10^2–10^4	5–50
Pharmaceuticals	10–10^3	25–100

molecular weight of the desired product divided by the total molecular weight of all products. For example, the conventional oxidation of a secondary alcohol

$$3\,C_6H_5 - CHOH - CH_3 + 2CrO_3 + 3H_2SO_4$$
$$\rightarrow 3\,C_6H_5 - CO - CH_3 + Cr_2(SO_4)_3 + 6H_2O$$

has an atom efficiency of 360/860 = 42%. By contrast, the catalytic route

$$C_6H_5 - CHOH - CH_3 + \frac{1}{2}O_2 \rightarrow C_6H_5 - CO - CH_3 + H_2O$$

offers an atom efficiency of 120/138 = 87%, with water as the only byproduct. The reverse step, a catalytic hydrogenation, proceeds with 100% atom efficiency:

$$C_6H_5 - CO - CH_3 + H_2 \rightarrow C_6H_5 - CHOH - CH_3$$

Another useful indicator of environmental acceptability is the *E factor* (Environmental Factor), defined as *the weight of waste or undesirable byproduct divided by the weight of the desired product*. As Table 2.6 shows, the production of fine chemicals and pharmaceuticals generate the highest amounts of waste per unit weight of product. Atom efficiencies and E factors can be calculated from each other, but in practice E factors can be higher due to yields being less than optimum and reagents that are used in excess. Also, loss of solvents should be included, and perhaps even the energy consumption with the associated generation of waste CO_2.

To express that it is not just the amount of waste but rather its environmental impact, Sheldon introduced the environmental quotient EQ as the E factor multiplied by an unfriendliness quotient, Q, which can be assigned a value to indicate how undesirable a byproduct is. For example, $Q = 0$ for clean water, one for a benign salt, NaCl, and 100–1000 for toxic compounds. Evidently, catalytic routes that avoid waste formation are highly desirable, and the more economic value that is placed on, for example, the unfriendliness quotient, the higher the motivation to work on catalytic alternatives. Waste prevention is far preferred over waste remediation.

2.5.3.3 *Definitions and Units*

- Activity
  A quantitative measure of how fast a catalyst works is its activity, which is usually defined as the reaction rate (or a reaction rate constant) for conversion of reactants into products. A high activity will result either in high productivity from relatively small reactors and catalyst volumes or in mild operating conditions, particularly temperature, that enhance selectivity and stability if the thermodynamics is more favorable.

- Selectivity
  Often products are formed in addition to those that are desired, and a catalyst has an activity for each particular reaction. A ratio of these catalytic activities is referred to as a selectivity, which is a measure of the catalyst's ability to direct conversion to the desired products. For a large-scale application, the selectivity of a catalyst may be even more important than its activity. High selectivity produces high yields of a desired product while suppressing undesirable competitive and consecutive reactions.
- Yield
  The yield of product A is the activity/conversion of the catalyst (conversion of the start product) multiplied by the selectivity toward product A.
  Suppose that reagent R is converted by a catalyst into A and B. The conversion is 60%, meaning that 60% of R is converted to either A or B. The selectivity of A is also 60%. This means that the yield of A $= 60\% \times 60\% = 36\%$.
- Turnover number
  The turnover number (TON) refers to the number of moles of substrate that a mole of catalyst can convert before becoming inactivated.
- Turnover frequency
  The turnover frequency (TOF) refers to the turnover per unit time. As the rate of a reaction (especially in a batch reactor, not so much in a plug flow reactor, see later) is not constant as a function of time (even for uncatalyzed reactions), the turnover frequency should be reported at maximum speed of the catalyst.
- Space velocity
  Space velocity refers to the quotient of the entering volumetric flow rate of the reactants divided by the reactor volume, which indicates how many reactor volumes of feed can be treated in a unit of time. For instance, a reactor with a space velocity of $5\,h^{-1}$ is able to process feed equivalent to five times the reactor volume each hour.
  Space velocity can be expressed mathematically as $SV = \frac{u_0}{V}$ with u_0 represents the volumetric flow rate of the reactants entering the reactor and V represents the volume of the reactor itself.
- Space time
  Space time is the reciprocal of the definition for the reactor space velocity: $\tau = \frac{1}{SV}$. Note that the space time is measured at the conditions of the reactor entrance while the space velocity is often measured at a set of standard conditions, so the reported space velocity may be different from the reciprocal of the measured space time.

2.5.4 Kinetics in a Heterogeneous Catalytic Reaction

(*Reworked, simplified and adapted from* [24]).

2.5.4.1 The Five Steps Determining the Kinetics of a Heterogeneous Reaction

A heterogeneous catalytic process consists typically of five steps, each of them might be the rate-determining step:

1. Transport of the substrate to the catalyst. Here, diffusion might play an important role.
2. Adsorption of the substrate at the surface of the catalyst; this is described by adsorption models.

3. Reaction of the adsorbed species at the surface of the catalyst.
4. Desorption of the reaction products off the surface, here catalyst poisoning should be considered.
5. Transport of the products out of the catalyst.

The steps 2–4 will be discussed in the kinetics in this chapter. The steps 1 and 5 will be discussed in the diffusion section of this chapter.

2.5.4.2 *Langmuir Model in Heterogeneous Catalysis*

In the previous section in this chapter, we derived the Langmuir equation (2.10), the Langmuir isotherm for the monolayer adsorption of one component on the surface of a material.

We will now expand this equation of *several* adsorbents:

In the case of several adsorbents, the Langmuir isotherm is easily expanded. The formula then is:

$$\theta_A = \frac{K_A P_A}{1 + \sum K_i P_i} \tag{2.64}$$

In heterogeneous catalysis, substrates often adsorb dissociatively (think of hydrogen sorption on Raney Nickel for dehydrogenation). In that case, with * representing an empty surface site or an adsorbed H-atom:

$$H_2 + 2(*) \rightleftharpoons 2H^*$$

$$k_{ads}.P_{H_2}.(\theta^*)^2 = k_{des}.(\theta_H)^2$$

$$\left(\frac{k_{ads}}{k_{des}} \right).P_{H_2}.(\theta^*)^2 = (\theta_H)^2; \text{ with } \left(\frac{k_{ads}}{k_{des}} \right) = K_H$$

$$\theta_H = \frac{\sqrt{K_H P_{H_2}}}{1 + \sqrt{K_H P_{H_2}}}$$

In general, the Langmuir isotherm for dissociative adsorption ($A_2 \rightarrow 2A$) can be written as:

$$\theta_A = \frac{\sqrt{K_{A2} P_{A2}}}{1 + \sqrt{K_{A2} P_{A2}}} \tag{2.65}$$

and

$$\theta^* = \frac{1}{1 + \sqrt{K_{A2} P_{A2}}} \tag{2.66}$$

We will now describe the kinetics of surface catalyzed reactions. There are several models, depending on the type of reaction. We will discuss the two most important ones for bimolecular reactions. We will start, however, with monomolecular reactions.

2.5.4.3 *Monomolecular Surface Reactions*

We assume that a reactant R adsorbs on the surface of the catalyst and is then reacted toward product P that immediately desorbs. If the product P does not desorb, the catalyst obviously loses its activity (the surface gets poisoned).

So schematically:

$$R + cat \rightleftharpoons R\,cat\ (\text{adsorption})$$
$$R\,cat \rightleftharpoons P + cat\ (\text{reaction and desorption})$$

The kinetics of this reaction are given by

$$v = k\,\theta_R = k\,\frac{K_R P_R}{1 + K_R P_R} \tag{2.67}$$

We have assumed in this model that the chemical reaction on the surface is the rate-limiting step, or (in the Henry regime):

$$v = k\,K_R P_R \tag{2.68}$$

$$\text{with } k = k_0 e^{\frac{-E_a}{RT}}\ (\text{Arrhenius}) \tag{2.69}$$

$$\text{and } K = K_0 e^{\frac{\Delta H_{ads}}{RT}}\ (\text{heat of adsorption}) \tag{2.70}$$

This means:

$$v = k' e^{-\frac{E_a - \Delta H_{ads}}{RT}}\,P_R \tag{2.71}$$

The temperature has an opposite impact on the adsorption coefficient and the rate. At high temperature, k will be increasing, while K will be decreasing and vice versa. This is illustrated in Figure 2.14.

Example 2.1: For the catalytic decomposition of PH_3 over a W-wire, one measures at low PH_3 pressure an activation energy of $100\,kJ\,mol^{-1}$, whereas at high PH_3 pressure an activation energy of $130\,kJ\,mol^{-1}$ is measured. Calculate the heat of adsorption.

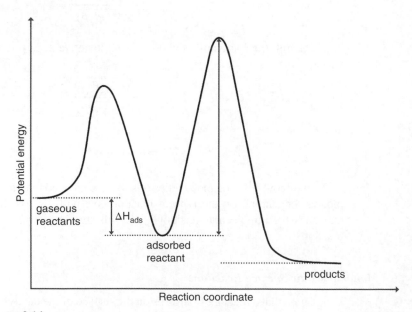

Figure 2.14
Energy profile of a heterogeneous catalysis.

Answer: This is a monomolecular problem, so

$$v = k\,\theta_R = k\,\frac{K_R P_R}{1 + K_R P_R}$$

At high pressure, this reduces to ($v = k$, zero order) and the heat of adsorption is irrelevant. At low pressure, the equation becomes ($v = kK_R P_R$, first-order). The heat of adsorption is the difference between both activation energies (see graph above), or $30\,\text{kJ mol}^{-1}$

An extra complication is expected when the products or the carrier gas (or solvent) adsorb competitively on the surface. In this case, the rate equation becomes (with K_s and P_s the adsorption constant and pressure of the solvents):

$$v = k\,\theta_R = k\,\frac{K_R P_R}{1 + K_R P_R + K_P P_P + K_s P_s} \tag{2.72}$$

2.5.4.4   *Bimolecular Reactions – Langmuir–Hinshelwood Kinetics*

For bimolecular reactions, the Langmuir–Hinshelwood model proposes that both molecules A and B are adsorbed on the surface and then react toward product P while on the surface.

$$A + cat \rightleftharpoons A - cat\,(\text{adsorption})$$

$$B + cat \rightleftharpoons B - cat\,(adsorption)$$

$$A - cat + B - cat \rightleftharpoons P - cat\,(reaction)$$

$$P - cat \rightleftharpoons P + cat\,(\text{desorption})$$

These kinetics are described as follows:

$$v = k\theta_A \theta_B = k\,\frac{(K_A P_A)(K_B P_B)}{\left(1 + K_A P_A + K_B P_B + \sum K_i P_i\right)^2} \tag{2.73}$$

The term $\sum K_i P_i$ represents adsorbed contaminations.

Various approximations can now be made, depending on the relative magnitudes of the two equilibrium constants. If one species, say A, is only weakly adsorbed, so that $K_A \ll K_B$, then for roughly equal pressures of the two species, $P_A = P_B$, the expression becomes

$$v = k\theta_A \theta_B = k\,\frac{(K_A P_A)(K_B P_B)}{(1 + K_B P_B)^2} \tag{2.74}$$

which reduces at high pressure to

$$v = k\,\frac{K_A P_A}{K_B P_B} \tag{2.75}$$

High pressures of B inhibit the reaction by saturating the surface.

At low pressures, when $K_B P_B \ll 1$, we obtain

$$v = kK_A K_B P_A P_B \tag{2.76}$$

and the rate is directly proportional to P_B. A maximum occurs in the rate at some point in between the high- and low-pressure cases as the pressure of B is increased. This optimum is typically plotted in a *volcano* plot and is based on the Sabatier effect.

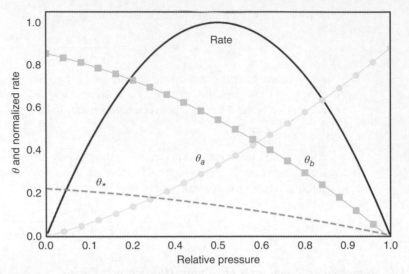

Figure 2.15
Pressure dependence of catalytic reaction.

The Sabatier effect is a qualitative concept in chemical catalysis named after the French chemist Paul Sabatier. It states that the interactions between the catalyst and the substrate should be "just right"; that is, neither too strong nor too weak. If the interaction is too weak, the substrate will fail to bind to the catalyst and no reaction will take place. On the other hand, if the interaction is too strong, the catalyst gets blocked by substrate or product that fails to dissociate (photograph: public domain).

This is further illustrated in following simulation, which shows that (obviously) the highest reaction rate A → B is obtained at intermediate coverages (Figure 2.15).

2.5.4.4.1   *Quasi-Equilibrium Approximation; Case: Heterogeneous CO Oxidation*

Let's introduce some more concepts of kinetics in practice, by following example. CO oxidation, an important step in automotive exhaust catalysis, is catalyzed by noble metals. The reaction mechanism is:

$$CO + (*) \xleftrightarrow{K_1} CO^* \qquad \text{(step 1)}$$

$$O_2 + 2(*) \xleftrightarrow{K_2} 2\,O^* \qquad \text{(step 2)}$$

$$CO^* + O^* \underset{k_3^-}{\overset{k_3^+}{\longleftarrow\!\!\!\longrightarrow}} CO_2^* + (*) \qquad \text{(step 3)}$$

$$CO_2^* \overset{K_4^{-1}}{\longleftrightarrow} CO_2 + (*) \qquad \text{(step 4)}$$

In this mechanism, step (3) is the rate-determining step. In *the quasi-equilibrium approximation* we assume that all other steps are so fast that they can be considered as equilibrated.

This means, that, for example,

$$v_1 \approx 0 \to k_1^+ P_A \, \theta^* = k_1^- \, \theta_A \to \theta_A = \frac{k_1^+}{k_1^-} P_A \to \theta_A = K_1 P_A \, \theta^*$$

Applying this to CO oxidation yields:

$$\theta_{CO} = K_1 P_{CO} \, \theta^* \qquad (2.77)$$

$$\theta_O = \sqrt{K_2 P_{O_2}} \, \theta^* \qquad (2.78)$$

$$\theta_{CO_2} = K_4^{-1} P_{CO_2} \, \theta^* \qquad (2.79)$$

and thus

$$\theta_{CO} + \theta_O + \theta_{CO_2} + \theta^* = 1 \to \theta^* = \frac{1}{1 + K_1 P_{CO} + \sqrt{K_2 P_{O_2}} + K_4^{-1} P_{CO_2}} \qquad (2.80)$$

The rate is that of the rate-determining step:

$$v = k_3^+ \theta_{CO} \theta_O - k_3^- \theta_{CO_2} \theta^* \qquad (2.81)$$

$$v = k_3^+ K_1 P_{CO} \theta^* \sqrt{K_2 P_{O_2}} \theta^* - k_3^- K_4^{-1} P_{CO_2} \theta^* \theta^*$$

$$v = k_3^+ K_1 \sqrt{K_2} P_{CO} \sqrt{P_{O_2}} \, \theta^{*2} \left(1 - \frac{k_3^- K_4^{-1} P_{CO_2} \theta^{*2}}{k_3^+ K_1 \sqrt{K_2} P_{CO} \sqrt{P_{O_2}} \, \theta^{*2}} \right)$$

$$v = k_3^+ K_1 \sqrt{K_2} P_{CO} \sqrt{P_{O_2}} \, \theta^{*2} \left(1 - \frac{K_3^{-1} K_4^{-1} P_{CO_2}}{K_1 \sqrt{K_2} P_{CO} \sqrt{P_{O_2}}} \right)$$

$$v = k_3^+ K_1 \sqrt{K_2} P_{CO} \sqrt{P_{O_2}} \, \theta^{*2} \left(1 - \frac{P_{CO_2}}{P_{CO} \sqrt{P_{O_2}} \, K_G} \right)$$

with

$$K_G = K_1 \sqrt{K_2} K_3 K_4$$

K_G is the equilibrium constant for the overall reaction: $CO + \tfrac{1}{2} O_2 \leftrightarrow CO_2$

CO_2 interacts so weakly with the surface that its presence at the surface may be neglected. All terms containing P_{CO2} become then zero.

$$v = k_3^+ K_1 \sqrt{K_2} P_{CO} \sqrt{P_{O_2}} \theta^{*^2} \tag{2.82}$$

At low temperatures, the surface is dominated by adsorbed CO. This is called MARI *(the Most Abundant Reaction Intermediate)*. In such a case, we may assume that

$$\theta_{MARI} = \theta_{CO} \quad and \quad \theta^* = 1 - \theta_{CO} = \frac{1}{1 + K_1 P_{CO}} \tag{2.83}$$

The rate then becomes:

$$v = k_3^+ K_1 \sqrt{K_2} P_{CO} \sqrt{P_{O_2}} \theta^{*^2}$$

$$= \frac{k_3^+ K_1 \sqrt{K_2} P_{CO} \sqrt{P_{O_2}}}{(1 + K_1 P_{CO})^2} \approx \frac{k_3^+ K_1 \sqrt{K_2} P_{CO} \sqrt{P_{O_2}}}{(K_1 P_{CO})^2} = \frac{k_3^+ \sqrt{K_2} \sqrt{P_{O_2}}}{K_1 P_{CO}} \tag{2.84}$$

The reaction orders are 0.5 for oxygen and -1 for CO in the low temperature region. The negative order of CO means that the surface is completely covered by CO. Any increase in CO pressure will slow down the reaction because CO will block the catalytic sites.

At high temperatures, desorption prevails, implying that the coverages of all species are small and the surface is almost empty. This is the *Nearly Empty Surface Approximation*. In this approximation, $\theta^* \approx 1$, and the rate equation is now simplified to:

$$v = k_3^+ K_1 \sqrt{K_2} P_{CO} \sqrt{P_{O_2}} \theta^{*^2} \approx k_3^+ K_1 \sqrt{K_2} P_{CO} \sqrt{P_{O_2}} \tag{2.85}$$

Now the reaction order is 0.5 in oxygen and $+1$ in CO! As the surface is now nearly empty, increasing the CO pressure will enhance the reaction rate. We see in this example that the reaction rate is not only strongly dependent on the partial pressures of the reactants, but also on the temperature. This is further illustrated in Figure 2.16.

2.5.4.5   *The Eley–Rideal Mechanism*

The Eley–Rideal mechanism assumes a different reaction path to the Langmuir–Hinschelwood approximation: in this mechanism only one molecule adsorbs on the catalyst surface, the other component reacts from the gas phase with the adsorbed molecule:

$$A + cat \rightleftharpoons A - cat$$
$$A - cat + B \rightleftharpoons AB + cat$$

This results in a simpler rate equation:

$$v = k\,\theta_A P_B = k\,P_B \frac{K_A P_A}{1 + K_A P_A} = k' \frac{P_A P_B}{1 + K_A P_A} \tag{2.86}$$

The Eley–Rideal mechanism does not lead to a maximum in the rate as P_A and/or P_B are changed, a property that can be used to distinguish this mechanism from the Langmuir–Hinshelwood mechanism (Figure 2.17). Surface combinations of atoms and free radicals generally occur by Eley–Rideal mechanisms.

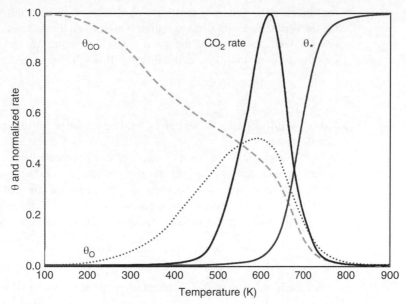

Figure 2.16
Coverages and reaction rate for the CO oxidation.

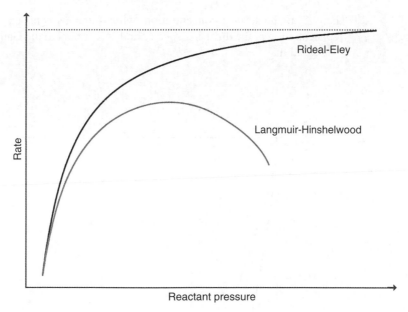

Figure 2.17
Rates of Langmuir–Hinshelwood and Rideal–Eley mechanisms (in the literature, there is no fixed position for the two names, so both Rideal–Eley and Eley–Rideal are used).

2.5.5 Diffusion Phenomena

2.5.5.1 *Types of Diffusion*

Diffusion phenomena can be subdivided into three types. For large pores, typically those of 100 nm are considered larger, the interactions between the diffusing molecules is the only rate-limiting step that needs to be taken into account. This is the *molecular diffusion* that is accurately described by the laws of Fick. The *first law of Fick* describes the so-called steady state diffusion. The flux does not change as function of time. In this case, the flux N_R (flux of reagent) is defined as:

$$N_R = -D_F \frac{dC_R}{dx}$$

(2.87)

D is the diffusion coefficient (m^2 s^{-1}). The subscript "F" refers to Fickian or molecular diffusion.

Fick's second law describes the more often encountered situation in which the diffusion is "non-steady-state," as the flux and the concentration gradient varies as a function of time. Typical concentration profiles for this time of diffusion are shown in Figure 2.18. The second law predicts how diffusion causes the concentration to change with time according to the following differential equation.

$$\frac{\partial C}{\partial t} = D \frac{\partial^2 C}{\partial x^2}$$

(2.88)

A general solution of this differential equation is given by (see also Figure 2.18):

$$\frac{C_x - C_0}{C_s - C_0} = 1 - erf\left(\frac{x}{2\sqrt{Dt}}\right)$$

(2.89)

here C_s is the surface concentration, which remains constant, C_0 is the final bulk concentration of the diffusing species and "erf" refers to the Gaussian error function. We recall that

$$erf(x) = \frac{2}{\sqrt{\pi}} \int_0^x e^{-t^2} dt$$

(2.90)

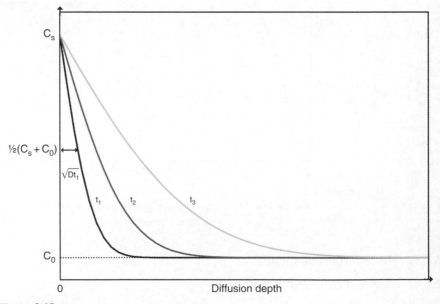

Figure 2.18
Concentration profiles in the second law of Fick.

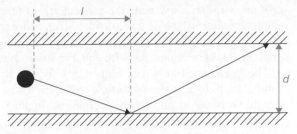

Figure 2.19
Knudsen diffusion.

$2\sqrt{Dt}$ is referred to as the "diffusion length" and provides a measure of how far the concentration has propagated in the x-direction by diffusion in time t. Figure 2.18 shows the resulting concentration profiles for three times ($t_1 < t_2 < t_3$); C_s the start concentration at the surface; x is the diffusion depth.

When the pores become smaller, typically between 1 and 100 nm, the interaction of the molecules with the pore walls becomes important. This type of diffusion is referred to as *Knudsen diffusion*.

Finally, when the pores are so small that the molecules can barely pass through them, the effective diffusion coefficient is strongly dependent on the aperture size. The term *configurational diffusion* is applied. In this case, when the pore size is about the same as the critical dimension of the molecule being transported, a decrease of only 0.1 nm in pore size can reduce the transport rate by orders of magnitude.

Figure 2.19 shows a typical case of Knudsen diffusion. Consider the diffusion of gas molecules through very small capillary pores. If the pore diameter (d) is smaller than the mean free path (l) of the diffusing gas molecules and the density of the gas is low, the gas molecules collide with the pore walls more frequently than with each other.

The Knudsen diffusion coefficient is therefore strongly dependent on the pore diameter d; and is given by the general formula (derived from kinetic gas theory):

$$D_K = \frac{d_{pore}}{3}\sqrt{\frac{8k_B N_A T}{\pi M}} \qquad (2.91)$$

Whereas Fickian diffusion coefficients have typical orders of magnitude of 1 cm^{2}s^{-1}, the Knudsen coefficients are typically around 10^{-2} to 10^{-5} cm^{2}s^{-1}, showing the significant transport limitations in these cases.

2.5.5.2 Thiele Modulus

Let's consider a single spherical particle of a catalyst having uniformly distributed catalytic groups. The molecules of a reactant molecule diffuse down a concentration gradient; the concentration of the reactant at the periphery of the particle is higher than the concentration at the center, not only because of diffusion phenomena, but also because the reactant will react at some point inside the pore toward the product and will be consumed.

So, the diffusion reactant molecule at any point in the catalytic pore has two options: either it undergoes reaction yielding the product, or it continues to diffuse further inside the catalyst pore toward the particle center. If the catalyst is very active, or if the pores are very small, the concentration of the reactant near the center of the catalytic center will be near zero with a steep concentration gradient. If the pores are much larger than the reactant and the catalytic activity is low, then the concentration gradient might be almost flat with equal reaction concentration all over the particle.

The *Thiele modulus* was developed by E.W. Thiele in his paper "Relation between catalytic activity and size of particle" in 1939. It was developed to describe the relationship between diffusion and reaction rate in porous catalyst pellets with no mass transfer limitations. This value is generally used in determining the effectiveness factor for catalyst pellets.

In its most general form, the Thiele modulus is defined as:

$$\phi = L\sqrt{\frac{kC_{bulk}^{n-1}}{D_{eff}}} \qquad (2.92)$$

n is the reaction order and L is the characteristic length of the catalyst model under consideration. For example, a spherical particle has its radius as a characteristic length, a cylindrical pore has its length. For other geometries a good estimation of the characteristic length is the volume divided by the surface area

For the diffusion coefficient we use the *effective* diffusion coefficient; for straight uniform pores, the value of D_{eff} can be estimated by following simple formula:

$$\frac{1}{D_{eff}} = \frac{1}{D_{Fick}} + \frac{1}{D_K} \qquad (2.93)$$

The methods to estimate the values for D_{Fick} and D_K in porous catalysts are empirical, and will not be discussed in detail here. They involve the parameters ε (void fraction of the particle) and τ (turtuosity, accounting for nonuniformity of pores).

$$\tau = \frac{\text{distance molecule travels between two point}}{\text{shortest distance between these two points}} \qquad (2.94)$$

$$\varepsilon = \frac{\text{void volume}}{\text{total volume}} \qquad (2.95)$$

In these cases

$$D_{K,eff} \text{ or } D_{Fick,eff} = \frac{D_{K \, or \, Fick}\cdot\varepsilon}{\tau} \qquad (2.96)$$

2.5.5.2.1 *Deriving the Thiele Modulus (First-Order Kinetics)*

The Thiele modulus can be derived by choosing a good model for a heterogeneous catalyst. Let us consider one pore of a porous catalyst and approximating it by a straight channel with length $2L$ between two infinite plates. At the entrance of the pores the concentration equals the bulk concentration. A reaction takes place inside the pores (on the catalytic centers) lowering the concentration inside the pores, reaching a minimum in the middle between the "entrance" and "exit" of a pore (Figures 2.20 and 2.21).

Now consider and infinitesimal element of the catalyst pore and write the mass balance for the reactant on that element (Figure 2.22).

Writing the mass balance (0 = in − reaction − out) gives:

$$0 = N_R - v \cdot dz - (N_R + dN_R) \text{ (flow in − reagent consumed − flow out)} \qquad (2.97)$$

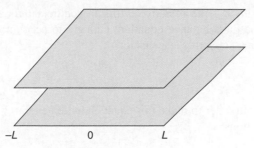

Figure 2.20
Infinite plates with length of 2 L.

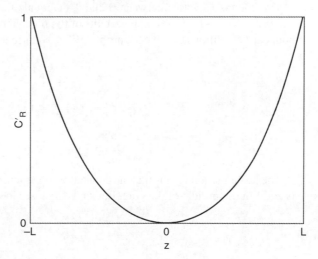

Figure 2.21
Concentration profile.

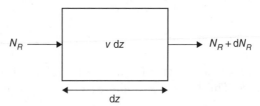

Figure 2.22
Infinitesimal element of the catalyst pore.

with N_R being the diffusion of the reactant R through the pore (in mol.m^{-2}.s^{-1}) and v the reaction rate per volume unit (assuming first-order, mol.m^{-3} s^{-1}). Writing this balance as a differential equation gives:

$$\frac{dN_R}{dz} = -v \tag{2.98}$$

The diffusion of the reactants through the pores follows Fick's first law:

$$N_R = -D_{eff}\frac{dC_R}{dz} \tag{2.99}$$

with D_{eff} the effective diffusion coefficient for the reactant. If we include this in the mass balance equation, taking into account the reaction is assumed first-order ($v = kC_R$), this leads us to:

$$D_{eff}\frac{d^2C_R}{dz^2} = v = kC_R \tag{2.100}$$

In order to make this equation more general, we will implement two more substitutions:

$$C'_R = \frac{C_R}{C_{R,bulk}} \text{ and } z' = \frac{z}{L} \tag{2.101}$$

This means C'_R varies between 0 and 1 (1 equals the bulk concentration) and z' varies between −1 and 1, being the exits of the pore. This also allows us to formulate boundary conditions to solve the problem. The equation and boundary conditions are:

$$\frac{d^2C'_R}{dz'^2} = \frac{L^2k}{D_{eff}}C'_R \tag{2.102}$$

$$C'_R = 1 \text{ at } z' = 1 \tag{2.103}$$

$$\frac{dC'_R}{dz'} = 0 \text{ at } z' = 0 \tag{2.104}$$

The last equation means that the concentration reaches a minimum (first derivative is zero) in the middle of the pore. Let's take a closer look at the prefactor:

$$\frac{L^2k}{D_{eff}}$$

This increases with increasing rate coefficient (faster reaction rate) and decreases with increasing diffusion coefficient (faster diffusion). We define the Thiele modulus as the square root of this prefactor:

$$\phi = \sqrt{\frac{L^2k}{D_{eff}}} = L\sqrt{\frac{k}{D_{eff}}} \tag{2.105}$$

This means that the Thiele modulus is a ratio between reaction speed and diffusion speed. This leads to the following differential equation:

$$\frac{d^2C'_R}{dz'^2} = \phi^2 C'_R \tag{2.106}$$

Solving the differential equation using the technique of Laplace transformations gives the following result:

$$C'_R = \frac{\cosh \phi z'}{\cosh \phi} \tag{2.107}$$

The concentration profile for different values of the Thiele modulus is plotted in Figure 2.23. At low Thiele moduli (slow reaction rates and fast diffusion) the concentration profile is almost straight, meaning there is only little conversion. At high Thiele moduli (fast reaction, slow diffusion) the reactants have completely disappeared at the beginning of the pore, this means any active sites further in the particle are not used. From a practical point of view both extreme cases are not wanted, the

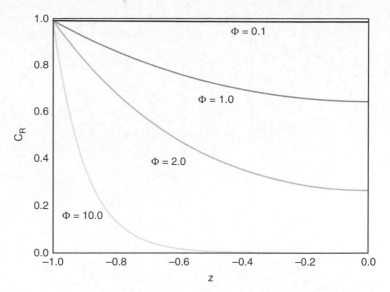

Figure 2.23
Concentration profile for different values of the Thiele modulus.

first one would mean there is almost no activity of the catalyst, in the second case a part of the (expensive) catalyst is never used.

2.5.5.3 Effectiveness Factor

Now, we can calculate a dimensionless *effectiveness factor*, η. This is the overall rate, divided by the maximum rate that would prevail if there would be no diffusion limitations at all. For an *infinite cylindrical pore*, the effectiveness factor is derived as:

$$\eta = \frac{v_{obs}}{v} = \frac{\tanh \phi}{\phi} \tag{2.108}$$

One can show that this equation transforms for *spherical particles* to:

$$\eta = \frac{v_{obs}}{v} = \frac{3}{\phi^2}[\phi \coth\phi - 1] \tag{2.109}$$

The dependence of the effectiveness factor on the Thiele modulus for a first-order reaction and spherical particles is shown in Figure 2.24. (The curves for other reaction orders and geometries are very similar.)

The shape of this curve is very important. There are two extreme cases.

- When the Thiele modulus is very small, typically below 1, then the effectiveness factor is close to 1; there are no diffusional limitations. The occurs when
  - The particle radius R (or length of pore L) approaches 0
  - The diffusivity D becomes very large
  - The catalytic activity, expressed in rate constant k becomes very small.
- When the Thiele modulus is very large, then

$$\eta \sim \frac{1}{\phi} \tag{2.110}$$

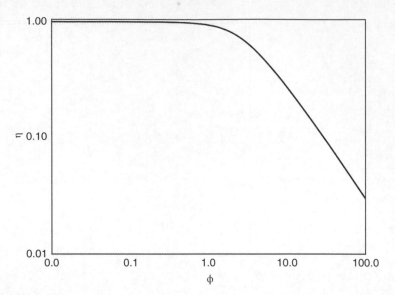

Figure 2.24
Effectiveness in terms of Thiele.

Diffusion limitations have an important influence on the observed rate. If we define the rate of the first-order reaction as

$$v = \eta k C_s \tag{2.111}$$

Then, for spherical particles at high diffusion limitations, where

$$\eta = \frac{3}{\phi} \tag{2.112}$$

$$v = \frac{3}{\phi} k\, C_s = \frac{3kC_s}{R} \sqrt{\frac{D_{eff}}{k}} = \frac{3C_s}{R} \sqrt{kD_{eff}} \tag{2.113}$$

This is an important result. It shows that the diffusion disguises the kinetics. In this case, the activation energy measured for the reaction is one-half that for the intrinsic reaction and one-half that for diffusion.

It is crucial in catalytic kinetics to resolve both the chemical and the physical processes. This is best done by a systematic variation of the Thiele modulus, which is most easily done by the systematic variation of the catalytic particle size R. When there is no effect of the particle size on the reaction rate, then the effectiveness factor is one and the measured kinetics are the intrinsic kinetics.

2.5.5.4   *Weisz–Prater Criterion*

The Weisz–Prater criterion uses measured values of the rate of reaction, k_{obs} to determine if internal diffusion is limiting the reaction. If C_{WP} is significantly larger than 1, there is serious diffusion limitation.

$$C_{WP} = \eta \phi^2 = 3(\phi \coth \phi - 1) \tag{2.114}$$

with

$$\eta\phi^2 = \frac{k_{obs}}{k}\left(L^2\frac{k}{D_{eff}}\right) = \frac{k_{obs}L^2}{D_{eff}} \text{ with } L = R_{pellet} \qquad (2.114)$$

This is an interesting formula since the intrinsic rate coefficient k (which is not directly measurable) is canceled out.

Example 2.2: A first-order reaction A → B was carried out over two different-sized pellets. The results of two experimental runs made under identical conditions are given as follows. Estimate the Thiele modulus and effectiveness factor for each pellet.

How small should the pellets be made to virtually eliminate all internal diffusion resistance?

Reaction	k_{obs} (10^{-5} mol g_{cat}^{-1} s^{-1})	R_p (m)
1	3.0	0.01
2	15.0	0.001

Solution: Make the ratio of the two Weisz–Prater criteria. The value D_{eff} cancels out (the same for both experiments):

$$\frac{k_{obs2}\,R_{p2}^2}{k_{obs1}\,R_{p1}^2} = \frac{\phi_2\,\coth\phi_2 - 1}{\phi_1\,\coth\phi_1 - 1}$$

Moreover, from the definition of the Thiele modulus, realizing that k *and* D_{eff} are the same for both experiments:

$$\frac{\phi_2}{\phi_1} = \frac{R_{p2}}{R_{p1}} \xrightarrow{yields} \phi_1 = 10\phi_2$$

Filling in the values:

$$\frac{15.10^{-5}.(0.001)2}{3.10^{-5}\,(0.01)^2} = \frac{\phi_2\,\coth\phi_2 - 1}{10\phi_2\,\coth(10\phi_2) - 1}$$

Solving this equation yields that $\phi_2 = 1.65$ and thus $\phi_1 = 16.5$. The effectiveness factors can now be calculated as:

$$\eta_2 = \frac{3}{(1.65)^2}[1.65\coth(1.65) - 1] = 0.85$$

$$\eta_1 = 0.18$$

Let us finally calculate the particle radius required to have an effectiveness factor of 0.95, at which diffusion limitations are negligible:

$$0.95 = \frac{3}{\phi^2}[\phi\coth\phi - 1] \xrightarrow{yields} \phi = 0.90$$

and

$$\frac{\phi_{perf}}{\phi_1} = \frac{R_{perf}}{R_{p1}} \xrightarrow{yields} \frac{0.90}{16.5} = \frac{R_{perf}}{0.01} \xrightarrow{yields} R_{perf} = 5.5\,10^{-4}m$$

2.A   Appendix

Solution of the differential equation for the flux in a catalytic particle via *the Laplace transformation technique.*

The Laplace transformation is a typical technique to solve differential equations, similar to the Fourier Transform technique. The Laplace transform is defined as:

$$\mathcal{L}\{f(t)\} = F(s) = \int_0^\infty e^{-st} f(t) dt \qquad (2.115)$$

Let's take, as an example, the Laplace transform of a function $sinh(at)$:

$$L\{\sinh at\} = \int_0^\infty e^{-st} \sinh(at) dt = \int_0^\infty e^{-st} \frac{e^{at} - e^{-at}}{2} dt$$

$$= \frac{1}{2} \int_0^\infty (e^{(a-s)t} - e^{(-a-s)t}) \, dt = \frac{1}{2} \left[\frac{1}{a-s} e^{at} e^{-st} + \frac{1}{a+s} e^{-at} e^{-st} \right]_0^\infty$$

$$= \frac{1}{2} \left[0 + 0 - \frac{1}{a-s} - \frac{1}{a+s} \right] = \frac{a}{s^2 - a^2}$$

In this way, the most common functions are tabulated (see Table 2.7).

Solving the differential equation with Laplace transformation includes transforming the equation to the Laplace domain, solving this equation, and then inverse transforming it back to the time domain.

Let us now solve our differential equation (2.106). We remove the primes for simplicity and to avoid confusion, the prime in the table and in following means: first derivative.

$$\frac{d^2 C_R}{dz^2} = \phi^2 C_R \qquad (2.116)$$

Using formula 36 from Table 2.7:

$$\mathcal{L}\left\{ \frac{d^2 C_R}{dz'^2} \right\} = s^2 F(s) - s\, C_R(0) - C_R'(0) = s^2 F(s) - s.C_R(0) - 0$$

The latter term $C'_R(0) = 0$, because this was our second boundary condition!

$$\mathcal{L}\{\phi^2 . C_R'\} = \phi^2 . F(s)$$

This gives:

$$s^2 F(s) - s\, C_R(0) - \phi^2 . F(s) = 0$$

$$F(s) = \frac{s}{s^2 - \phi^2}\, C_R(0)$$

Transform back (inverse Laplace, using formula 18 from Table 2.7):

$$C_R(z) = C_R(0) \cdot \cosh(\phi z)$$

We can now find the unknown $C'_R(0)$ from the boundary condition that we set: $C'_R(1) = 1$. (concentration = bulk concentration at the surface of the pellet).

$$C_R(1) = 1 = C_R(0) \cdot \cosh(\phi \cdot 1)$$

Table 2.7 List of Laplace transforms.

$f(t) = \mathcal{L}^{-1}\{F(s)\}$	$F(s) = \mathcal{L}\{f(t)\}$	$f(t) = \mathcal{L}^{-1}\{F(s)\}$	$F(s) = \mathcal{L}\{f(t)\}$
1	$\dfrac{1}{s}$	e^{at}	$\dfrac{1}{s-a}$
$t^n, n = 1, 2, 3, \ldots$	$\dfrac{n!}{s^{n+1}}$	$t^p, p > -1$	$\dfrac{\Gamma(p+1)}{s^{p+1}}$
$\sqrt{t}$	$\dfrac{\sqrt{\pi}}{2s^{\frac{3}{2}}}$	$t^{n-\frac{1}{2}}, n = 1, 2, 3, \ldots$	$\dfrac{1 \cdot 3 \cdot 5 \ldots (2n-1)\sqrt{\pi}}{2^n s^{n+\frac{1}{2}}}$
$\sin(at)$	$\dfrac{a}{s^2 + a^2}$	$\cos(at)$	$\dfrac{s}{s^2 + a^2}$
$t\sin(at)$	$\dfrac{2as}{(s^2 + a^2)^2}$	$t\cos(at)$	$\dfrac{s^2 - a^2}{(s^2 + a^2)^2}$
$\sin(at) - at\cos(at)$	$\dfrac{2a^3}{(s^2 + a^2)^2}$	$\sin(at) + at\cos(at)$	$\dfrac{2as^2}{(s^2 + a^2)^2}$
$\cos(at) - at\sin(at)$	$\dfrac{s(s^2 - a^2)}{(s^2 + a^2)^2}$	$\cos(at) + at\sin(at)$	$\dfrac{s(s^2 + 3a^2)}{(s^2 + a^2)^2}$
$\sin(at + b)$	$\dfrac{s\sin(b) + a\cos(b)}{s^2 + a^2}$	$\cos(at + b)$	$\dfrac{s\cos(b) - a\sin(b)}{s^2 + a^2}$
$\sinh(at)$	$\dfrac{a}{s^2 - a^2}$	$\cosh(at)$	$\dfrac{s}{s^2 - a^2}$
$e^{at}\sin(bt)$	$\dfrac{b}{(s-a)^2 + b^2}$	$e^{at}\cos(bt)$	$\dfrac{s-a}{(s-a)^2 + b^2}$
$e^{at}\sinh(bt)$	$\dfrac{b}{(s-a)^2 - b^2}$	$e^{at}\cosh(bt)$	$\dfrac{s-a}{(s-a)^2 - b^2}$
$t^n e^{at}, n = 1, 2, 3, \ldots$	$\dfrac{n!}{(s-a)^{n+1}}$	$f(ct)$	$\dfrac{1}{c}F\left(\dfrac{s}{c}\right)$
$u_c(t) = u(t - c)$	$\dfrac{e^{-cs}}{s}$	$\delta(t - c)$	e^{-cs}
$u_c f(t - c)$	$e^{-cs}F(s)$	$u_c(t)g(t)$	$e^{-cs}\mathcal{L}\{g(t + c)\}$
$e^{ct}f(t)$	$F(s - c)$	$t^n f(t), n = 1, 2, 3, \ldots$	$(-1)^n F^{(n)}(s)$
$\dfrac{1}{t}f(t)$	$\displaystyle\int_s^\infty F(u)\,du$	$\displaystyle\int_0^t f(v)\,dv$	$\dfrac{F(s)}{s}$
$\displaystyle\int_0^t f(t - \tau)g(\tau)\,d\tau$	$F(s)G(s)$	$f(t + T) = f(t)$	$\dfrac{\int_0^T e^{-st}f(t)\,dt}{1 - e^{-sT}}$
$f'(t)$	$sF(s) - f(0)$	$f''(t)$	$s^2 F(s) - sf(0) - f'(0)$
$f^{(n)}(t)$	$s^n F(s) - s^{n-1}f(0)$ $\quad - s^{n-2}f'(0) \ldots$ $\quad - sf^{(n-2)}(0)$ $\quad - f^{(n-1)}(0)$		

$$C_R(0) = \frac{1}{\cosh(\phi)}$$

And the final solution then becomes:

$$C_R(z) = C_R(0) \cdot \cosh(\phi z) = \frac{\cosh(\phi z)}{\cosh(\phi)} \qquad (2.117)$$

Exercises

1 Consider the nitrogen adsorption isotherm at 77 K of two porous materials. The first
   two columns are isotherm A and the last two columns are isotherm B. Calculate,
   based on these data, for both isotherms:
   a. The BET surface area;
   b. Pore size distribution (PSD) based on the adsorption data;
   c. PSD based on the desorption data;
   d. Discuss the difference between the PSD derived from the adsorption and des-
      orption data;
   e. Discuss the hysteresis;
   f. Calculate the total pore volume;
   g. Calculate the Wheeler pore radius and explain how you derived this formula;
   h. Determine micropore volume using the *t*-plot method.

P/P_0	Volume (cm³ g⁻¹)	P/P_0	Volume (cm³ g⁻¹)
0.010742	232.5873	0.010283	197.2427
0.031572	276.5159	0.020022	225.4825
0.051733	299.6349	0.03027	245.4357
0.085447	326.2857	0.039687	260.3304
0.1027	337.4207	0.050078	274.1725
0.12631	351.4365	0.060557	286.5175
0.1515	364.754	0.069071	295.5702
0.17691	377.0952	0.080985	307.2953
0.20212	388.5476	0.089542	315.4357
0.25584	410.7619	0.10154	325.7339
0.29887	427.619	0.10817	331.7134
0.35562	449.4921	0.12407	345.2281
0.3783	458.1429	0.13628	355.3275
0.40231	468.0556	0.15002	367.1959
0.42723	478.3333	0.1611	376.7865
0.45221	488.5079	0.17067	386.9971
0.47708	498.9048	0.18008	397.4064
0.50192	509.5159	0.189	408.9035
0.52655	520.6031	0.20014	425.7339
0.55109	532.1032	0.22324	493.3977
0.57585	544.4207	0.24715	582.2397
0.60015	557.3174	0.27582	631.9444
0.62429	571.254	0.30186	641.6812
0.6478	587.3096	0.32435	646.3743

(continued overleaf)

P/P_0	Volume (cm^3 g^{-1})	P/P_0	Volume (cm^3 g^{-1})
0.67791	609.4921	0.35026	650.614
0.70222	632.1746	0.37643	654.1842
0.72236	792.7698	0.40195	657.3157
0.75178	1013.349	0.42727	660.1169
0.77387	1027.508	0.45311	662.6345
0.80208	1037.452	0.47825	664.9269
0.85087	1048.54	0.50344	667.0818
0.90113	1059.968	0.54901	671.8362
0.95002	1072.746	0.59994	675.924
0.98039	1091.151	0.65068	679.7661
0.99371	1140.143	0.70115	683.3538
0.98206	1099.849	0.75157	686.883
0.94423	1073.706	0.8007	690.4796
0.89776	1061.571	0.85129	694.3508
0.86034	1053.825	0.89886	698.8508
0.82952	1047.357	0.95091	706.7485
0.79521	1039.762	0.98061	725.5789
0.76088	1031.325	0.99405	884.8216
0.73065	1022.651	0.9829	800.7075
0.6983	1009.873	0.94898	713.231
0.66585	959.4841	0.89582	701.8918
0.63601	605.9762	0.84872	696.6432
0.59491	557.8174	0.79547	693.6024
0.56521	542.127	0.74665	690.2514
0.53478	526.9921	0.695	686.7953
0.50024	510.8889	0.64584	683.307
0.46533	494.7064	0.59679	679.5643
0.43379	480.7064	0.56393	677.0468
0.3988	466.1508	0.52534	673.9005
0.34324	443.9286		
0.29941	427.2857		
0.2429	404.9365		
0.20013	386.6984		
0.14366	359.3651		

2 The following reaction mechanism for a global A → B reaction takes place on a solid catalyst:

$$A + (*) \rightleftharpoons A^*$$
$$A^* \rightleftharpoons B^*$$
$$B^* \rightleftharpoons B + (*)$$

   a. Determine the rate law of the reaction, using the pseudo-stationary-state approximation and assuming the product desorption is rate-limiting. All components are in the gas phase ($K = K_1 \cdot K_2 / K_3$).

   b. Write the rate law in the proper form, as to denote the kinetic factor, thermodynamic driving force and the adsorption term.

   c. What information does the kinetic factor hold?

3 A Pd@C catalyst is used to mediate a first-order heterogeneous irreversible reaction. The catalyst consists of microporous spherical particles. The effectiveness factor η was found to be only 14.3%. $D_{eff} = 2.10^{-6}$ m.s^{-1} and $k = 12.105$ s^{-1}.

  a. What is the Thiele modulus for this catalyst?

  b. What is the particle radius?

  c. State two possible reasons why the effectiveness factor is so low. Plot (sketch) the current concentration profile throughout a pore to indicate this problem.

  d. If the particle size may not be altered (reactor design), how could you still improve the effectiveness factor when looking at the catalyst design?

4 The oxidation of naphthalene occurs through the following mechanism:

Adsorption of naphthalene (A).

Adsorption of oxygen (B).

Surface reaction of A and B into the product AB.

Desorption of AB from the catalyst.

  The catalyst is a cobalt Schiff base complex, supported on silica.

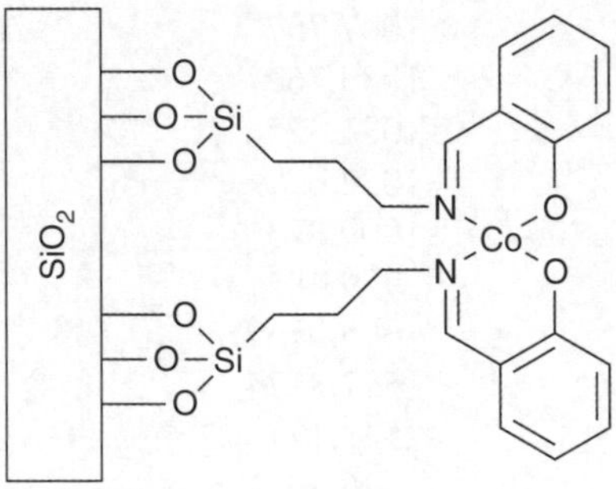

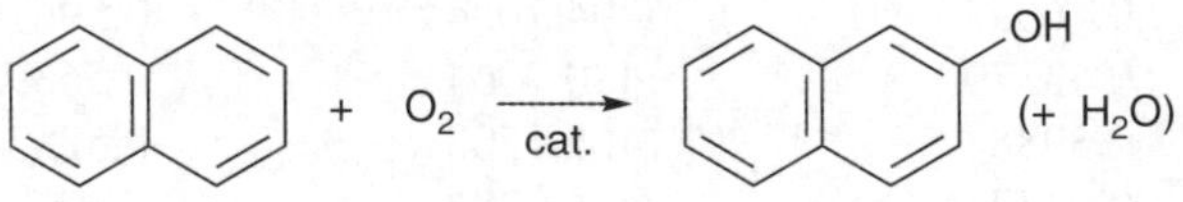

  a. The results of CHNO analysis for this catalyst are provided. Estimate the cobalt-loading (mmol g^{-1}) based on these results. Assume a 1 : 1 metal : complex ratio (meaning that each complex contains a cobalt center). Assume a perfectly pure silica support (no impurities).

  (H = 5.2 w%, C = 2.4 w%, N = 0.28 w%, O = 49 w%)

  b. Derive the rate law by using pseudo-steady-state approximation (PSSA) if the surface reaction of A and B into AB is the rate-limiting step (RLS/SBS). Assume A, B, and AB as gas phase components. Write the rate law properly, so as to denote the kinetic factor, the thermodynamic driving force, and the number of active sites required in the RLS.

  c. The reaction is performed in 1200 ml of solvent. The initial naphthalene concentration is 2 mol l^{-1}. 1 g of this catalyst is added. The oxygen is bubbled through the reaction system. The conversion of naphthalene in time is plotted in the following graph. Calculate the turnover number (TON) and turnover frequency (TOF) for this reaction. Keep in mind that the active sites are the cobalt sites of the catalyst.

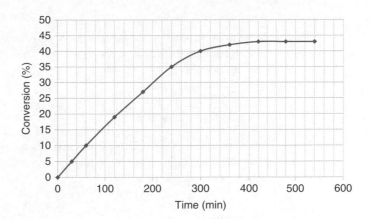

Answers to the Problems

1 a. $1358\,\text{m}^2\,\text{g}^{-1}$ // $1542\,\text{m}^2\,\text{g}^{-1}$

f. $1.76\,\text{ml}\,\text{g}^{-1}$ // $1.37\,\text{ml}\,\text{g}^{-1}$

g. $2.28\,\text{nm}$ // $1.77\,\text{nm}$

2 a. Product desorption is rate limiting: $r_3 = k_{+3}\theta_B - k_{-3}P_B\theta^*$ Adsorption of A and surface reaction are in pseudo-equilibrium. Solve for θ_B and θ^*

$$r_1 : \frac{\theta_A}{P_A\theta^*}\quad or\quad \theta^* = K_1 P_A \theta^*$$

$$r_2 : \frac{\theta_B}{\theta_A}\quad or\; \theta_B = K_2\theta_A = K_1 K_2 P_A \theta^*$$

$$1 = \theta^* + \theta_A + \theta_B\; or\; 1 = \theta^* + K_1 P_A \theta^* + K_2 K_1 P_A \theta^*$$

$$\theta^* = \frac{1}{1 + K_1 P_A + K_2 K_1 P_A}$$

$$r = \frac{k_{+3} K_1 K_2 P_A - k_{-3}P_B}{1/\theta^*}$$

a. Write the rate law in the proper form, as to denote the kinetic factor, thermodynamic driving force and the adsorption term.

$$\frac{k_{+3} K_1 K_2 \left(P_A - \dfrac{P_B}{K}\right)}{1/\theta^*}$$

a. The term before the brackets in (b) is the kinetic factor. It contains the rate constant of the rate-determining step and the equilibrium constants of the kinetically significant steps.

3 a. We are dealing with spherical particles, so the formula:

$$\eta = \frac{v_{obs}}{v} = \frac{3}{\phi^2}(\phi\coth\phi - 1)$$

can be used. Instead of solving this for ϕ, you can assume the modulus is very high, since the effectiveness is so low. The effectiveness becomes: $\eta = \frac{3}{\phi}$ or $\phi = 21$.

b.
$$\phi = \sqrt{\frac{L^2 k}{D_{eff}}} = L\sqrt{\frac{k}{D_{eff}}} \text{ with } L = R \text{ for spherical particles} \approx 26\ \mu m$$

c. Catalyst could be very reactive (reactant does not reach the core)
   • Particle size could be too high (takes a long time to reach the core)
   • Diffusion limitations

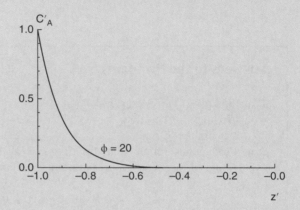

d. Instead of a microporous material, use a material with broader pores. Hence, the diffusion will increase, the Thiele modulus will decrease and the effectiveness will rise. Or: Make a catalyst with a lower Pd loading.

4 a. Calculate Co loading based on nitrogen content or carbon content. Not via H or O, because you do not know how the support (surface) looks like. But you are sure that no C or N is present in the support (pure).

Nitrogen: 0.28 wt.% = 0.0028 g per 1 g material
            $MW(N) = 14\ \text{g mol}^{-1} \rightarrow 0.0028\ \text{g}/(14\ \text{g mol}^{-1}) = 0.00020\ \text{mol g}^{-1}$
            $1 \times$ Co per $2 \times$ N in complex $\rightarrow 0.00010$ mol Co/g or
            0.10 mmol Co/g
Carbon: 2.4 wt.% = 0.024 g per g material
            $MW(C) = 12\ \text{g mol}^{-1} = 0.002\ \text{mol g}^{-1}$
            $1 \times$ Co per $20 \times$ C in complex (include alkyl tails!)
            $0.002\ \text{mol g}^{-1} / 20 = 0.1\ \text{mmol g}^{-1}$

---

b.
$$A + (*) \rightleftharpoons A^*$$
$$B + (*) \rightleftharpoons B^*$$
$$A^* + B^* \rightleftharpoons (AB)^* + (*)$$
$$(AB)^* \rightleftharpoons AB + (*)$$

$$RLS \text{ } or \text{ } SBS = step \text{ } 3 \implies r = r_3 = r_{+3} - r_{-3} = k_{+3}\theta_A\theta_B - k_{-3}\theta\theta_{AB}\theta^*$$

$$r = \frac{k_{+3}K_1K_2\left(P_AP_B - \frac{P_{AB}}{K}\right)}{(1 + K_1P_A + K_2P_B + K_4P_{AB})^2}$$

$$= \frac{(kinetic \text{ } factor)(thermodynamic \text{ } drive \text{ } force)}{square: \text{ } two \text{ } sites \text{ } involved}$$

c. TON: At the end of the reaction (equilibrium)! Conversion at equilibrium: ~43%

43% conversion = 0.43×2.4 mol = ~1 mol converted (1.032 mol)

Active sites: 0.1 mmol g^{-1} = 0.1 mmol (1 g cat)

TON = 1 mol converted per 0.1 mmol sites = 10 000

TOF: In the beginning of the reaction (unhindered catalysis)

Example: 60 min: Conversion = 10%

Conversion 10% = 0.1×2.4 mol = 0.24 mol converted after 60 min

TOF = (0.24 mol converted) / (0.0001 mol cat $\times$ 3600 s) = 0.667 s^{-1}

Other example: 120 min = 7200 s, conversion = 19% so 0.46 mol converted. TOF = 0.64 s^{-1}

References

1. Langmuir, I. (1918). *J. Am. Chem. Soc.* 40: 1361–1403.
2. Lowell, S. (1979). *Introduction to Powder Surface Area*. Wiley.
3. Wang, J.L. and Chen, C. (2009). *Biotechnol. Adv.* 27: 195–226.
4. Sing, K.S.W., Everett, D.H., Haul, R.A.W. et al. (1985). *Pure Appl. Chem.* 57: 603–619.
5. Thommes, M., Kaneko, K., Neimark, A.V. et al. (2015). *Pure Appl. Chem.* 87: 1051–1069.
6. (a) Van Der Voort, P., Ravikovitch, P.I., De Jong, K.P. et al. (2002). *J. Phys. Chem. B* 106: 5873–5877. (b) Van Der Voort, P., Ravikovitch, P.I., De Jong, K.P. et al. (2002). *Chem. Commun.* 1010–1011.
7. (a) Fomina, M. and Gadd, G.M. (2014). *Bioresour. Technol.* 160: 3–14. (b) Tchobanoglous, G., Burton, F.L., Stensel, H.D., and Metcalf, I.E. (2003). *Wastewater Engineering: Treatment and Reuse*. McGraw-Hill Education.
8. Volesky, B. (2001). *Hydrometallurgy* 59: 203–216.
9. Park, J., Won, S.W., Mao, J. et al. (2010). *J. Hazard. Mater.* 181: 794–800.
10. Tran, H.N., You, S.J., and Chao, H.P. (2016). *Waste Manag. Res.* 34: 129–138.
11. Tran, H.N., You, S.J., Hosseini-Bandegharaei, A., and Chao, H.P. (2017). *Water Res.* 120: 88–116.
12. Lagergren, S. (1898). *Kungliga Svenska Vetenskapsakademiens Handlingar*, vol. 24, 1–39.
13. Blanchard, G., Maunaye, M., and Martin, G. (1984). *Water Res.* 18: 1501–1507.
14. Ho, Y.S. and McKay, G. (1999). *Process Biochem.* 34: 451–465.
15. Ho, Y.S. and McKay, G. (1998). *Process Saf. Environ.* 76: 332–340.
16. Roginsky, S.Z. and Zeldovich, J. (1934). *Acta Phys. Chem. USSR* 1: 554.
17. Chien, S.H. and Clayton, W.R. (1980). *Soil Sci. Soc. Am. J.* 44: 265–268.
18. Weber, W.J. and Morris, J.C. (1963). *J. Sanit. Eng. Div.* 89: 31–60.
19. Weber, W.J. and Smith, E.H. (1987). *Environ. Sci. Technol.* 21: 1040–1050.
20. Walter, W.J. (1984). *J. Environ. Eng.* 110: 899–917.
21. Hall, K.R., Eagleton, L.C., Acrivos, A., and Vermeulen, T. (1966). *Ind. Eng. Chem. Fundam.* 5: 212.
22. Freundlich, H. (1906). *Z. Phys. Chem.* 57: 385–471.
23. Julkapli, N.M. and Bagheri, S. (2015). *Int. J. Hydrogen Energy* 40: 948–979.
24. Chorkendorff, I. and Niemantsverdriet, J.W. (2003). *Concepts of Modern Catalysis and Kinetics*. Weinheim: Wiley-VCH Verlag GmbH & Co. KGaA. ISBN: *3-527-30574-2*.

3 Zeolites and Zeotypes

3.1 Crystallographic Directions and Planes

As zeolites are crystalline materials (and many other materials discussed in this book are also crystalline materials), we will first briefly introduce some conventions on crystallographic directions and planes.

Directions and planes are described by three indices. The base for these indices is the so-called *unit cell*. A unit cell is the smallest section of the crystal structure that contains all symmetry elements; in other words, the crystal lattice can be expanded by multiplying the number of unit cells.

3.1.1 Crystallographic Directions

A *crystallographic direction* is defined as a vector. The three *directional indices* that describe this vector are determined as follows

(1) Draw a vector that starts at the origin of the coordinate system. This vector may be translated to other positions, as long as the parallelism is maintained.
(2) The lengths of the vector projections on the a-, b-, and c-axes of the unit cell are determined. The numbers are then manipulated to reduce them to the smallest integers. These integers are represented between square brackets without commas; for example, [uvw]. They are referred to as Miller indices.

As an example, we show in Figure 3.1a the often encountered [100, 110], and [111] directions in a unit cell. In Figure 3.1b, we show a given direction. Let us determine this direction.

(1) The vector that is drawn passes through the origin of the coordinate system; so no translation is necessary.
(2) The projections on the coordinate axes are, respectively, 1/2a, 1b, and 0c. Reduction of the smallest set of integers (multiply by 2) yields a crystallographic direction of [1 2 0].

Indices can also be negative. This is shown as an upper score in the integers, for instance [$a\bar{b}c$]. In Figure 3.1c we draw another crystallographic direction. Let's determine the direction again.

(1) The vector that is drawn passes through the origin of the coordinate system, so no translation is necessary.
(2) The projections on the coordinate axes are, respectively, 1a, −1b, and 0c. Reduction of the smallest set of integers (multiply by 1) yields a crystallographic direction of [1 $\bar{1}$ 0].

Introduction to Porous Materials, First Edition.
Pascal Van Der Voort, Karen Leus and Els De Canck.
© 2019 John Wiley & Sons Ltd. Published 2019 by John Wiley & Sons Ltd.

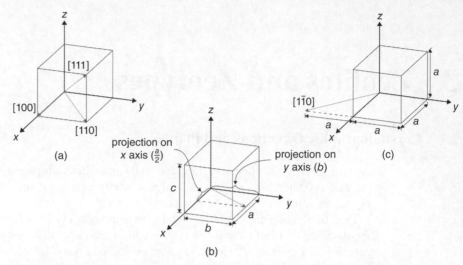

Figure 3.1

(a) Common crystallographic directions (b) determine the shown crystallographic direction (c).

Hexagonal crystals require special attention as their unit cell does not correspond to Cartesian coordinates. This is solved by using a four-axis coordinate axis system (the *Miller–Bravais* coordinate system) as shown in Figure 3.2a.

The three axes a_1, a_2, and a_3 are in the same plane (the *basal plane*) and are at 120° to one another. The z-axis is then perpendicular to the basal plane.

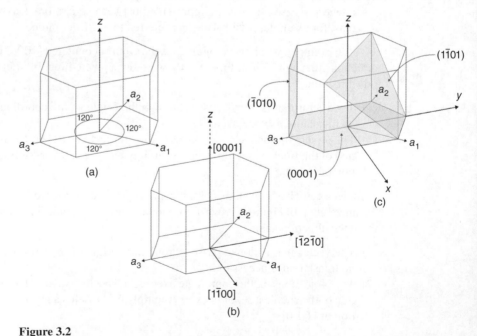

Figure 3.2

(a) The coordinate axis system for a hexagonal unit (Miller–Bravais). (b) Orthogonal directions in hexagonal system and (c) planes in hexagonal system (see later).

In order to express a crystallographic direction in a hexagonal system, the three-index system is converted to a four-index system by the following transformations:

$$[u'v'w'] \rightarrow [uvtw]$$

$$u = \frac{n}{3}(2u' - v')$$

$$v = \frac{n}{3}(2v' - u')$$

$$t = -(u + v)$$

$$w = nw'$$

with n being the factor to reduce the indices to the lowest possible integers.

An example is shown in Figure 3.2b. After transformation, the [010] direction in a cubic crystal becomes the [$\bar{1}2\bar{1}0$] direction in a hexagonal lattice.

$$u = \frac{1}{3}(0 - 1) = -\frac{1}{3}$$

$$v = \frac{1}{3}(2 - 0) = \frac{2}{3}$$

$$t = -\left(-\frac{1}{3} + \frac{2}{3}\right) = -\frac{1}{3}$$

$$w = 0$$

$$n = 3$$

$$[\bar{1}2\bar{1}0]$$

3.1.2 Crystallographic Planes

Crystallographic planes are represented by the three (four) *Miller indices*, called (hkl). Note that here that round brackets (parentheses) are used contrary to the crystallographic directions.

The methodology to obtain the Miller indices of a plane is similar to that explained in Section 3.1.1. In brief:

(1) If the plane passes through the origin, a new origin must be chosen at the corner of an adjacent cell, still within the unit cell.
(2) The plane now either parallels or intersects the three axes, the lengths of the planar intercepts are determined. If the plane is parallel, the length is ∞.
(3) The reciprocal of these numbers is taken, a plane that runs parallel to an axis has an infinite intercept and a zero index.
(4) The numbers are changed into the smallest set of integers and represented without commas within parentheses (round brackets) as (hkl).

Figure 3.3 shows examples of (100), (110), and (111) planes. Note that there are endless possibilities for drawing such planes in any given crystal.

Let us determine the crystal plane that is shown in Figure 3.4.

(a)

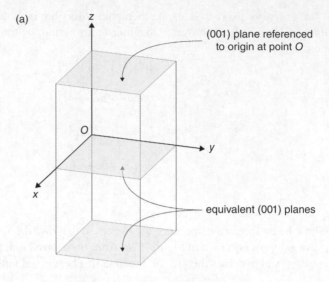

(b)

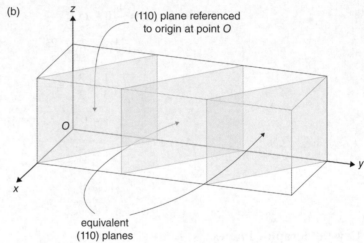

(c)

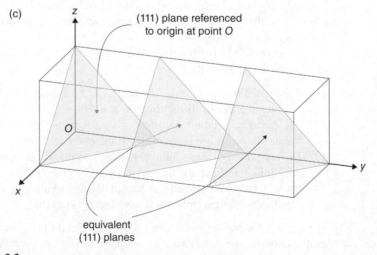

Figure 3.3
Representation of the (endless) set of (001), (110), and (111) planes in a cubic reference system. Source:
Reproduced with permission of John Wiley & Sons, Ltd. Redrawn from Ref. [1].

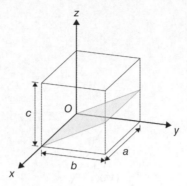

Figure 3.4
Determine the crystal plane shown in this figure. Source: Reproduced with permission of John Wiley & Sons, Ltd. Redrawn from Ref. [1].

As the plane goes through the origin, we need to create a new origin (O′) at the corner of an adjacent unit cell, but still part of the original unit cell. We thus translate the origin from O to O′ as shown in Figure 3.5.

We now notice that, in the new coordination system, the plane runs parallel to the *x*-axis, so the a-value is ∞. The b-value of the plane in the new coordination system is −1 and the c-value is 1/2. In Miller indices, this gives the reciprocal of these values, being $(0\bar{1}2)$.

For hexagonal crystals the same transformations apply. We have shown in Figure 3.2c some common planes in the hexagonal crystal system. To determine these planes, we use a four-index (*hkil*) scheme with

$$i = -(h + k)$$

Otherwise, the three *h, k,* and *l* indices are the same for both indexing mechanisms. This is exemplified in Figure 3.6.

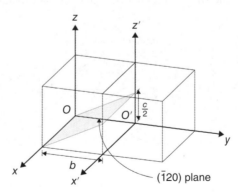

Figure 3.5
Translation of the origin to the nearest adjacent cell. Source: Reproduced with permission of John Wiley & Sons, Ltd. Redrawn from Ref. [1].

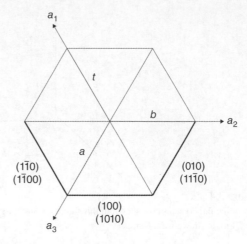

Figure 3.6
Relation between Miller indices and Miller–Bravais indices for indicated planes in a hexagonal symmetry (2D view, top view, z-axis is omitted for clarity). The three planes are crystallographically equivalent.

3.2 X-Ray Diffraction

Crystallinity (position of crystallographic planes) is typically determined by X-ray diffraction (XRD), either as powder (lower resolution) or as a single crystal (higher resolution). X-rays have a wavelength in the order of nanometers and below, which makes them very suitable for determining the distances between crystallographic planes by constructive interference of the diffracted waves.

In 1913, Bragg found that crystalline solids have remarkably characteristic patterns of reflected X-ray radiation. In these crystals, at certain specific wavelengths and incident angles intense peaks of scattered radiation were observed. We must envision the crystal having parallel crystallographic planes of atoms, separated by the interplanar distance d. The conditions for a sharp peak in the intensity of the scattered radiation are: (i) the X-rays should be specularly reflected by the atoms in one plane and (ii) the reflected rays from the successive planes interfere constructively.

Figure 3.7 shows reflected X-rays from adjacent crystalline planes. The two planes AA′ and BB′ are parallel with an interplane distance of d. The X-rays enter the planes

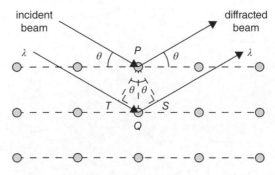

Figure 3.7
Bragg diffraction. Two beams with identical wavelengths and phases approach a crystalline solid and are scattered off two different crystallographic planes.

with a wavelength λ at an angle θ with the planes. As we explained previously, both planes have the same (hkl) indices. Two rays in the X-ray beam are scattered by atoms P and Q, respectively. The X-ray scattered in the second plane travels a distance $\overline{SQ} + \overline{QT}$ longer than the other beam. Simple trigonometric observation shows us that

$$\overline{SQ} = \overline{QT} = d \sin \theta \qquad (3.1)$$

In order to interfere constructively (and show a diffraction peak), the distance of $\overline{SQ} + \overline{QT}$ needs to be equal to a whole multiple of λ, or

$$2d \sin \theta = n\lambda \qquad (3.2)$$

This equation is known as *Bragg's Law*.

William Laurence Bragg (1890–1971) was an Australian born British physicist and X-ray crystallographer, and the discover of Bragg's law of X-ray diffraction in 1912 (at the age of 22!).

His father, W. H. Bragg, built an apparatus in which a crystal could be rotated to precise angles while measuring the energy of reflections. This enabled father and son to measure the distances between the atomic sheets in a number of simple crystals. W. H. Bragg reported their results at meetings and in a paper, giving credit to "his son" (unnamed) for the equation, but not as a co-author, which strongly impacted their relationship.

Together with his father, he received the Nobel Prize for Physics in 1915, "for their services in the analysis of crystal structure by means of X-rays." Even today, he is the youngest Nobel Laureate in Physics ever; he received the prize at the age of 25. Photo: https://en.wikipedia.org/wiki/Lawrence_Bragg#/media/File:Wl-bragg.jpg, *in the public domain.*

Using XRD it is possible to identify the crystal structure. In a common way, the wavelength of radiation is fixed and the angle of incidence is allowed to vary in practice to observe diffraction peaks corresponding to reflections from different crystallographic planes. Hereafter, the distance between the planes can be determined using Bragg's law.

The XRPD (X-Ray Powder Diffraction) pattern is achieved by recording the intensities of X-rays scattered by the sample at various angles between the incident beam on the sample and the beam scattered by it. The set of rotational displacements of the detector and sample (or of the X-ray source) together with the recording of scattered X-ray intensity at each displacement is called a diffractogram.

3.3 Zeolite Structures

Zeolites are "crystalline aluminosilicates with fully cross-linked open framework structures made up of corner-sharing $[SiO_4]^{4-}$ and $[AlO_4]^{5-}$ tetrahedra." They contain pores (cavities) in the order of 0.3–1.5 nm, and are therefore usually microporous. For every Al^{3+} that is introduced in the network, virtually replacing a Si^{4+} node, a negative charge is introduced in the network that is compensated by "exchangeable cations," typically Na^+ or H^+. Zeolites are typically labeled as MZ, where M is the exchangeable cation and Z refers to the topology of the framework. We will explain in more detail later (Figure 3.8).

The first zeolite, stilbite, was discovered by Cronstedt in 1756, who found that the mineral loses water rapidly on heating and thus seems to boil. The name "zeolite" comes from the Greek words $\zeta\varepsilon\acute{\omega}$ (*zeo*, onomatopoeia, to boil) and $\lambda\acute{\iota}\theta o\varsigma$ (*lithos*, stone) [4]. A representative empirical formula of a zeolite is $M_{2/n}O.Al_2O_3.xSiO_2.yH_2O$.

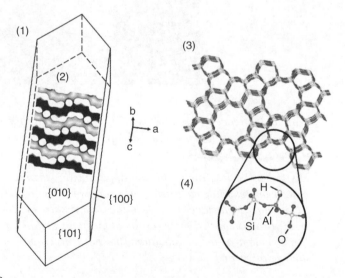

Figure 3.8
Schematic representation of a zeolite. (1) The single crystal with its corresponding surface planes (indicated with [2] by these authors), (2) the internal pore system and voids, (3) zoomed in part of the crystal structure, showing the SBUs and the pore structure, and (4) further zoomed in, the elemental composition, and individual bonds in the zeolites. Source: Reproduced with permission ACS [3].

The $[SiO_4]^{4-}$ and $[AlO_4]^{5-}$ tetrahedra are arranged in such a way that zeolite crystals contain well-defined networks of channels and cavities or molecular dimensions. The cavities can be tuned accurately by changing the type of the zeolite structure by choosing its chemical composition and synthesis conditions. The network of channels and cavities in zeolites is so vast that the internal surface of zeolites notably exceeds their external surface and can be as big as $600 \, m^2 \, g^{-1}$.

Natural zeolites occur in Australia, New Zealand, India, the Eiffel area, Iceland, Northern Ireland, Faeroe Islands, Russia, and northwestern USA. They have different colors, due to transition metals in the structure. Some natural occurring crystals are shown in Figure 3.9. We refer the reader to the website of the International Zeolite Association for a full database of structures and compositions of both naturally occurring and synthetic zeolites. The current URL is www.iza-online.org.

Although natural zeolites find some application in agriculture (soil enrichment) and construction, most zeolites that are commercially used are synthetic zeolites.

The classification of the hundreds of types of zeolites is based on their "secondary building units" (SBUs). There are dozens of types of SBU, going from a simple ring to a complex polyhedron. These SBUs link together to form a complex three-dimensional structure. Figure 3.10 presents the most important SBUs [5] that build up zeolite structures. In this notation, the cornerpoints are either a Si or an Al atom, the lines between the cornerpoints represent an oxygen bonding.

The SBUs can be arranged in various ways, in order to make up a crystalline topology of zeolites.

As an example, the well-known and naturally occurring structure faujasite (FAU) is formed using the sodalite cage (or β-cage) consisting of 24 tetrahedral (T) atoms (six 4-rings, four 6-rings, three 6-2 units, or four 1-4-1 units) shown in Figure 3.11. The two-dimensional Periodic Building Unit (PerBU) is obtained when β-cages are linked through double 6-rings (D6Rs) into the hexagonal faujasite layer depicted in Figure 3.12.

Figure 3.9

Some examples of naturally occurring zeolites. Source: Courtesy of Bert Weckhuysen.

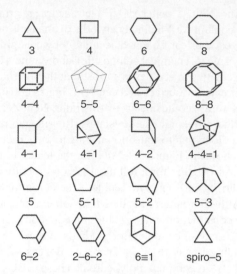

Figure 3.10
Examples of SBUs currently known to be found in zeolites. Source: Reproduced with permission of ACS [5].

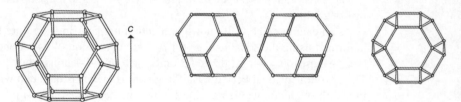

Figure 3.11
The sodalite cage. From left to right: perspective view perpendicular to c; two parallel projections (different scale), related by a rotation of +30° and −30° about c; and parallel projection down c. Source: Taken from www.iza-structure.org/databases/ModelBuilding/EMT.pdf.

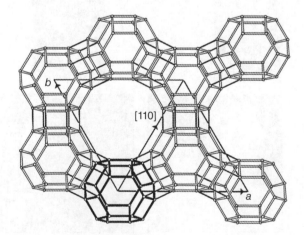

Figure 3.12
PerBU in faujasite, built from β-cages (one cage in bold), viewed along the c-axis. Source: Taken from www.iza-structure.org/databases/ModelBuilding/EMT.pdf.

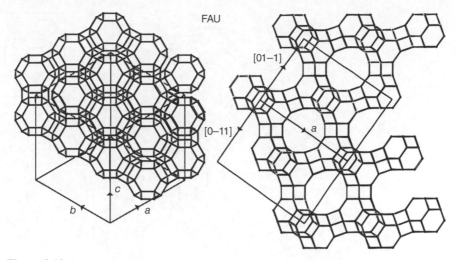

Figure 3.13
Unit cell content in faujasite viewed along the cubic axes [111] (left) and [011] (right). Source: Taken from www.iza-structure.org/databases/ModelBuilding/EMT.pdf.

These PerBUs can still form two zeolite topologies. When the neighboring Per-BUs are exclusively related by reflection axes, then the topology EMT is formed (not shown). When the neighboring PerBUs are exclusively related by inversion, then the structure of the faujasite is formed. See Figure 3.13.

Hundreds of zeolite structures have been described, most of them are synthetic zeolites. The first ones were the faujasite structures that we described above (Linde X, Linde Y). The synthetic zeolites Na-X and the H-Y belong to this family. They are used in large quantities for ion-exchange and catalytic cracking (see later).

Another industrially important family that is also a natural zeolite is the family of the mordernite (framework structure MOR). These structures are shown in Figure 3.14. Note in the mordenite structure the long straight pores through the material that should facilitate mass transport through them and decrease diffusional limitations.

3.4 Applications of Zeolites

Zeolites have three major industrial applications that we will discuss in some detail. Due to their exchangeable cations as charge compensators, they are widely used as ion-exchangers and in particular as water softeners in washing powders. When the exchangeable cation is an H^+-ion, zeolites become solid acids, and for that reason they are widely used in industry as cracking catalysts. Finally, due to their perfectly defined micropores, which has granted them their alternative name of *molecular sieves*, they are used as gas purification agents.

3.4.1 Ion-Exchange, Water Softening

In many areas in the world, the water contains too high concentrations of Ca^{2+} and/or Mg^{2+}, which render washing products ineffective, as, together with the detergents,

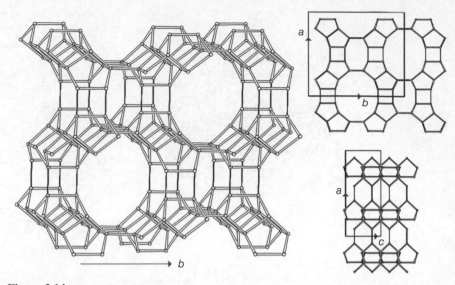

Figure 3.14
Connection mode in MOR viewed along c (left) and projection of the unit cell content along c (top right) and along b (bottom right). Source: Taken from www.iza-structure.org/databases/ModelBuilding/MOR .pdf.

they form an insoluble suspension that renders water and clothes gray (the so-called grime). In the 1950s to 1970s, abundant amounts of phosphates (mostly pentasodium triphosphate, $Na_5P_3O_{10}$) were added to the washing products in order to prevent the grime, as phosphates would bond to these ions. However, as phosphates are also fertilizers, in many areas, rivers and streams were turned green by the unlimited growth of algae. This in turn led to very high oxygen consumption leading to the death of fish and other living organisms in these waters.

Therefore, in the 1970s phosphates were largely replaced by aluminosilicates. Amorphous aluminosilicates (by sol-gel processes) are also widely used, the crystalline variants of the aluminosilicates are then often the corresponding zeolites. In the case of zeolite substitution, the zeolite *NaA* is often used.

The *zeolite A* was first prepared by the company Linde (Union Carbide). It is one of the three famous zeolites: Linde type A, X, and Y. The abbreviation **LTA** stands for Linde type A. All three zeolites are built up by the sodalite cage, as presented in Figure 3.11. But in zeolite A, the sodalite cages are built in a different pattern to zeolite X and Y. Both *zeolites X and Y* are *faujasites* as already discussed and represented in Figure 3.13. The difference between zeolite X and zeolite Y is the Al content. In X-zeolites, the ratio of silica to alumina is between 2 and 3, in Y-zeolites it is higher than 3. The faujasite zeolite is a naturally occurring structure, described in 1842 and named after Berthélemy Faujas de Saint-Fond (1741–1819), a French geologist.

These structures are compared in Figure 3.15. The zeolite A in its sodium form has a *cation-exchange capacity* (CEC) of 3–5 mmol Na^+-ions per gram. This is a huge amount. In washing products this cheap zeolite is very effective in exchanging the hardeners Ca^{2+} and Mg^{2+} for "innocent" Na^+ ions.

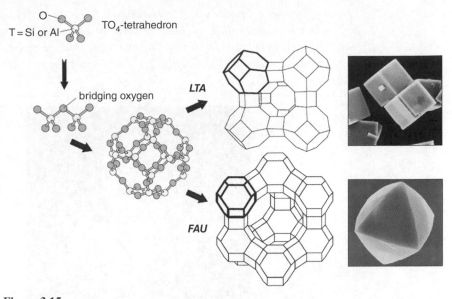

Figure 3.15
The framework structure of LTA (zeolite A) and faujasite (zeolite X or Y), together with a SEM image of their crystals. Both types are built from the sodalite cage. Source: Courtesy of Bert Weckhuysen.

The two "founding fathers" of the synthetic zeolites are Richard M. Barrer (1910–1996) (right) and Robert M. Milton (1920–2000). Photograph ACS Symposium in Los Angeles, September 22, 1988. (c) ACS Symp. Ser. 1989, 398, 11.

Richard Barrer was the "academic," coming from New Zealand, and worked at Cambridge, London, and Aberdeen. He spent the largest part of his career (1954–1976) at London Imperial College. He was one of the first to create synthetic zeolites (a.o. Barrerite). Also, a unit of gas permeability, the Barrer, was named after him.

Robert Milton was the "industrial." He worked at Linde, a German founded company, which was, and still is, the largest industrial gas company. Linde Air Products since 1917 has been a part of Union Carbide. Here, Milton produced synthetic zeolites, still used today; the Linde types A, X, and Y are the best known ones among many others.

Bert Weckhuysen is a specialist in zeolites and heterogeneous catalysis, focusing on the operando (in situ) spectroscopy of the working catalyst. He obtained his Ph.D. at the KULeuven (Belgium) in 1995 and after a postdoctoral stay he was tenured at the University of Utrecht (The Netherlands) in 2000. He is currently a "university professor" in Utrecht and was knighted in the Netherlands in 2015.

3.4.2   Catalysis

Zeolites in their acidic form are widely used in the petrochemical industry for cracking and isomerization reactions. A full overview of all reactions and all zeolites used in this industry goes beyond the scope of this book. We will give a few examples for some large industrial processes.

3.4.2.1   *Fluid Catalytic Cracking (FCC) – USY Zeolite*

FCC is one of the most widespread techniques in petrochemistry used to convert the high boiling hydrocarbon fractions of crude oils into valuable fuels. Next to the cracking of the alkanes producing smaller ones, it produces alkenes as high value byproducts.

3.4.2.1.1   *Acidity in Zeolites – Cracking Mechanisms*
Acidity in Zeolites
Actually, the *acidity in zeolites* is rather complex. As in all aluminosilicate zeolitic structures discussed so far, they have Al^{3+}-ions and in their acidic form also exchangeable H^+-ions, so they possess both a Lewis acidity and a Brønsted acidity.The *Brønsted acid sites* (the H^+-ions) are located on the bridging oxygen atoms, usually in a Si—O—Al bridge. In cation-exchanged zeolites, a second source of Brønsted sites are the dehydrated metal nodes.

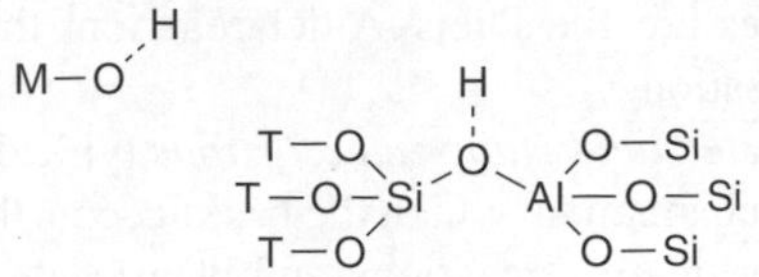

$$M-O\overset{\cdot\cdot H}{}$$

$$
\begin{array}{cc}
T-O & \quad \overset{H}{\underset{|}{O}} \quad O-Si \\
T-O-Si & \searrow \quad \diagup \\
 & O \cdots Al-O-Si \\
T-O & \diagup \quad \nwarrow \\
 & O-Si
\end{array}
$$

According to the *Löwenstein rule* [6] (aluminum avoidance rule), in a zeolite it is forbidden to form an Al—O—Al bond, but all other combinations are possible. This means that following combinations are possible: (i) Al_3Si—OH—$AlSi_3$, (ii) Al_2SiSi—OH—$AlSi_3$, (iii) $AlSi_2Si$—OH—$AlSi_3$ and (iv) Si_3Si—OH—$AlSi_3$. As for the dehydrated metal sites, according to the *Gutmann rule* [7], if the surrounding atoms have a higher electronegativity, there will be an electron shift from the less electronegative H to the more electronegative O, resulting in a weakening of the O—H bond rendering it more acidic. The terminal OH bonds on the exchanged metals are much stronger than the bridging OH bond, and therefore the terminal OH bonds are much less acidic in general than the bridging ones. The actual acid-catalytic behavior of the zeolite is predominantly determined by the Brønsted acid sites.

However, the zeolite framework also contains *Lewis acid sites*. These are the electron deficient sites (a double bond containing an unoccupied orbital) exhibiting the ability to accept electrons during interaction with molecules. Historically, it was suggested by Uytterhoeven et al. [8] that Lewis acidity was created by the formation of a trigonally coordinated (Al^- —O— Si^+) group when two hydroxyl groups are removed from the zeolite framework (Figure 3.16a).

This mechanism was corrected later on as it became clear that the dihydroxylation of hydrogen forms of faujasites resulted in the loss of aluminum (dealumination) instead of the formation of lattice Lewis sites. It is currently believed that undefined species, such as AlO^+ or charged $Al_xO_y^{n+}$ or tiny nanoparticles inside or outside the pore system, created during the pretreatment, activation, and dihydroxylation of zeolites, are responsible for the Lewis acidity. The corrected model is shown in Figure 3.16b.

Exact knowledge about the number of acid sites and the strength of the Lewis and Brønsted sites is very important to tune the acid catalyst toward the right application.

Lewis acid catalyzed reactions typically act as electron acceptors for a lone pair bearing an electronegative atom in the substrate; typically, oxygen, sulfur, and halogens. In this way, they withdraw electron density from the substrate, facilitating heterolytic bond cleavage or activating the substrate toward nucleophilic attack. Typical

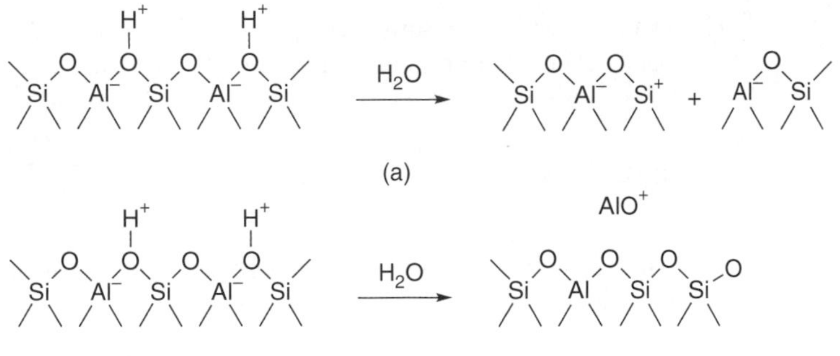

(a)

(b)

Figure 3.16

Lewis acid sites in zeolites; (a) an old model by Uytterhoeven and (b) the current model.

examples are the Diels–Alder reaction, the Friedel–Crafts reaction, and the aldol condensation.

Brønsted acid catalyzed reactions typically follow the Brønsted–Lowry acid–base pair mechanism, in which the base accepts the proton from the acid. Typical examples are esterification reactions and clearly also alkane cracking. In that case, one can operate at lower temperatures and pressures compared to thermal cracking, and it is easier to tune the end products toward high value products.

Catalytic cracking reactions can be subdivided into two types of reaction. The primary reactions involve the active participation of the catalyst and are the reactions that initially convert the hydrocarbon molecule into a positively charged carbonium ion. The most important *primary reactions* are:

(1) The protonation of an alkene in the feedstock toward its carbonium ion:

$$C_nH_{2n} \xrightarrow{H^+} C_n^+H_{2n+1}$$

(2) Abstraction of a hydride by an acid catalyst:

$$C_nH_{2n+2} \xrightarrow{cat} C_n^+H_{2n+1} + Cat.H^-$$

(3) The abstraction of a hydride by an already-formed carbocation:

$$C_nH_{2n+2} + R^+ \rightarrow C_mH_{2m+1} - C^+H - C_kH_{2k+1} + RH$$

(4) Fission (β-fission) of a carbonium ion in an adjacent bond:

(5) Fission of branched alkanes, cyclic alkanes, and aromatics substituted with one or more side chains, after which the carbonium ion loses its proton again to form an alkene:

The *secondary reactions* are plentiful and involve all sorts of structural rearrangements and conversions. Here are just a few examples:

(1) Double bonds will shift from *end standing* toward "middle of chain" double bonds. The latter are more stable and also have a better octane number:

(2) Alkenes will further break down with carbocations as intermediates. In this way, mostly small hydrocarbons are formed:

(3) Cycloalkanes (mostly cyclohexanes) are further broken into alkenes and aromatic hydrocarbons by a series of deprotonation reactions:

(4) Aromatics are formed from alkenes:

3.4.2.1.2 The Catalyst – Ultra Stable Y Zeolite (USY)

The cracking of alkanes is Brønsted-acid catalyzed. Sulfuric acid was used as a catalyst in the very early days, but the Frenchman Eugene Houdry was the first to use a solid acid catalyst, a natural clay (bentonite) that was washed with sulfuric acid to remove the Na^+ and Mg^{2+} ions and to replace them with acid, exchangeable H^+ ions.

Eugene Houdry, born in France (Domont) in 1892 was a mechanical engineer. He moved to the USA in 1930, where he died in Pennsylvania in 1962.

*The first Houdry Unit for FCC was built in Pennsylvania in 1937 at Sun Oil (Sunoco).
Later, Standard Oil (ExxonMobil) improved the system by introducing the continuous
FCC.*

*But he also invented the catalytic converter for cars (see the photograph, where he
holds his prototype of such a convertor). It was not used at the time, as the tetraethyl
lead that was heavily used in the 1950s and 1960s to improve the RON (octane number) poisoned the catalyst. Finally, he was also the inventor of the butadiene process,
where butane is converted into butadiene for synthetic rubber fabrication*

(*Photograph:* http://www.sciencehistory.org/historical-profile/eugene-houdry).

The current catalyst in the FCC process is called USY, or Ultra Stable Y. It is the
Linde type Y zeolite that is further processed to remove some of the Al^{3+}, rendering
the catalyst less active and more stable in the harsh conditions of the process.

One of the possible syntheses routes is to start with the NaY zeolite and exchange
the Na^+-ions with NH_4^+-ions to form a NH_4Y zeolite. This zeolite is then further
steamed. During this process, two phenomena occur: the NH_4^+ ions decompose into
NH_3-gas that leaves the zeolite and H^+-ions that are now the exchangeable ions. The
NH_4Y has thus turned into a HY zeolite. But during these steaming conditions many
of the Al^{3+}-ions in the framework migrate to the surface of the crystal, forming the
so-called Extra Framework Alumina (EFAL) there. Most of this EFAL is removed
by acid washing. This migration of the Al^{3+} yields a structure that is higher in silicon content and thus also more stable (remember, the more Al^{3+}, the less stable the
framework) but it simultaneously creates several defects in the structure that produce
larger pores or even mesopores.

While the debate is still ongoing as to the exact processes that occur during steaming, we now have a clear view on what the USY looks like. In a seminal paper in
Angewandte Chemie, Krijn de Jong and coworkers [9] were able to visualize the internal pores in the USY crystal using 3D-electron-tomography, a Transmission Electron
Microscopy (TEM) technique that is able to show 3D pictures of the materials by
measuring the same samples multiple times at different angles. The result is shown
in Figure 3.17, which clearly shows the difference in a normal 2D-picture (where all

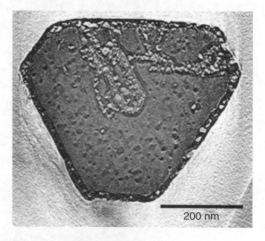

Figure 3.17

A reconstructed slice of 1.7 nm through 3D-TEM (tomography) of a USY. Source: Reproduced with
permission of John Wiley & Sons, Ltd. Ref. [9].

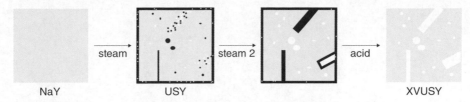

Figure 3.18
Model of the generation of mesopores in zeolite Y. The zeolite is gray, the amorphous alumina is black, and the empty mesopores are white. Source: Reproduced with permission of John Wiley & Sons, Ltd. Redrawn from Ref. [9].

details are lost due to the thickness of the sample) and the 3D-tomography picture, in which the authors virtually "cut" a slice of 1.7 nm.

From these experiments, the authors could present a possible scheme for the changes in the zeolite that occur during the steaming toward an USY. In Figure 3.18, the NaY zeolite is exchanged to the NH_4Y zeolite that is subsequently steamed into the USY zeolite. During this process, two phenomena occur: many of the Al^{3+} sites migrate, mostly to the outer surface but also partially to some nests of alumina within the framework. Also, mesoporous are formed in the structure and the authors established that these were not cylindrical pores (as is often believed) but clear spherical pores. An extra second steaming stem enlarges these effects, more alumina is built inside and outside the structure and the spherical mesoporous become larger. A final acid treatment removes all the EFAL, leaving a high silica zeolite with large mesopores in the structure and a very high stability. The acronym XVUSY stands for "extra very ultra-stable Y".

3.4.2.1.3   *Reactant Shape Selectivity*

The USY zeolite has another important advantage. It shows *Reactant Shape Selectivity*. When cracking alkanes, it is important to mainly crack the unbranched alkanes, as these form waxes with high boiling points. The branched alkanes (even with simple methyl branches) have much lower boiling points and are in fact an excellent ingredient in of the final fuel. USY zeolites have such a pore opening that branched alkanes cannot enter the pore system, so predominantly unbranched alkanes will enter the zeolite pores and will be cracked (Figure 3.19).

The actual FCC catalysts consists of multiple components. Typically, zeolite particles (5 µm), alumina (200 nm), a binder (5 nm), and clay (2 µm) are suspended and then spray dried into tiny particles of a size of about 75 µm (Figure 3.20).

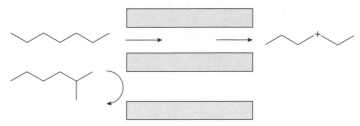

Figure 3.19
Reactant shape selectivity in the cracking of alkanes.

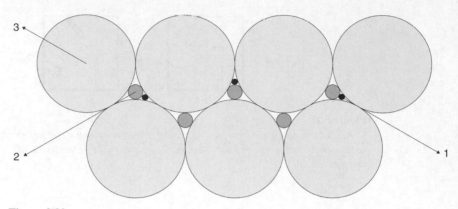

Figure 3.20
FCC particle: 1 zeolite, 2 alumina and 3 binder. Source: Reproduced with permission of Elsevier. Redrawn from Ref. [10].

3.4.2.1.4   Fluid Catalytic Cracking – Fluidized Bed Technology

The feedstock for the FCC is the fraction of the oil that has a boiling point >340 °C and an average molecular weight of 200–600. In the industry, this fraction is called heavy gas oil (HVGO). In the FCC, this feedstock is heated to high temperatures when contacting the cracking catalyst. The catalyst breaks the long alkanes down into smaller fragments that have much lower boiling points and are collected as a vapor.

However, there are several issues with this seemingly simple process. First of all, the *coking* of the catalyst reduces its catalytic activity in a matter of seconds. Coking means that heavy weight (usually aromatic) compounds deposit on the catalyst. A second problem is that cracking is an endothermic process, which requires massive heating.

Both problems were solved by an ingenious reactor design. The FCC unit consists of two units: an active catalytic part and a regenerator (see Figure 3.21).

The hot, high-boiling oil feedstock is injected in the catalyst riser (Figure 3.21(7)) where the oil is vaporized and encounters the tiny catalyst particles. The hot vapors drive the catalyst upward. In engineering terms, this is called a *fluidized bed reactor*. The fluid can be either a gas or a liquid and it goes upward at a high enough velocity to suspend the powder and cause it to behave like a fluid.

The actual cracking takes place in the riser in a few seconds. The mixture of the (cracked) hydrocarbon vapors together with the catalyst flow further upward to enter the reactor. Several cyclones (5) separate the catalyst particles from the vapors. While the vapors are collected for further processing, the catalyst is sent to the regenerator via a slide valve (9).

The catalyst at this point, after a very short time in the riser, is deactivated by coking and needs to be regenerated. This is done in the regenerator. The regenerator operates at 715 °C, and the combustion of coke on the catalyst particles is strongly exothermic and produces a large amount of heat. This heat is adsorbed (at least partially) by the catalyst particles, providing some of the energy for the endothermic cracking in the riser. The flue gases emerging from the coke combustion are recovered, the catalyst is ready for another cycle, and enters the riser again via the "catalyst withdrawal well" (6) and another slide valve (8).

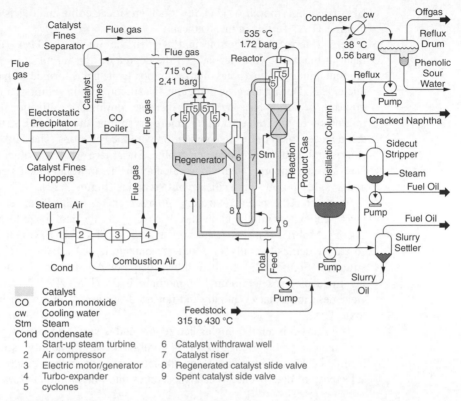

Figure 3.21
Schematic overview of a FCC unit. Source: Courtesy: https://upload.wikimedia.org/wikipedia/commons/9/95/FCC.png (public domain).

The amount of catalyst circulating in an FCC plant is around 5 kg per kg of feedstock, an incredibly high amount! Luckily, the system can operate for six months to a few years without major maintenance. As the catalytic particles break up gradually (catalyst fines), a new catalyst has to be inserted regularly.

3.4.2.2 Isomerization – Templated Zeolites – ZSM-5

Researchers in zeolite development are continuously looking for novel zeolites and two properties are highly desired: the zeolite should have a high silicon content (higher stability, but much more difficult to synthesize) and should preferably have larger pores.

Both issues have been resolved simultaneously by adding small organic ammonium salts to the synthesis mixture, typically triethylammonium hydroxide (TEA). In such a synthesis, next to the Al- and Si-sources and the bases (NaOH, KOH), these ammonium salts are also added that act as a small template to occupy space in the crystal (before calcination) stabilizing the synthesis gel and crystal growth.

Researchers have been experimenting with ammonium salts in synthesis for a long time, and reports on the synthesis of the first "templated" zeolite in 1967 took some time to be appreciated. The material was called zeolite β, and now has the acronym

BEA (Beta polymorph A) [11]. It was a high silica zeolite and this is one of the roles that the TEA$^+$ fulfills in the synthesis.

Peter (Pierre) Jacobs wrote in 1987 that the amount of TEA$^+$ that is incorporated in the framework depends on the degree of crystallinity: first, basically only Al-containing sodium (potassium) aluminosilicate gel is formed with only very small nuclei [12]. During the growth regime, the silicon (still in solution) migrates from the liquid to the solid phase and further increases during crystallization. Then the TEA$^+$ migrates into the frameworks; the more crystalline the material is, the more TEA$^+$ can be incorporated. It is in competition with the Na$^+$-ions. The TEA$^+$ incorporation is most efficient for a silicon-rich zeolite; for a more Al-rich zeolite, more and more Na$^+$-ions are incorporated. Upon calcination the ammonium salts are decomposed and a highly porous, high silica zeolite remains (Figure 3.22).

Soon after the patent of zeolite β, research on high silica, templated zeolites increased exponentially upon the discovery of the zeolite ZSM-5 [13] (Zeolite Socony Mobil – 5).[1] Its framework type is MFI (*ZSM-FI*ve). It belongs to the so-called pentasil family of zeolites, characterized by five- and 10-rings in the structure (Figure 3.23). In the original recipe or Argauer and Landolt, the templating molecule was tetrapropyl ammonium hydroxide (TPA). Later, milder routes were described that eventually did not need the use of the expensive, smelly, and toxic TPA.

The ZSM-5 is another high silica zeolite and is very acidic in its protic form. The very regular pore system of ZSM-5, combined with its acidity, is used in industry for hydrocarbon isomerization and alkylation of hydrocarbons. Here another type of selectivity of heterogeneous catalysis comes into play: *product shape selectivity*.

3.4.2.2.1   Product Shape Selectivity

A nice example of product shape selectivity is shown in Figure 3.24. The alkylation of toluene with methanol is acid catalyzed and produces a thermodynamical equilibrium of approximately equal concentrations of *ortho-*, *meta-*, and *para*-xylene. Only the *para*-xylene has a commercial value because it is a raw material in the large-scale synthesis of various polymers, such as a component in the production of terephthalic acid for polyesters and terephthalates.

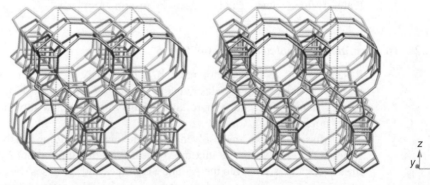

Figure 3.22
Zeolite beta, viewed along [010]. Source: Taken from www.iza-structure.org/databases/ModelBuilding/EMT.pdf.

---

[1] Mobil was previously known as Socony Vacuum Oil Co. Mobil then merged in 1999 with Exxon into ExxonMobil.

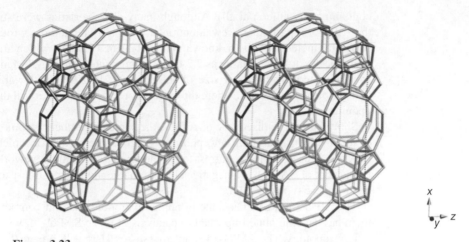

Figure 3.23
ZSM-5, MFI framework, viewed along [010] and clearly showing the pentagons that build the structure.
Source: Taken from www.iza-structure.org/databases/ModelBuilding/EMT.pdf.

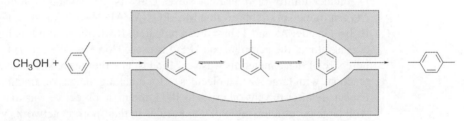

Figure 3.24
Product shape selectivity in the alkylation of toluene toward *para*-xylene.

The *ortho-* and *meta*-xylenes that are formed within the pores of the catalyst are unable to leave the pore, as these branched molecules are too bulky. Only the slim *para*-xylene is able to leave to pore system, each time causing a re-equilibration in the pore system and a theoretical yield of 100% *para*-xylene. It is a nice example of how heterogeneous catalysts can *seemingly* change the thermodynamics of a system, at least at the macro-scale. Such phenomena, where the pore systems play a decisive role in catalytic yield and selectivity, is often referred to as *catalysis in confined spaces*.

In 1978, an Al-free ZSM-5 was reported in *Nature*. This material was called silicalite [14].

3.4.2.3 *Methanol to Olefins (MTO) – Zeotypes*

3.4.2.3.1 *Zeotypes*

In the 1980s, the quest for "larger pores" continued. While bouncing at the limits of the classic (Al,Si,O) containing zeolites, researchers started to also try other (tetrahedral) ions in the synthesis of crystalline porous materials.

The concept that microporous materials do not necessarily need to contain silica was proven by the discovery of the $AlPO_4$ materials in 1982 [15]. Edith Flanigen and coworkers (from the Union Carbide Corporation) reported 20 new three-dimensional frameworks that consisted of the atoms Al, P, and O (aluminophosphates, not containing any silicon at all). Although many characteristics were still unexplored, the AlPO-5 in particular drew attention at the time. The authors wrote: "Large-pore structures include A1PO$_4$-5, known to have 12-ring channels which allow adsorption of molecules at least as large as 2,2-dimethylpropane (0.62 nm)." Many more AlPOs were reported. The critical size of the channels in AlPO with various structures varies from 0.3 to 1.2 nm, where the largest pore apertures were still the 12-ring pore openings.

Shortly after the discovery of AlPOs, the same researchers at Union Carbide discovered silicon-aluminum-phosphorus-oxygen frameworks, for which the acronym "SAPO" is used [16]. In contrast to the electrically neutral framework of AlPO, the framework of SAPO has a negative electric charge inducing cation-exchange properties for these types of solid.

The AlPOs and SAPOs came in numerous topologies and compositions. One of them has become quite important in catalysis, it is the SAPO-34, with the chabazite (CHA) topology. The SAPO-34 zeolite possesses a large CHA cage (0.94 nm in diameter) and a small eight-ring pore (0.38 nm) opening, as well as moderate acidity, rendering it the ideal catalyst for the methanol-to-olefin process (see the next section) (Figure 3.25).

Another family of molecular sieves consists of a metal-aluminum-phosphorus-oxygen framework (abbreviated MeAPO). MeAPOs with Me=Co, Fe, Mo, Mn, Zn, B, Be, Ga, Ge, As and Ti have been prepared. If silicon is additionally present in the crystal lattice, the acronym MeAPSO is used. However, the increase in the average pore size of these materials was very modest and they were still not very effective in reactions with large sized molecules or for cracking the gas oil fraction of petrolum. Especially after the gas oil crisis in 1973, the main objective was to obtain materials with large pores allowing bulky molecules in their porous network. VPI-5 (from the acronym for the Virginia Polytechnic Institute), was the first zeotype in which the classic barrier of only 12 atoms in a ring cavity was broken. Its framework topology has been named VFI (*Virginia Polytechnic Institute FIve*). VPI-5 contains an 18-ring as a cavity. This cavity, however, remains too small to allow bulky organics to enter the pore system (Figure 3.26).

Only one zeotype has been prepared with a large pore opening. The rather exotic gallophosphate cloverite (CLO) was synthesized in 1991 with a structural formula $|(C_7H_{14}N^+)_{24}|_8$ [F$_{24}$ Ga$_{96}$P$_{96}$ O$_{372}$(OH)$_{24}$]$_8$ [17]. The name cloverite comes from

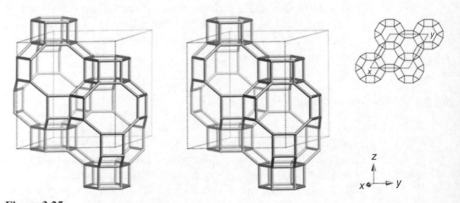

Figure 3.25
Chabazite (CHA) framework viewed normal to [001] (upper right: projection down [001]). SAPO-34 has the CHA topology. Source: Taken from www.iza-structure.org/databases/ModelBuilding/EMT.pdf.

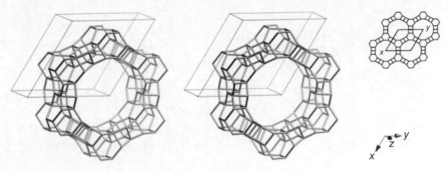

Figure 3.26
Framework of VPI-5 (VFI) viewed along [001] (upper right: projection down [001]). Note the 18-ring pore openings. Source: Taken from www.iza-structure.org/databases/ModelBuilding/EMT.pdf.

"clover" as the pore opening resembles a four-leafed clover. This pore opening was a 20-ring with a maximum diameter of 1.32 nm (Figure 3.27).

3.4.2.3.2 The Methanol-to-Olefin Process

The more general methanol-to-hydrocarbon process was discovered by Mobil Oil (Exxon Mobil) in 1977. As the name suggests, it converts methanol into more useful products such as gasoline and olefins. The MTO process then narrows this down to the production of ethylene and propylene (preferably propylene) from methanol. These olefins are the basis for the formation of many polyolefins, such as polyethylene and polypropylene.

Critical for this process are acidic zeolites that have the correct pore size and pore entry diameters to allow for these reactions. The most commonly used *zeotypes* for this process are the H-SAPO-34 (CHA) and the H-ZSM-5 (MFI), together with the *zeolite β*. Important for this process, as we will discuss later, is to have zeolites or zeotypes with a well-defined pore diameter and a well-defined pore entrance and exit (Figure 3.28).

In the MTO process, methanol is converted into olefins using zeolites and zeotypes as acid catalysts. The general process (methanol-to-hydrocarbons) can be tuned to either predominantly produce olefins (MTO) or gasoline (MTG).

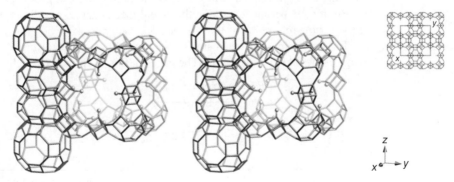

Figure 3.27
Structure of cloverite (CLO), viewed along [001] (upper right: projection down [001]). Notice the small eight-ring pore openings and the large 20-ring "clover-like" pore opening. Source: Taken from www.iza-structure.org/databases/ModelBuilding/EMT.pdf.

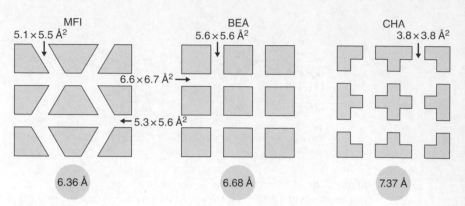

Figure 3.28

Illustration of some zeolite topologies relevant for MTO chemistry, with indication of relevant pore sizes and pore openings. In gray, the maximum diameter of a sphere that can be included is given. Source: Reproduced with permission of the RSC. Adapted from Ref. [18].

The methanol can be made from synthesis gas (mixture of CO and H_2). Synthesis gas can be produced from almost any gasifiable carbonaceous species, such as natural gas, coal, biomass, and waste.

In an acid catalyst, the methanol will first form dimethylether (DME) and water, together with some methoxy species that are bound to the surface of the catalyst. In a second step, the DME will dehydrate to form ethylene and propylene. Depending on the reaction circumstances, the reactions will continue to form higher olefins (by oligomerization of the lower olefins) and aromatics. For example, propene may trimerize, and then crack into butenes and pentenes. Disproportionation to aromatics and alkanes is highly undesirable; in the 1980s, this step was essential to the MTG process, but it must be greatly suppressed if the goal is olefin synthesis [19].

Figure 3.29 shows a schematic overview of the processes.

Researchers have noted that, in the 10-ring zeotype H-ZSM-5, a lot of methylbenzene and alkanes are produced, whereas the eight-ring H-SAPO-34 shows much more of the desired low alkenes. The reason was found to be twofold: first of all, the H-SAPO-34 has a more moderate acidity, reducing the secondary olefin reactions that require a high acid strength. But, secondly, and maybe more importantly, the branched benzenes that were formed were unable to leave the zeolite pores through the smaller eight-ring windows. This is another example of *product shape selectivity*.

But it soon became clear that the actual mechanism is very complex; this mechanism is now referred to as the *carbon pool mechanism*.

The carbon pool mechanism is schematically outlined in Figure 3.30. First, the cage of the SAPO-34 is drawn. The cages of this material are roughly 1 nm long, and

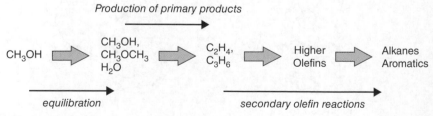

Figure 3.29

Schematic illustration of the conversion of methanol to olefins over a solid acid zeolite catalyst. Source: Reproduced with permission of the RSC. Adapted from Ref. [18].

each cage is connected to its six nearest neighbors through windows roughly 0.38 nm in diameter. So, we have a large pore with narrow entrances.

As shown in Figure 3.30, cycloalkanes formed by propene trimerization can transfer hydrogen to other propenes, forming the small amounts of propane seen during MTO and leaving a methylbenzene molecule trapped in a cage. As we will see, this methylbenzene molecule is needed to make the cage an active site for MTO catalysis; it can age into other, less active, aromatic species, but once a cage contains an aromatic species it will continue to do so until the catalyst is regenerated by combustion.

A key scientific and technological issue in MTO catalysis is understanding and controlling the relative rates at which an active cage produces ethylene versus propene. In either case, a cage containing pentamethylbenzene is formed by olefin elimination and re-alkylation completes the catalytic cycle.

However, the catalyst deactivates as a function of time-on-stream. The methylbenzene components of active H-SAPO-34 catalyst cages age into polyaromatic species. The exact mechanism is not known, but as suggested in Figure 3.31 it may begin when a linear butene molecule (formed by oligomerization and cracking of propene, or some other undesirable secondary reaction) alkylates a methylbenzene, followed by hydrogen loss and ring closure.

With time-on-stream, methylnaphthalene (which is a mildly active reaction center) ages to phenanthrene, which in turn can age to pyrene. Both phenanthrene and pyrene are inactive as reaction centers, not participating in the extension of side chains and elimination of olefins, though they can be methylated to some extent depending on topology of the catalyst.

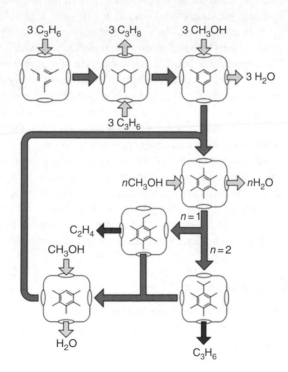

Figure 3.30

Schematic overview of the "carbon pool mechanism." Source: Reproduced with permission of Springer Nature [19].

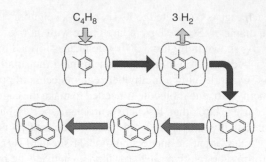

Figure 3.31
Schematic of the aging of the organic reaction centers of the hydrocarbon pool. Source: Reproduced with permission of Springer Nature [19].

Figure 3.32 shows the evolution of the lifecycle of the catalyst. Each cage is around 1 nm and a typical crystallite can vary from one to several hundred nanometers in diameter. Early in its lifetime, a few percent (or less) of the catalyst cages contain methylbenzenes and the occasional methylnaphthalene (situation (a)). In "middle age," the catalyst crystallite has a larger number of both methylbenzene and methyl-naphthalene sites, but the organics in some cages have aged further to phenanthrene or pyrene and lost activity (situation (b)). Finally, at the end of life, more than half of all cages contain polycyclic aromatic hydrocarbons in a deactivated catalyst and mass transport is severely restricted (situation (c)).

This lifecycle is also reflected in the product that a catalyst forms. In the beginning (the so-called *induction* period), no aromatic species are formed at all and the catalyst only produces DME and water (Figure 3.33). As soon as some methylbenzene species are formed, the catalyst is in the *initiation* phase and the catalyst starts to produce ethylene and propylene, together still with some DME. When the catalyst is "in regime," the band of active catalyst moves downward (sideways), producing the desired products and leaving the deactivated catalyst behind, like ash from a cigar. Finally, when all the active sites in the catalyst are deactivated by coke, the catalyst is dead and needs to be reactivated by combustion of the coke.

In fact, this process very much resembles the catalytic cracking process we described earlier and also here is a technological solution for some of these issues, a fluidized bed reactor. In the cigar model (a prop-flow reactor or fixed bed reactor), the deactivated parts of the catalyst will cause very serious transport limitation and diffusional barriers for methanol to diffuse to the active part of the catalyst.

A lot of research is ongoing today to better understand the complex mechanisms that occur in the carbon pool model and to further tune selectivity toward propylene, as this commodity has a much higher economic value than ethylene.

3.4.2.4   *Zeolites in a Biorefinery*

3.4.2.4.1   *The Nature of the Second Generation Biomass*

The previous examples of the use of zeolites in catalysis (zeolite for FCC, templated zeolite for alkylation and isomerization, zeotype for MTO, all in the acid form) are big processes run by big oil companies.

However, oil companies and many small emerging companies feel the urgency to reduce our fossil oil consumption to achieve a more sustainable, circular way to fabricate chemicals and fuels.

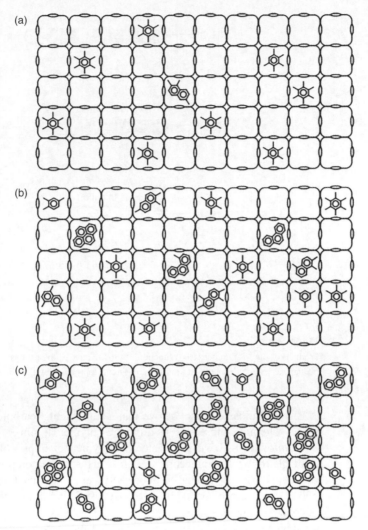

Figure 3.32
Evolution of the lifecycle of H-SAPO-34 in the MTO process. Source: Reproduced with permission of Springer Nature [19].

The use of biomass has become an important source of chemicals and fuels. It soon became clear in the 2000s that the sources of biomass for chemical production should not compete with food production and no acres of land for food production should be sacrificed for biomass.

This complicates of course the biomass conversion process, as the "simple" feedstocks, such as sugar (sucrose, production of bioethanol) or vegetable oils (polymers) are in direct competition with the food industry because the raw materials are sugar beets, soy beans, and so on.

A biomass source that is far less in competition with food production comes from *lignocellulose*. Lignocellulose is dry plant matter, in our case not related to edible plants. High energy feedstocks are, for example, elephant grass and fast growing trees such as poplars.

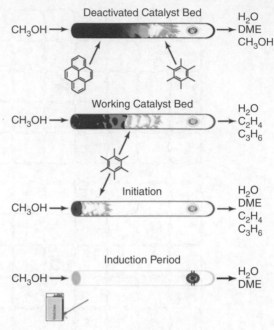

Figure 3.33
Products of the MTO process as a function of the lifecycle of the catalyst, authors represent the catalytic bed here as a cigar that slowly burns away. Source: Reproduced with permission of Springer Nature [19].

Lignocellulose consists of two major ingredients; carbohydrate polymers (cellulose and hemicellulose) and an aromatic polymer (lignin).

Cellulose (40–50 wt.%) and starch are typical examples of carbohydrate polymers available from wheat and corn. Cellulose is the major component in the fibrous materials of cell walls. A typical cellulose polymer is shown in Figure 3.34. It is basically a polymer of sucrose bonded by β-1,4-glycosidic bonds. As cellulose occurs as dense, microcrystalline domains, access to ether bonds is difficult (contrary to starch) and the molecule is therefore not so easy to depolymerize (Figure 3.34).

Hemicellulose (25–40 wt.%) is the glue in the cellulose fibers, it has short and branched polymer chains built up by several hexoses (six-ring sugars) and pentoses (five-ring sugars). Hemicellulose is – contrary to cellulose – relatively easy to hydrolyze (depolymerize).

Lignin (10–25 wt.%) is the third ingredient. It provides rigidity to the plant. It is a complex polyaromate with a three-dimensional structure built by propyl phenol alcohols, bonded both by ether and C—C links. The large heterogeneity of the molecular and its inert character make this ingredient the toughest challenge in a biorefinery (Figure 3.35).

Clearly, the use of this *second-generation biomass* (contrary to first generation biomass: sugars and vegetable oils) is not straightforward. In any application, before even thinking of using a zeolitic catalyst, the biomass needs to be broken down into smaller molecules. This can be done by *fast pyrolysis*. When bringing the biomass for a very short time to 500 °C, a black liquid called *bio-oil* is produced. Unfortunately, the bio-oil consists of more than 400 different molecules that are difficult to separate. While it is a valuable source for energy and for syngas, it is not useful for the production of chemicals.

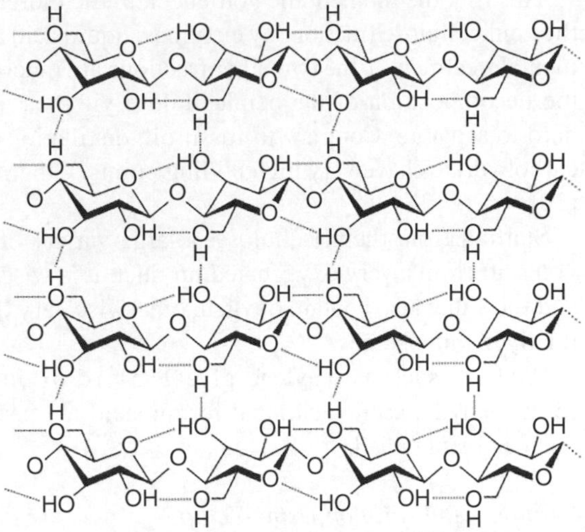

Figure 3.34
Crystalline patches of cellulose. Source: Taken from https://commons.wikimedia.org/wiki/File:Cellulose_
strand.svg, CC BY-SA 3.0, https://commons.wikimedia.org/w/index.php?curid=26213703.

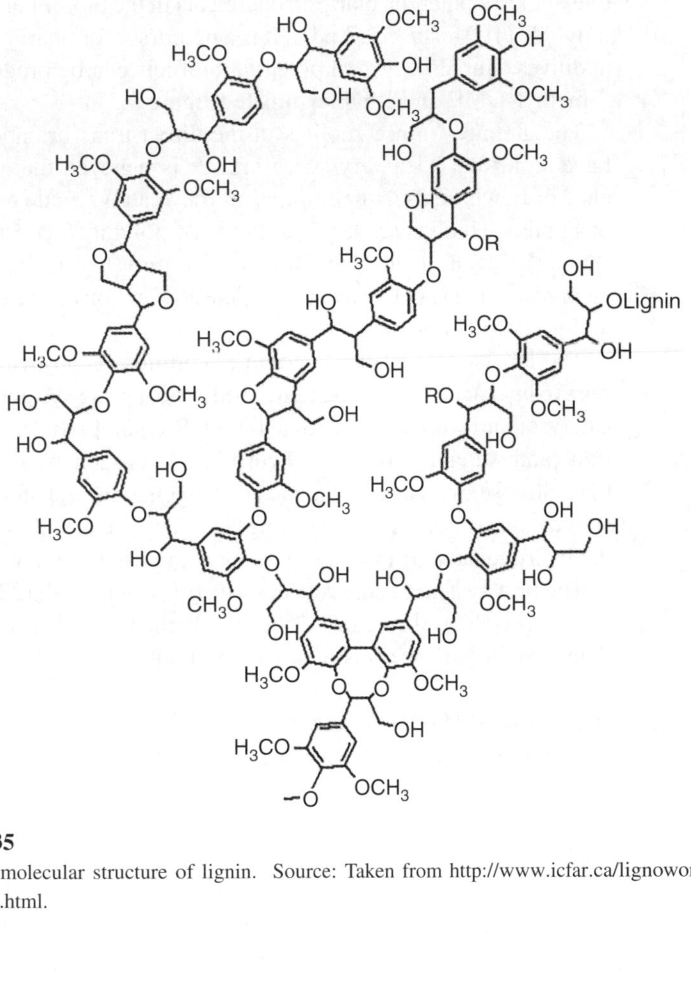

Figure 3.35
A possible molecular structure of lignin. Source: Taken from http://www.icfar.ca/lignoworks/content/
what-lignin.html.

This is at the moment the bottleneck in the biorefinery. While it is easy to degrade the hemicellulose fraction (by enzymes, liquid acid and bases, and many other experimental treatments), the two other fractions are much harder to decompose. Even then, the decomposition of the hemicellulose yields a number of products that are very hard to separate. Contrary to fossil oil, distillation is not possible as the sugars and carbohydrates have very high boiling points. Other techniques need to be established [20].

Starting from (hemi)cellulose, a large variety of chemicals can be produced. As you start from a very oxygenated product, it is wise to produce high value chemicals and fuels that also contain oxygen, otherwise very deep deoxygenation is required to finally obtain alkanes.

Products such as ethylene glycol, 5-hydroxymethylfurfural (HMF) (see later), formaldehyde, acetic acid, and formic acid are just some examples of the products that can be obtained.

3.4.2.4.2 Dehydration – Rehydration Using Zeolites

The molecule 5-HMF (5-hydroxymethyl furfural) is an important platform molecule as it can be obtained from cellulose and can be converted into formic acid and levulinic acid (Figure 3.36).

This levulinic acid is been coined by the U.S. Department of Energy (U.S. DoE) as one of the 12 potential platform chemicals in the biorefinery concept (http://www.nrel.gov/docs/fy04osti/35523.pdf). It is a precursor for pharmaceuticals, plasticizers, and additives. But also, several potential biofuels can be produced, like γ-valerolactone, 2-methyl-DMF, and ethyl levulinate (Figure 3.37).

The attempts to use zeolites in the direct transformation of cellulose to 5-HMF have so far not been very successful. It is believed that this is caused by the presence of Lewis acids in the zeolites, as they catalyze side reactions with the formation of humins. Humins are large dark-brown polymers, consisting of a soluble fraction (humic acid) and an insoluble fraction (humin) (Figure 3.36). However, using furfuryl alcohol as a feedstock, almost complete conversion can be obtained using ZSM-5 with an Si/Al ratio of 30 [21].

The large water-insoluble polymer cellulose can obviously not enter the zeolitic pore to be cracked and converted. The best catalytic system at the moment is a H-USY catalyst that is further impregnated with Ru nanoparticles. Ru is a known hydrogenation catalyst, and in this case hydrolytic hydrogenation is required: the splitting of the cellulose, with a simultaneous hydrogenation, transforming the sugars into more stable sugar alcohols and avoiding the side reactions to humins. A sugar alcohol is the hydrogenated form of a sugar; some are used as sweeteners (Figure 3.38).

The best system to date was described Ennaert et al. [22], using an H-USY zeolite with a very low Si/Al ratio (2:6), and loaded with Ru nanoparticles yielding more than 95% of pure hexitols (C6-alcohols) out of the cellulose feedstock.

3.4.2.4.3 Zeolite-Assisted Hydrodeoxygenation of Bio-Oils

Bio-oils that come from the fast pyrolysis of biomass contain a large number of phenolic compounds that originate from the lignin fraction. While these phenolic compounds are highly valuable, it is almost impossible to purify them from this complex "soup" of molecules.

Hydrodeoxygenation is the process by which aromatic compounds are hydrogenated (turned in non-aromatic compounds, e.g. phenol to cyclohexanol) and are

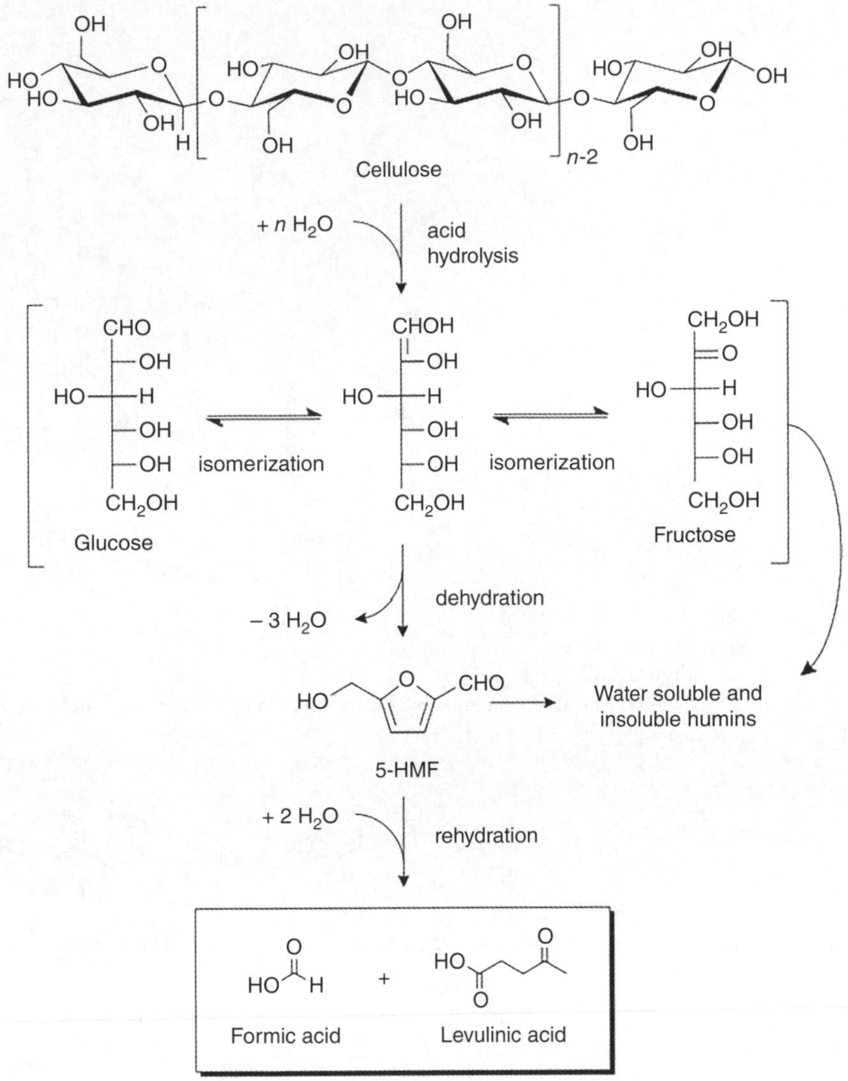

Figure 3.36

Proposed reaction pathway for the acid catalyzed hydrolysis of cellulose to levulinic acid. Source: Reproduced with permission of Elsevier. Adapted from Ref. [10].

subsequently dehydrated to lose oxygen and turn into a cyclic alkane. The cyclic alkanes can be used as fuels. To create such a catalyst, you need two catalytic functions: a hydrogenation catalyst (typically Ni, Pd, or Pt nanoparticles) and an acid catalyst for the dehydration reaction.

Jones and coworkers [23] developed a HY zeolite loaded with Pt nanoparticles in a fixed bed reactor under a 40 bar H_2 pressure at 200 °C. They were able to produce large amounts of cyclohexane using this zeolite. Since then, many different catalytic systems have been explored, for example Ni@H-ZSM-5, Pd@H-ZSM-5, Pd@β, and so on (Figure 3.39).

In many cases, the product shape selectivity of the catalyst plays a major role. In the beta zeolite and in HY (or USY) zeolites, polymerization products are also formed,

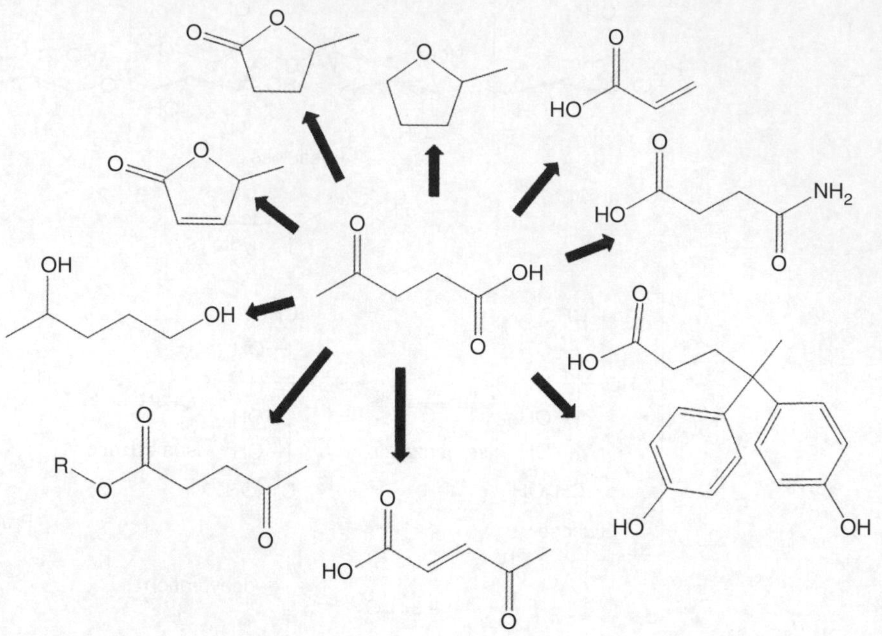
O
O
O
O
O
HO
O
O
O
NH2
HO
O
OH
OH
OH
O
O
HO
O
R
O
O
O
HO
OH
HO
O
O

Figure 3.37
Useful chemicals produced from levulinic acid. Source: Taken from https://commons.wikimedia.org/w/index.php?curid=44513671.

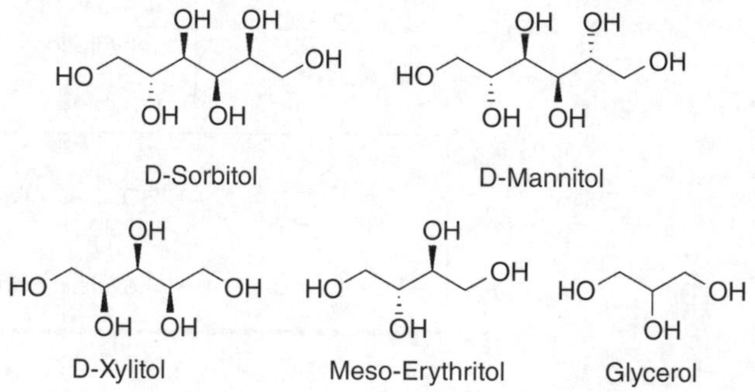
D-Sorbitol
D-Mannitol
D-Xylitol
Meso-Erythritol
Glycerol

Figure 3.38
Some common sugar alcohols (polyols).

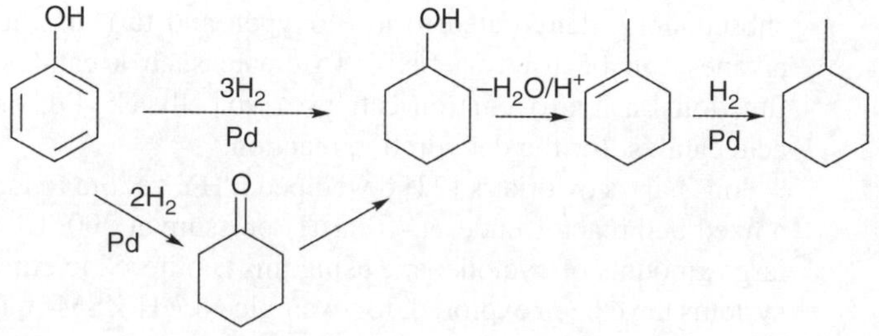
OH
3H2
Pd
OH
–H2O/H+
H2
Pd
2H2
Pd
O

Figure 3.39
Hydrodeoxygenation of phenol. Source: Reproduced with permission of Elsevier. Adapted from Ref. [20].

such as bi(cyclohexane) and oxygenated derivatives thereof. However, when using the smaller pore windows of ZSM-5, only the monomers are formed because only these compounds can leave the pore system.

3.4.3 Gas Sorption and Purification

A lot of literature over several decades is available on the use of zeolites in gas purification and gas separation technology. Remember that zeolites have the nickname "molecular sieves," although gas separation is not only based on size but also on interaction affinity. Just as we did in the catalysis, we will limit ourselves to a few relevant examples and, in this case, we will focus on the zeolitic membranes.

The mechanism of gas separation by a zeolitic membrane follows these steps:

3.4.3.1 Adsorption

The first step is obviously the adsorption of the gases onto the zeolite's surface. Especially at low temperatures, the *affinity* of the gas molecular for the surface is much more important than the size. Molecules with a higher polarizability, and dipole or quadrupole moments adsorb stronger on the surface, in many cases simply blocking the adsorption sites and the pores for other molecules. CO_2 and H_2O are very strong adsorbers, especially on the strongly polar surfaces of zeolites. CH_4 and H_2 are weak adsorbers (Table 3.1).

The zeolites can be tuned for optimal adsorption. Their properties as polarity, topology, and flexibility of the framework, the type of counter cation and the zeolite pore size determine adsorption behavior. One of the most important parameters is polarity, which in turn depends on the chemical composition of the zeolite. With increasing Al content, the framework becomes more polar. In general, the more polar the zeolite framework, the stronger it adsorbs various molecules through enhanced interactions. An example of this effect is the CO_2/H_2 separation. CO_2 is a strong adsorber and blocks the pores for H_2.

Table 3.1 Properties of some gas molecules.

Molecule	Kinetic diameter (Å)	Polarizability (Å^3)	Dipole moment (D)	Quadrupole moment (D Å)
H_2O	2.65	1.450	1.870	2.30
H_2	2.89	0.80	0.000	0.66
CO_2	3.30	2.650	0.000	4.30
O_2	3.47	1.600	0.000	0.39
N_2	3.64	1.760	0.000	1.52
CO	3.69	1.95	0.112	2.50
CH_4	3.76	2.600	0.000	0.02
C_2H_4	4.16	4.260	0.000	1.50
C_2H_6	4.44	4.470	0.000	0.65
$n-C_4H_{10}$	4.69	8.20	0.050	–
$i-C_4H_{10}$	5.28	8.29	0.132	–
SF_6	5.50	6.54	0.000	0.00

Source: Reproduced with permission of Elsevier. Adapted from Ref. [24].

3.4.3.2 *Diffusion*

Diffusion selectivity becomes important when the components have significant size differences. This is the true *molecular sieving* effect. The window size of the zeolite is very important in these cases. An example is the H_2/CH_4 separation (both weak adsorbers, but with a size difference).

The most extreme case of molecular sieving is the *size exclusion*, when one or more components are too bulky to enter the zeolites pore system. An example is the $H_2/$iso-butane separation on eight-ring zeolites.

It should be noted that these considerations are an idealized view of 100% defect free zeolites. If many non-zeolitic pores of much larger sizes are present in the zeolite film, contributions of Knudsen diffusion and molecular diffusion can be significant, and even become the dominant diffusion pathways. Knudsen diffusivity is proportional to the pore diameter, so the size of defects plays an important role. Based on these results, the fraction of defective areas should constitute less than 10 ppm for a zeolite membrane to render it operational in the molecular sieving regime [24].

The zeolite films are then typically cast onto a macroporous support. It is important that the support is macroporous (not interfering with the diffusion process) and inert (not changing the gas composition after the zeolite membrane). Macroporous anodized alumina and mesoporous silica are typical supports.

By far the most studied zeolite structure for preparation of membranes is ZSM-5 (MFI) named silicalite-1 in its pure silica form. MFI possesses a three-dimensional pore network consisting of intersecting sinusoidal (a-direction) and straight (b-direction) channels; the 10MR pores of MFI are around 0.55 nm in size. The popularity of this structure is explained by the relative ease of preparation. However, from the perspective of gas separation, it should be noted that MFI pores are larger than the kinetic diameter of most permanent gases. Thus, MFI and other 10- or 12-ring zeolite membranes can only be efficient for adsorption-controlled gas separations or separation of larger hydrocarbon molecules.

Zeolites with smaller pores can offer real molecular sieving for separation of permanent gases, as typical pore size of these zeolites is below 4 Å, which is close to the kinetic diameter of many permanent molecules. To date, several eight-ring structures have been utilized for the preparation of gas selective zeolite membranes. Table 3.2 gives an overview of some already applied and other promising 8MR structures.

Table 3.2 Pore apertures of common zeolites, compared to the kinetic diameter of molecules.

Molecule	Kinetic diameter (nm)	Zeolite	Pore size (nm)
He	0.25	K-A	0.30
NH_3	0.26	Li-A	0.40
H_2O	0.28	Na-A	0.41
N_2	0.36	Ca-A	0.50
SO_2	0.36	Na-ZSM-5	0.54
Propane	0.43	Na-Y	0.74
Benzene	0.53	Na-mordenite	0.67
CCl_4	0.59	AlPO-5	0.80
Cyclohexane	0.62	VPI-5	1.20

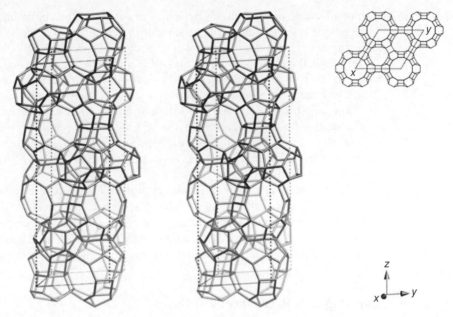

Figure 3.40

Framework viewed normal to [001] (upper right: projection down [001]). Source: Taken from http://europe
.iza-structure.org/IZA-SC/Atlas_pdf/DDR.pdf.

Where the LTA zeolites and even the pure silica LTA (named ITQ-29, after
ITQ = Instituto de Tecnología Química, Valencia, Spain), the performance is
relatively poor, because the pores are just a bit too big for, as an example, effective
CH_4/H_2 separation, and there difficulties in making pure and defect free samples.

A promising zeolite for methane/hydrogen separation is the *D*eca-*D*odecil, 3*R*
(*DDR*) zeolite. DDR is an eight-ring zeolite, originally prepared by Hermann Gies
in 1986 [25]. The acronym (the same as the framework code) comes from deca- and
dedecahedra, three layers, and rhombohedral (the description of the topology), abbreviated to DDR. The topology of DDR is shown in Figure 3.40.

Main applications of these membranes are CO_2 and H_2 separations. For instance,
selectivities of the order of 10^3 and 10^2 have been reported for CO_2/CH_4 and CO_2/air
separations, respectively [26].

So, lately, CHA membranes and zeotype SAPO-34 are currently considered the
most promising candidates for light gas separations. Especially SAPO-34 membranes
have shown excellent performance. They are highly selective and permeable for CO_2
in various mixtures as well as for H_2 and other small molecules [27].

3.5 Solid-State NMR

3.5.1 Introduction to the Technique NMR

Nuclear Magnetic Resonance or NMR spectroscopy has grown during the recent
decades as an invaluable tool for chemical analysis. Currently, it is mostly used on
samples in the liquid state in a wide variety of applications; for example, for structure
determination of organic and inorganic compounds, such as proteins, catalysts, and

biomolecules, or the study of dynamics in solution. The applicability of the NMR technique is without any doubt undeniable in very different fields. In most occasions, a complete analysis and interpretation of the full spectrum can be achieved with a very limited amount of sample. It is considered to be the spectroscopic technique of choice by many researchers and additionally is also non-destructive. This probably makes it the most powerful technique for structure analysis.

On the other hand, the use of NMR on solids, such as zeolites and silicas, is more challenging but has progressed by the development of specific techniques that also now allow measuring samples in the solid state. During recent years, the advances made in this field have created many opportunities to also employ NMR for structural analysis of solids.

For the interpretation of NMR spectra, it is necessary to have some understanding of the basic physical principles of NMR spectroscopy. However, a detailed background on the mathematics and physics of NMR is beyond the scope of this chapter. For more details, we refer to some very interesting reviews and books on the principles of liquid- and solid-state NMR spectroscopy [28].

3.5.2 Nuclear Magnetic Resonance: The Basics

NMR, also called Magnetic Resonance Imaging (MRI) in a medical environment, considers nuclei of atoms that are brought into an external magnetic field (B_0). The nuclei of several elemental isotopes possess a spin (I), which is characteristic for a certain isotope. The spin can be integral ($I = 1, 2, 3, \dots$) or fractional ($I = 1/2, 3/2, 5/2, \dots$), but it can also be equal to zero for certain isotopes (e.g. $I = 0$ for ^{16}O, ^{32}S, $\dots$). A selection of isotopes and their spins is presented in Table 3.3. The isotopes with $I = 1/2$ are most often used in NMR, such as ^1H, ^{13}C, ^{19}F, and ^{31}P, as these are the most suitable for this technique. The following explanation of the theoretical background of NMR will be based on 1/2 spins.

The spin of a nucleus creates a magnetic field (B_x). This "magnet" has a magnetic moment (μ), which is proportional to the spin of the nucleus. When this spin is brought into an external magnetic field (B_0), two spin states will occur: $+1/2$ (α) and $-1/2$ (β) (Figure 3.41). The lower energy state will align itself with the external field, while the higher energy $-1/2$ spin state will align itself against B_0. It is clear that an energy difference between both spin states exists at that moment and this difference is dependent on the external magnetic field and also on the type of nucleus

Table 3.3 A selection of isotopes together with their natural abundance and spin.

Isotope	Natural abundance (%)	Spin (I)	Magnetic moment (μ)[a]
^1H	99.9844	1/2	2.7927
^2H	0.0156	1	0.8574
^{11}B	81.17	3/2	2.6880
^{13}C	1.108	1/2	0.7022
^{17}O	0.037	5/2	−1.8930
^{19}F	100	1/2	2.6273
^{29}Si	4.7	1/2	−0.5555
^{31}P	100	1/2	1.1305

[a]In units of nuclear magnetons = 5.05078×10^{-27} J T^{-1}.

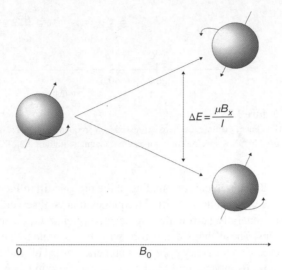

Figure 3.41
When applying an external magnetic field (B_0) to a nucleus with spin, excitation into a $+1/2$ and $-1/2$ spin state will occur. The energy difference between both states is dependent on the spin (I), the magnetic moment (μ), and the magnetic field created by the spin (B_x).

(thus μ and I) that is brought into that magnetic field. The energy difference (ΔE) is generally very small, but can be enlarged by increasing B_0 (Figure 3.42). It is usually expressed in units of MHz and corresponds to radio frequency energy. This means that radio frequency irradiation, which is energetically the mildest type, can excite nuclei to the $+1/2$ and $-1/2$ spin state. When an external magnetic field is applied, each frequency will correspond to a certain nucleus (Figure 3.43). Thus, a clear distinction of the nucleus can be made by choosing the right magnetic field.

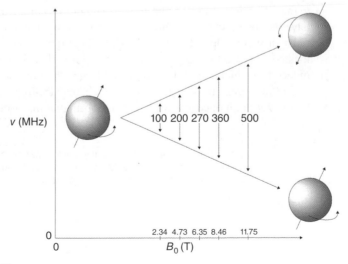

Figure 3.42
The energy difference for a proton spin is depicted in the presence of different external magnetic fields.

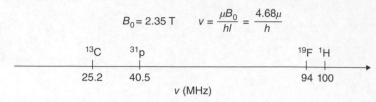

Figure 3.43

The energy difference, here expressed as a frequency in MHz, is shown for different nuclei commonly used in NMR spectroscopy. An external magnetic field of 2.35 T is applied.

The chemical surrounding of the proton will influence the energy difference as well and can result in a shift in frequency that is observed. For example, a proton ($I = 1/2$) is generally surrounded by negatively charged electrons in covalent compounds or ions. These charged electrons will also respond to the external magnetic field and will create a secondary *opposite* magnetic field. This induced field will shield the nucleus from B_0, which implies that a larger B_0 field must be applied to achieve excitation. In brief, proton nuclei that are highly shielded require a high field. This principle allows us to distinguish between different protons via electron shielding, which depends on the chemical environment (Figure 3.44). In brief, the position of the signal in the spectrum will depend on the surrounding atoms and functional groups, and allows us to perform structure analysis.

The nuclei ^1H, ^{19}F, and ^{31}P are highly abundant (Table 3.3) and give strong signals in NMR spectroscopy. ^{13}C, ^{15}N, and ^{29}Si are less abundant and normally give weak signals. For example, long acquisition times can be used to increase the signal to noise ratio.

Frequency as such is, however, not used in practice. It is not an easy unit to work with and the location of the NMR signal is dependent on the external magnetic field, which will always differ a small amount. This can be solved by transforming the frequency into *chemical shift* or δ, with parts per million, or ppm as units. This is more elaborately explained in reference [28a]. A theoretical example of a ^1H NMR spectrum is presented in Figure 3.45. Three groups of signals can be identified due to the different chemical environment. The protons of the methyl group ($—CH_3$) will have the lowest chemical shift. The protons of the methylene group ($—CH_2—$) and the hydroxyl group (HO—) will be shifted to higher chemical shift values as they are located closer to the oxygen atom of the hydroxyl. This electronegative element will induce a de-shielding of the proton and larger chemical shift values.

Furthermore, it must be noted that the chemical shift is influenced by many effects: hybridization, inductive effects of neighboring atoms, aromaticity, and so on. NMR

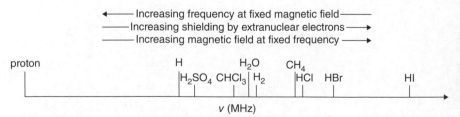

Figure 3.44

The signal of a certain proton will shift according to its chemical environment.

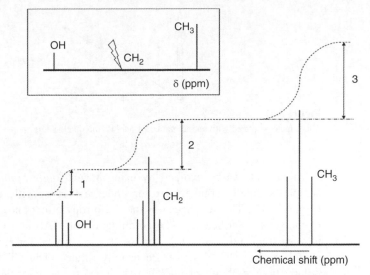

Figure 3.45
The ^{1}H NMR spectrum of ethanol (HO—CH$_2$CH$_3$).

gives us the opportunity to perform different types of experiment, depending on the nuclei, and obtain a complete structural analysis. It has evolved into a very versatile technique, but the interpretation of the (sometimes complicated) spectra requires some skill and practice.

3.5.3 Solid-State NMR: The Challenges

It is generally known that NMR on solid materials is more difficult than on liquids or compounds that can be dissolved in an appropriated solvent. In liquid state NMR, the molecules can move around and anisotropic interactions are averaged out, leading to sharp and resolved NMR signals or peaks in a spectrum. In solids, this mobility is absent, and anisotropic interactions and different orientations of, for example, crystals and large aggregates of molecules will lead to a broadening of signals (Figure 3.46).

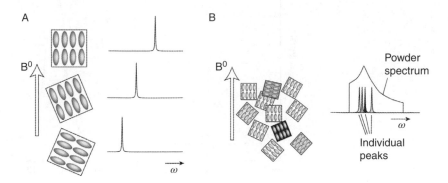

Figure 3.46
A different orientation of certain crystals will result in slightly shifted signal. A powder spectrum is the sum of all individual peaks. Source: Reproduced with permission of John Wiley & Sons, Ltd. Redrawn from Ref. [29].

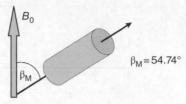

Figure 3.47
The sample is rotated with an inclination of 54.74°; the magic angle.

This effect can be avoided by using *Magic-Angle Spinning* (MAS). The solid sample is rapidly spun in a rotor under a very specific magic angle of 54.74° (Figure 3.47). This averages out the anisotropy effect and improves the resolution of the peaks significantly. The combination of MAS and *Cross Polarization* (CP) is very frequently used to increase the sensitivity for nuclei that are not very abundant (e.g. ^{13}C, ^{29}Si) or are heavily diluted. The CP technique transfers the polarization of abundant nuclei such as ^1H to the rare nuclei. This enhances the signal to noise ratio and reduces the normally long acquisition times. A typical experiment is noted down as {^1H}-^{29}Si CP-MAS-NMR. This is basically a Silicon NMR experiment where the sample is rotated under the magic angle and CP with the aid of protons is used to further boost the signal. The CP MAS technique is often used in ^{13}C and ^{29}Si solid-state NMR. Other techniques also exist, such as multiple-pulse sequences, homo- and heteronuclear decoupling, and recoupling [28c].

The development of these special techniques for solid-state NMR has transformed it into a very valuable tool and it is becoming equally important as liquid NMR in this field for (bio)-organic chemistry and pharmaceutical sciences.

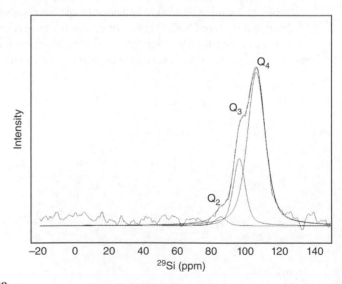

Figure 3.48
NMR Q-sites. Source: Taken from PhD dissertation by Matthias Ide, promoter P. Van Der Voort, Ordered Mesoporous Silica Materials in Liquid Chromatography, 2012, UGent.

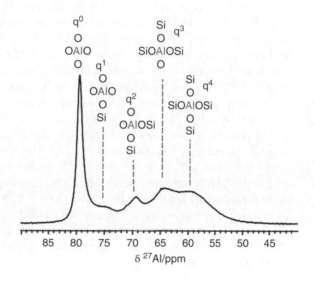

Figure 3.49
Different Q-sites.

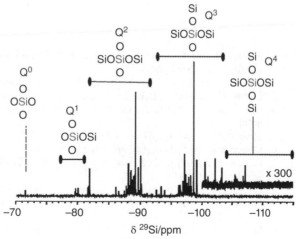

Figure 3.50
The ^{29}Si and ^{27}Al NMR spectrum of silicate and aluminosilicate solutions. Clear similarities can be seen between the two compounds. Source: Reproduced with permission of Elsevier [31].

3.5.4   The Application of Solid-State NMR

Specifically, for the study of the surface of silica, ^{29}Si-NMR, often with ^{1}H cross-coupling, is a very powerful technique. Of the naturally occurring isotopes ^{28}Si (92.21%), ^{29}Si (4.70%), and ^{30}Si (3.09%), ^{29}Si only has a 1/2 spin and therefore a magnetic moment. This puts it in the same league as the other elements of group 14 in the periodic table, such as carbon, tin, and lead.

In the study of many silicas, silicates, zeolites, and other silica-based materials, $\{^{1}H\}$-^{29}Si CP-MAS-NMR is a very useful technique as it discerns the number of siloxane and silanol groups, referred to as Q-sites. A Q^4-site represents a silicon with four siloxane bonds, representative of the bulk of the silica material. A Q^3-site represents a silicon atom with three siloxane linkages and one silanol linkage, and therefore represents a typically free (isolated) or vicinal (hydrogen bridges) silanol group. A Q^2 site then typically represents a silicon with two silanols, the geminal silanols.

FTIR and ^{29}Si-MAS-NMR are therefore very complimentary techniques in the study of silica: the FTIR can discern between the isolated sites (free and geminal) and the hydrogen-interacting silanols (vicinal). NMR, on the other hand, shows the difference between a Q^3 (one silanol per silicon) and Q^2-site (two silanols per silicon) (Figure 3.48). The combination of the two techniques leads to a full understanding of the hydroxyl population on the silica surface. A recent example of the quantification of Q-sites by $\{^{1}H\}$-^{29}Si CP-MAS-NMR is shown in Figure 3.49 [30].

The possibility to measure ^{27}Al NMR is also important in the field of materials science, especially when discussing zeolites. The element aluminum has a nuclear spin $I = 5/2$ and is NMR active. As $I > 1/2$, broader signals are mostly observed and overlap of NMR resonances can occur. Although the resolution is lower, it normally has a higher signal to noise ratio. Similar to the ^{29}Si NMR spectrum, the ^{27}Al NMR spectrum can provide valuable information on the local chemical environment of the nucleus [31]. Figure 3.50 shows an alkaline silicate and an aluminosilicate solution. Resonances of the Q^n sites for silicon ($SiO_{4-n}(OSi)_n$) and q^n sites for aluminum ($AlO_{4-n}(OSi)_n$) can be clearly seen.

References

1. W. D. Callister, D. G. Rethwisch, Materials Science and Engineering – An Introduction. 9, Wiley, New York, 2014.
2. Abe, E., Pennycook, S.J., and Tsai, A.P. (2003). *Nature* 421: 347–350.
3. Cundy, C.S. and Cox, P.A. (2003). *Chem. Rev.* 103: 663–701.
4. Cronstedt, A.F. (1756). *Kongl Vetenskaps Academiens Handlingar*, vol. 17, 120. Stockholm.
5. Morris, R.E. (2005). *J. Mater. Chem.* 15: 931–938.
6. Löwenstein, W. (1954). *Am. Mineral.* 39: 92–96.
7. Gutmann, V. (1978). *The Donor-Acceptor Approach to Molecular Interactions*. New York: Springer US.
8. Uytterhoeven, J.B., Christner, L.G., and Hall, W.K. (1965). *J. Phys. Chem.* 69: 2117.
9. Janssen, A.H., Koster, A.J., and de Jong, K.P. (2001). *Angew. Chem. Int. Ed.* 40: 1102–1104.
10. Sels, B. and Kustov, L. (2016). *Zeolites and Zeolite-like Materials*. Amsterdam: Elsevier.
11. R. L. Wadlinger, G. T. Kerr, E. J. Rosinski, U.S. Patent 3.308.069, 1967.
12. Perezpariente, J., Martens, J.A., and Jacobs, P.A. (1987). *Appl. Catal.* 31: 35–64.
13. R. J. Argauer, G. R. Landolt, U.S. Patent 3.702.886, 1972.
14. Flanigen, E.M., Bennett, J.M., Grose, R.W. et al. (1978). *Nature* 271: 512–516.
15. Wilson, S.T., Lok, B.M., Messina, C.A. et al. (1982). *J. Am. Chem. Soc.* 104: 1146–1147.
16. Lok, B.M., Messina, C.A., Patton, R.L. et al. (1984). *J. Am. Chem. Soc.* 106: 6092–6093.

17. Estermann, M., Mccusker, L.B., Baerlocher, C. et al. (1991). *Nature* 352: 320–323.
18. Van Speybroeck, V., De Wispelaere, K., Van der Mynsbrugge, J. et al. (2014). *Chem. Soc. Rev.* 43: 7326–7357.
19. Haw, J.F. and Marcus, D.M. (2005). *Top. Catal.* 34: 41–48.
20. Ennaert, T., Schutyser, W., Dijkmans, J. et al. (2016). *Zeolites and Zeolite-Like Materials*, 1e. Elsevier.
21. Lange, J.P., van de Graaf, W.D., and Haan, R.J. (2009). *Chem. Sus. Chem.* 2: 437–441.
22. Ennaert, T., Geboers, J., Gobechiya, E. et al. (2015). *ACS Catal.* 5: 754–768.
23. Hong, D.Y., Miller, S.J., Agrawal, P.K., and Jones, C.W. (2010). *Chem. Commun.* 46: 1038–1040.
24. Kosinov, N., Gascon, J., Kapteijn, F., and Hensen, E.J.M. (2016). *J. Membr. Sci.* 499: 65–79.
25. Gies, H. (1986). *Z. Kristallogr.* 175: 93–104.
26. Van Den Bergh, J., Zhu, W.D., Kapteijn, F. et al. (2008). *Res. Chem. Intermed.* 34: 467–474.
27. (a) Li, S.G., Zong, Z.W., Zhou, S.J. et al. (2015). *J. Membr. Sci.* 487: 141–151; (b) Poshusta, J.C., Tuan, V.A., Pape, E.A. et al. (2000). *AlChE J.* 46: 779–789.
28. (a) Claridge, T.D.W. (1999). *High-Resolution NMR Techniques in Organic Chemistry*, Tetrahedron Organic Chemistry Series, vol. 19. Elsevier; (b) Duer, M.J. (2007). *Solid State NMR Spectroscopy: Principles and Applications*. Wiley; (c) Laws, D.D., Bitter, H.M.L., and Jerschow, A. (2002). *Angew. Chem. Int. Ed.* 41: 3096–3129; (d) Dybowski, C. and Bal, S. (2008). *Anal. Chem.* 80: 4295–4300; (e) Carbajo, R.J. and Neira, J.L. (2013). *NMR for Chemists and Biologists*. Springer.
29. Levitt, M.H. (2008). *Spin Dynamics: Basics of Nuclear Magnetic Resonance*. New York: Wiley.
30. Ide, M., El-Roz, M., De Canck, E. et al. (2013). *Phys. Chem. Chem. Phys.* 15: 642–650.
31. Neuhaus, D., Bodenhausen, G., Gadian, D. et al. (2016). *Prog. Nucl. Magn. Reson. Spectrosc.* 94–95: A2–A2.

4 Silica, A Simple Oxide – A Case Study for FT–IR Spectroscopy

4.1 Different Methods to Synthesize Silica

The name silica comprises a large class of products with the general formula SiO_2 or $SiO_2.xH_2O$. Silica is an abundantly naturally occurring material in the Earth's crust; over 90% of which is composed of silicate minerals, making silicon the second most abundant element (about 28% by mass) after oxygen. But it is also widely present in plants such as bamboo, rice, and barley, and often has beautiful structures. In Chapter 1, we showed some nice examples of beautiful forms of silica, in the form of frustules and extracted bamboo leaves.

4.1.1 Silica Gels and Sols

The most documented method for the preparation of silica is the *sol-gel route*. Sol-gel processing is used not only for the preparation of silica gels, but also for the synthesis of ceramic products, ranging from thin films and coatings over porous membranes to composite bodies [1].

The sol-gel process is a wet chemical method involving hydrolysis and condensation of metal alkoxides and inorganic salts. The synthesis of silica gel using sodium silicate as a starting product was well documented by Iler, [2] Unger [3], and Barby [4] at the end of the 1970s.

Stöber et al. [5] reported the controlled synthesis of spherical silica powder from tetraethoxysilane (TEOS) using ammonia as a catalyst in 1968. These particles are widely used in chromatography, the synthesis method and the particles are usually referred to as the Stöber method and Stöber particles.

Activity in the research on alkoxysilanes as precursors was initiated by the work of Yoldas [6] and Yamane et al. [7], who demonstrated the preparation of silica monoliths by careful drying of gels. Since the 1980s and up to now, the sol-gel process using alkoxysilane precursors has been a major research topic, aiming for a thorough understanding of the various process steps and a broadening of the field of applications. The huge amount of literature in this context has been excellently reviewed and synthesized by Brinker and Scherer, [8] Hench and West [9], and others.

Introduction to Porous Materials, First Edition.
Pascal Van Der Voort, Karen Leus and Els De Canck.
© 2019 John Wiley & Sons Ltd. Published 2019 by John Wiley & Sons Ltd.

4.1.1.1   Step 1: Hydrolysis

Sol-gel silica synthesis is based on the controlled condensation of $Si(OH)_4$ entities. These may be formed by hydrolysis of soluble alkali metal silicates or of alkoxysilanes. The commonly used compounds are sodium silicates and $Si(OEt)_4$, commonly referred to as TEOS. The $Si(OH)_4$ (silicic acid) cannot be bought, as it is not stable. The general class of $Si(OR)_4$ compounds are called alkoxysilanes. Next to TEOS, tetramethylsilane (TMOS) is often used. Its hydrolysis rate is faster.

To get a rapid and complete hydrolysis, an acid or a base catalyst may be used. In both cases the reaction occurs by a nucleophilic attack of the oxygen contained in water to the silicon atom. Hydrolysis mechanisms were studied by Osterholz and Pohl [10] using alkyltrialkoxysilanes instead of tetra-alkoxysilanes. Both types of compound follow the mechanism described next.

Base-catalyzed hydrolysis (Scheme 4.1) is a two-step process with formation of a pentacoordinate intermediate. Acid catalyzed hydrolysis (Scheme 4.2) proceeds by an S_N2-type mechanism. The leaving alkoxy group is rapidly protonated and a water molecule performs a nucleophilic attack on the central silicon atom.

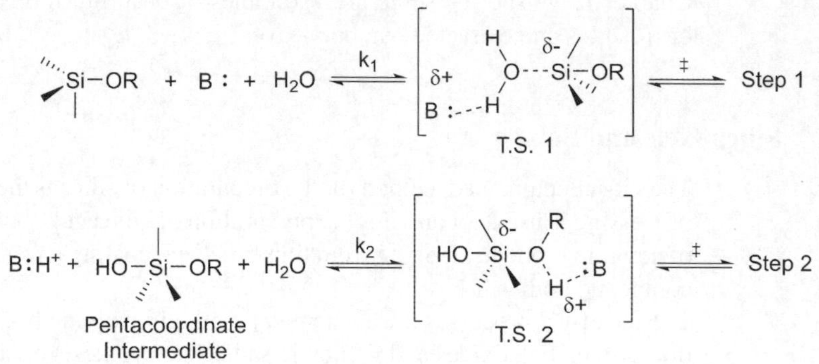

Scheme 4.1

Base catalyzed hydrolysis. Source: Redrawn from [11] with permission.

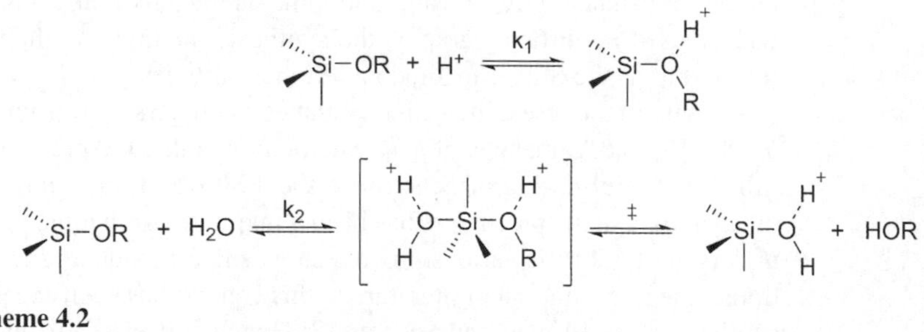

Scheme 4.2

Acid catalyzed hydrolysis. Source: Redrawn from [11] with permission.

4.1.1.2   Step 2: Condensation

The formed silicic acid molecules condense to form siloxane bonds with release of water:

$$\equiv Si-OH \quad + \quad HO-Si \quad \longrightarrow \quad \equiv Si-O-Si \equiv \quad + \quad H_2O$$

Condensation may also proceed by the reaction of the alkoxysilane with a silanol group, releasing an alcohol:

$$\equiv \text{Si}-\text{OH} \quad + \quad \text{RO}-\text{Si} \quad \longrightarrow \quad \equiv \text{Si}-\text{O}-\text{Si} \equiv \quad + \quad \text{ROH}$$

The condensation may be acid or base catalyzed. Hydrolysis and condensation occur simultaneously. The relative rate of both processes determines the sol structure. As discussed by Ying et al. [12], in acidic conditions, hydrolysis is faster than condensation. The rate of condensation slows down with an increasing number of siloxane linkages around a central silicon atom. This leads to weakly branched polymeric networks. On the contrary, under basic conditions, condensation is accelerated relative to hydrolysis. The rate of condensation increases with an increasing number of siloxane linkages. Thus, highly branched networks with ring structures are formed. This generates large, bulkier, more ramified polymers.

Three stages are recognized in silicic acid polymerization:

1. polymerization of monomers to form small primary particles;
2. growth of primary particles;
3. linking of particles into branched chains, then networks, finally extending throughout the liquid medium, thickening it to a gel.

By controlling the pH and with the addition of electrolytes that induce flocculation, the relative importance of steps 2 and 3 may vary. A control of both factors will either favor the growing of the particles or the linkage of particles to form chains, as illustrated in Figures 4.1 and 4.2.

Solid silica spheres are prepared by the Stöber method. This involves the condensation of TEOS in alcoholic solution of water and ammonia. The mechanism of this synthesis was elucidated by Van Blaaderen et al. [14]. The particle growth is rate-limited by the production of hydrolyzed monomer molecules. It involves a surface condensation of monomers and oligomers, while aggregation of particles only occurs in the early stages of the condensation.

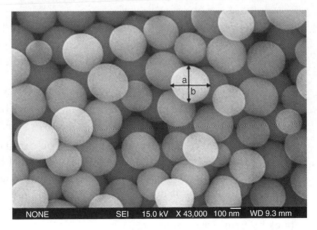

Figure 4.1

SiO_2 particles of average size 324.1 ± 15.7 nm under experimental conditions 8 M ethanol, 0.045 M TEOS, and 14 M NH_3 at 70 °C. Average $d = (a + b)/2$ (nm). Source: Reproduced with permission of Elsevier [13].

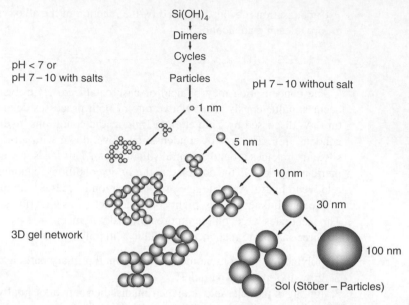

Figure 4.2
Growth of silica particles as a function of time and pH. Source: Reproduced with permission of Elsevier [8].

4.1.1.3 Step 3: Gelation

As the polymeric network extends throughout the total volume, the sol thickens to a gel. The sol-to-gel conversion is a gradual process, which is easily observed qualitatively but difficult to measure analytically. The gelation point t_{gel} is defined as the point where elastic stress is supported. The t_{gel} is not an intrinsic property of a sol. It is influenced by the size of the container, the solution pH, the nature of the salt concentration, the anion and solvent, the type of initial alkoxy group, and the amount of water. In simpler words, the gel is formed when the silica polymer extents from one side of the beaker to the other side, resulting in a stiff product with a viscosity near infinity.

The formed gel is termed a *hydrogel* when water is the solvent, or an *alcogel* if an alcohol is used as solvent. The hydrogel structure is controlled by temperature, the pH of the medium, the nature of the solvent, the nature of the added electrolyte, and the type of the starting salt or alkoxide.

Fast drying of the hydrogel results in a loss of the pore-filling liquid and the material densifies. The pores are narrowed by capillary forces exerted by this liquid. A *xerogel* is formed, with little to no porosity. (ξηρός, xērós, "dry")

If a gel is dried carefully (usually under supercritical conditions), the pore narrowing by capillary attraction is avoided. The formed gel has a very large pore volume (up to 98% of the total volume) and is therefore named an *aerogel*. This broad success of the sol-gel method is due to its ability to form pure and homogeneous products at very low temperatures. An overview is presented in Figure 4.3, taken from the famous book by Brinker and Scherer [8].

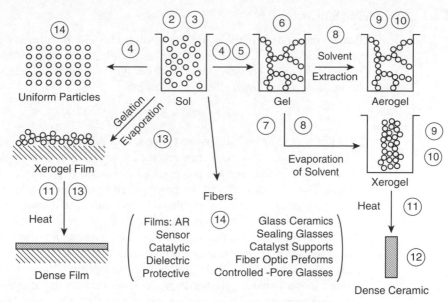

Figure 4.3
Aerosols and gels. Source: Reproduced with permission of Elsevier [8].

4.1.1.4 Step 4: Aging

When the gel is kept in contact with the pore-filling liquid, its structure and properties keep changing as a function of time. This process is called aging. During the aging period, four processes affect the porous structure and surface area of the silica gel. These are polycondensation, syneresis, coarsening, and phase transformation.

Polycondensation is the further reaction of silanols and alkoxy groups in the gel structure to form siloxane bonds by reactions 3 and 4. These reactions result in a densification and stiffening of the siloxane network.

Syneresis is the shrinkage of the gel network, resulting in the expulsion of the pore liquid. This shrinkage is caused by the condensation of surface groups inside the pores, resulting in pore narrowing.

Coarsening or *Ostwald ripening* is the solution and redeposition of small particles. The cause of this process is the higher solubility of convex surfaces compared to concave surfaces. Therefore, the small particles are dissolved and redeposition on larger particles occurs. Also, necks between particles will grow and small pores may be filled in. This results in an increase in the average pore size and a decrease in the specific surface area.

To summarize, during aging the pore structure, surface area, and stiffness of the gel network are changed and controlled by the following parameters: time, temperature, pH, added electrolyte, and pore fluid.

In contrast to the crystalline silica, the synthetic amorphous silicas have a high surface area. The silica gels consist of connected particles with irregular pores. Their surface area can go up to $800 \, m^2 \, g^{-1}$ with pore sizes in the micro-, meso-, and macro-ranges.

4.1.2   Pyrogenic Silicas

Besides preparation in the liquid phase, silica may also be formed with high temperature processes, such as using a flame, arc, or plasma. One of the most widely used methods to synthesize pure silicas is the burning of $SiCl_4$ with hydrogen and oxygen. In the flame, the following reaction takes place:

$$2H_2 + O_2 + SiCl_4 \rightarrow SiO_2 + 4HCl$$

In this way, "fumed" silicas are formed. This process was developed by Degussa and the thus formed silicas were marketed as Aerosil. Now many other brand names are available. In the reaction process HCl is formed, which is evacuated from the system. The characteristics of the produced silica may be controlled by a variation of the reagent concentrations, the flame temperature, and the time of presence in the combustion chamber. Thus, the specific surface area, particle size, and particle size distribution may vary. In this type of silica, primary particles are linked into linear chains and a non-porous structure is produced (Figure 4.4).

Arc silicas are formed by the reduction of high-purity sand in a furnace. This type of silica shows a greater variation in particle size. Primary particles do not form chains but form dense, non-microporous secondary particles. The volatilization of sand in a plasma jet produces plasma silicas. These are ultrafine silica powders.

For fumed silicas, the surface area can be theoretically predicted as follows (Figure 4.5).

The surface area and volume of the fumed silica is given by the following formulas, where n is the number of silica particles:

$$S = n \, 4\pi r^2 \tag{4.1}$$

$$V = n \frac{4}{3} \pi r^3 \tag{4.2}$$

Often, it is easier to work with density than with the volume. One can go from one to the other with the following relationship

$$V = \frac{1}{\rho} \text{ (for 1 g).} \tag{4.3}$$

Figure 4.4
Fumed silica when agitated easily forms a smoke.

Figure 4.5

SEM pictures of AeroPerl 300. Source: taken from http://www.aerosil.com/product/aerosil/downloads/technical-overview-aerosil-fumed-silica-en.pdf.

The substitution of Eqs. (4.2) and (4.3) into (4.1) yields the following relationship between the density and surface area.

$$S = \frac{3}{\rho r} \qquad (4.4)$$

For a typical surface of 400 m^2 g^{-1} and a density of silica of $2.2 \cdot 10^6$g m^{-3}, the radius of particles is 3.4 nm.

4.1.3 Precipitated Silicas

The precipitated silicas include a wide range of silicas with a variety of structural characteristics. Most of the preparation methods are patented. In general, the formation involves a coagulation and precipitation from silica solutions. Properties are therefore supposed to be like those of the gels. For these silicas, however, preparation conditions aim to avoid gel growth and stimulate precipitation.

An overview of the different types of synthetic silicas is presented in Table 4.1.

4.2 The Surface of Silica

As silica consists only of silicon and oxygen, the main bonds in silica are *siloxane* bonds (Si-O-Si). Every silicon is tetrahedrally surrounded by four oxygens. Only at the surface can the oxygen can be bonded to a hydrogen to create surface hydroxyl groups, referred to as *silanols*.

The silanols can be divided into *isolated* groups (or free silanols), where the surface silicon atom has three bonds into the bulk structure and the fourth bond attached to a single OH group, and *vicinal* silanols (or bridged silanols), where two single silanol groups attached to different silicon atoms are close enough to make a hydrogen bond. The word "vicinal" stems from the Latin word *vicinus* (neighbor). A third type of silanol, the *geminal* silanol, consists of two hydroxyl groups that are attached to one silicon atom. "Geminal" stems from the Latin word *gemini* (twins). The geminal silanols are too close to hydrogen bond with each other, whereas the free hydroxyl groups are too far apart. Figure 4.6 shows the different types of silanols in a graphical way.

Table 4.1   Types of silica.

Type of silica	Subtype	Description
Synthetic silicas		Surface area, pore volume, pore size and particle size are to some extent independently controllable. (commercially interesting)
Colloidal silicas		Stable dispersions or sols of discrete particles of amorphous silica.
Silica gels		Coherent, rigid, 3D network of contiguous particles of colloidal silica.
	Hydrogels	Silica gel, in which the pores are filled with the corresponding liquid (water).
	Xerogels	A gel from which the liquid medium has been removed, resulting in a compressed structure and a reduced porosity.
	Aerogels	A special form of xerogel, from which the liquid has been removed in such a way as to prevent any collapse or change in the structure as liquid is removed (Iler). A gel with water-collapsible macropores as judged by direct porosimetry (Barby).
Pyrogenic silicas		Made at high temperature.
	Aerosils	Flame hydrolysis products of $SiCl_4$; very pure materials.
	Arc silicas	Made by the reduction of high-purity sand.
	Plasma silicas	Ultra-fine silica powders, made by the direct volatilization of sand in a plasma jet.
Precipitates		Made by the precipitation of the silicic acid solution.

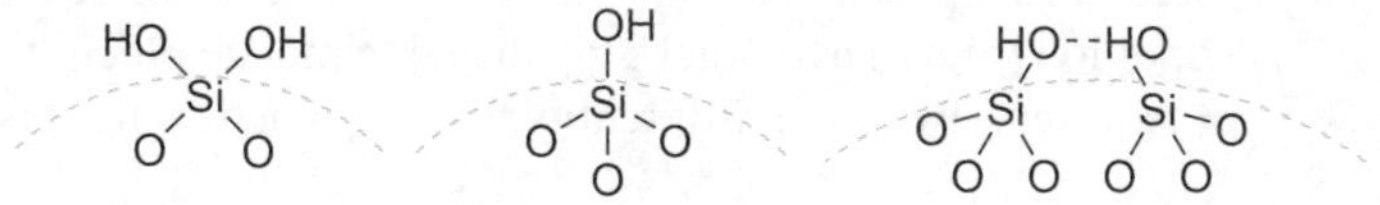

Figure 4.6

Types of silanols. The gray line represents the surface.

The surface structure of amorphous silica is highly disordered, so one cannot expect a regular arrangement of hydroxyl groups. Hence the surface of amorphous silica gel may be covered by isolated as well as vicinal hydroxyl groups. Irrespective of whether a surface contains both types or only isolated hydroxyl groups (as in crystalline silica), complete surface coverage with silanols can be achieved. In this case, the surface is fully *hydroxylated*.

On exposing the hydroxylated silica further to water, it is further able to adsorb water physically by means of hydrogen bonding. In fully hydroxylated non-porous silica species, a multilayer of adsorbed water is built up by increasing the partial pressure. In fully hydroxylated porous silica species, additional capillary condensation takes place on the adsorbed multilayer: on increasing the partial pressure, the pore volume is gradually filled with liquid water. The uptake of physically adsorbed water

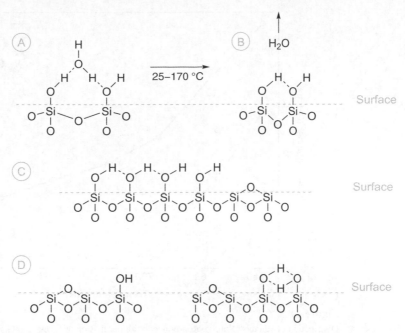

Figure 4.7
Hydrolysis reaction scheme. (A) fully hydrated silica, water is adsorbed physically and through hydrogen bonding; (B) fully hydroxylated silica, the surface is fully covered with silanols, but there is no physically adsorbed water; (C) partially hydroxylated silica, some silanols have condensed to water and siloxane bridges are present at the surface. The silanols are still mainly present as vicinal silanols also called bridged silanols; and (D) almost completely dehydroxylated silica: there are no more hydrogen bond interacting silanols, only free and geminal silanols remain. When further increasing the temperature, these silanols will also disappear to leave a complete siloxane surface.

is termed *hydration*. The terms hydration and hydroxylation are often confused in literature, which is not surprising since it is very difficult to separate the two processes. Both processes can be reversed. *Dehydration* is the loss of physisorbed water as a function of increasing temperature, whereas *dehydroxylation* stands for the condensation of hydroxyl groups to form siloxane bonds. Figure 4.7 depicts these processes.

The silanol number (α_{OH}) is a temperature dependent constant that contains the number of silanol groups expressed as groups per square nanometer. The conversion from mmoles per gram to groups per square nanometer is relatively straightforward, and is left as an exercise. The distribution of silanol numbers as a function of temperature is shown in Figure 4.8, as published by Van Der Voort.

The silanols that are present at the surface of the silica render it possible to functionalize the material. In such cases, we call the silica itself the *support* material. Functionalized silicas have a wide range of applications, in chromatography, catalysis, sorption, sensing, and many more.

4.3 Fourier Transform Infrared Spectroscopy

Fourier transform infrared (FT–IR) spectroscopy is one of the most frequently used techniques to study silica, and – in our case – solid materials in general. In this section, we give an introduction to this important technique using silica as a case study.

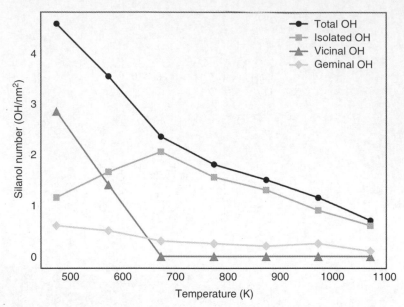

Figure 4.8
Distribution of silanols. Source: Reproduced with permission of Elsevier [11].

4.3.1   Principles of Infrared Spectroscopy

Atoms in molecules and solids do not remain in fixed relative positions, but vibrate about their mean position. This vibrational motion is quantized and at room temperature most of the molecules in a given sample are in their lowest vibrational state. Absorption of electromagnetic radiation with the appropriate energy allows the molecules to become excited to a higher vibrational level. The required energy for this transition comes from the infrared region of the electromagnetic light. The absorption of infrared light as a function of wavelength gives rise to an infrared spectrum with specific spectroscopic fingerprints, which can be assigned to certain molecular entities.

When two independently moving atoms, each with three degrees of freedom (x, y, z), form a diatomic molecule, the six degrees of freedom are transformed into three degrees of freedom for translational motion of the molecule in space, two degrees of freedom for rotational motion, and one for vibrational motion of the atoms in the molecule. The two atoms vibrate in phase around their equilibrium position in the molecule in opposite directions with an amplitude that is inversely proportional to their masses. The classical mathematical approach of the atomic vibrations in a molecule is that of the harmonic oscillator. The chemical bond is considered to be a spring that exerts a force F on the atom given by:

$$F = -k(r - r_{eq}) \qquad (4.5)$$

where k is the force constant in Nm^{-1} and r_{eq} the equilibrium bond distance in m. The minus sign indicates that the force is opposite to the direction of movement of the atom. This is schematically illustrated in Figure 4.9.

The resulting vibrational energy is

$$E_{vib} = \frac{1}{2}k(r - r_{eq})^2 \qquad (4.6)$$

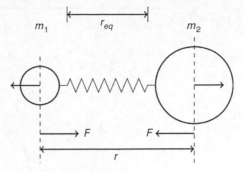

Figure 4.9
The (harmonic) oscillator.

and the frequency

$$\omega_{vib} = 2\pi v_{vib} = \sqrt{\frac{k}{\mu}} \tag{4.7}$$

with μ = reduced mass = $\dfrac{m_1 \cdot m_2}{(m_1 + m_2)}$. Thus, the harmonic vibration of a diatomic molecule can be viewed as the vibration of a spring with mass equal to the reduced mass of the molecule and force constant k. The latter is a measure of the chemical bond in Hz. In vibrational spectroscopy, one uses wavenumbers instead of frequencies. Making use of the fundamental relation, $c = v\lambda$, with λ = wavelength in m and c = light velocity in vacuum ($2.99 \cdot 10^8$ m $\cdot$ s^{-1}), one obtains the unit of the *wavenumber*.

This wavenumber is expressed in m^{-1}, with c in m $\cdot$ s^{-1} and k in N $\cdot$ m^{-1} and μ in kg. To obtain the often-used cm^{-1} (previously called the Kaisar), one has to divide by a factor of 100.

From this equation, one can conclude that the vibrational frequency of a chemical bond is related to the masses of the vibrating atoms and the force constant. The larger the force constant, the higher the vibration frequency. On the other hand, vibration frequencies relate inversely to the masses of the vibrating atoms: a light atom oscillates faster than a heavy one.

Quantum mechanically, the harmonic vibration of a diatomic molecule is treated by solving the corresponding time-independent Schrodinger equation. A more sophisticated approach uses the model of the anharmonic oscillator as shown in Figure 4.10.

In this case, the corresponding quantum mechanical expression of the energy is:

$$E_n = \left(n + \frac{1}{2}\right) h v_{vib} - \left(n + \frac{1}{2}\right)^2 x_a h v_{vib} \tag{4.8}$$

x_a is called the anharmonicity constant, which is always small but positive;, and v_{vib} is the frequency of the harmonic vibration. The transition from $n = 0$ to $n = 1$ is obtained by the absorption of a photon with frequency v_a such that

$$h v_a = E_1 - E_0 = (1 - 2x_a) h v_{vib} \tag{4.9}$$

One observes that the vibration frequency of the anharmonic oscillator, v_a, is lower that that of the harmonic vibration, v_{vib}. The difference between the energy levels of the harmonic and anharmonic vibrations increases with quantum number n. The second consequence of anharmonicity is that besides $\Delta n = \pm 1$; also higher order

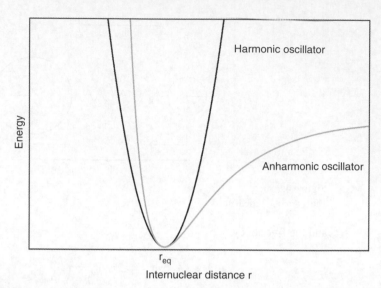

Figure 4.10
Anharmonic oscillator approximation to a vibration.

transitions are allowed, $\Delta n = \pm 2, \pm 3, \ldots$ and visible in the spectra. These bonds are called *overtones* and are usually much less intense than the fundamental modes. These overtones are shown in Figure 4.11.

When N independently moving atoms are combined in a molecule, the $3N$ degrees of freedom are transformed into three degrees of translational freedom, three for rotation (two for linear molecules, because the molecular axis is not a rotational axis) and $3N - 6$ ($3N - 5$) for vibrations.

In a good approximation, and in practice, groups of atoms in a molecule, such as the functional groups in organic molecules, vibrate independently of the rest of the molecule.

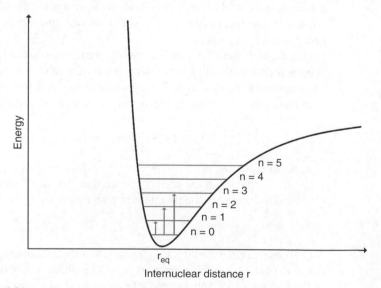

Figure 4.11
Fundamental and overtone vibrations.

Figure 4.12
Symmetric and asymmetric stretching vibrations of water.

They give rise to the so-called group frequencies. These group frequencies are tabulated and form the basis of the interpretation of any vibrational spectrum. The $3N - 6$ ($3N - 5$) fundamental vibrations can also be subdivides according to the type of movement of the atoms in a molecule. If the vibration is a bond lengthening/shortening, such as for the diatomic molecule, one has a stretching vibration designated by the symbol v. One adds the subscripts a and s to indicate, respectively, antisymmetric and symmetric stretching. In the last case, the lengthening/shortening of the bonds occurs in phase, in the first case out of phase, as illustrated in Figure 4.12. The number of stretching vibrations is equal to the number of bonds ($N - 1$), except for rings. The remaining $2N - 5$ vibrations give a change in bond angles and do not affect the bond lengths. They are called bending vibrations and are designated with the symbol δ. It is much easier to change a bond angle than a bond length. In vibrational terms, the force constant of a stretching vibration is larger than that of a bending vibration; therefore, the frequencies (wavenumbers) of a stretching vibration are always higher than those of bending vibrations.

In summary, there are several criteria for bond assignment in vibrational spectroscopy:

1. Tables of group frequencies
2. Strengths of the bonds
3. Stretching versus bending vibrations
4. Mass of the atoms

Finally, the number of vibrations increases rapidly with the number of atoms. Thus, while one has three normal vibrations in water, this is 30 for benzene. Experimentally, one does not observe all 30 vibrations for two reasons:

1. Normal vibrations might have the same frequency; they are degenerate.
2. Some vibrations are forbidding according to Laporte's selection rule. Here, symmetry and group theory play an important role. In general, vibrations that induce a change in dipole moment are allowed in infrared spectroscopy; vibrations that induce a change in polarizability of the molecule are allowed in Raman spectroscopy.

4.3.2 Principles of FT–IR

FT–IR has become an economic and multidisciplinary analysis tool to be found in a broad range of research and application laboratories. Unlike, for instance, Raman spectroscopy where the dispersive and the Fourier transform instrument co-exist and have unique (often complementary) properties, the FT–IR machine has completely overtaken the dispersive instruments, which are no longer produced. Surely, the most important advantage of Fourier transform spectroscopy is known as the Fellgett or multiplex advantage: modern instruments can now take a complete spectrum in less than 1/50 of a second. This has allowed numerous new possibilities:

1. The rapid collection time allows fast hyphenated techniques, such as GC–FT–IR or TGA–FT–IR.
2. The rapid collection time also allows the use of sampling techniques that intrinsically produce a low signal to noise ratio. The signal to noise ratio improves as the square root of the number of collected scans. This multiplex advantage has stimulated the use of several sampling accessories, such as diffuse reflectance cells, photo-acoustic cells, and attenuated total reflection cells, which require 32 to 1000 scans to produce a high-quality spectrum of even the most difficult samples.

The fundamental difference between a dispersive and a FT–IR instrument consists of the method of scanning the sample. In a dispersive instrument, the polychromatic source is monochromatized by a prism or a grating. These separated frequencies are measured independently. In a FT instrument, the polychromatic source is modulated into an interferogram that contains the entire frequency region of the source. Therefore, all frequencies are measured simultaneously.

4.3.2.1   *The Interferogram*

The heart of an FT–IR machine is the interferometer. The Michelson interferometer (Figure 4.13) consists of a beamsplitter, to divide the infrared radiation into two equal beams, and a fixed and a moving mirror. The beamsplitter is designed to transmit *exactly* 50% of the infrared light onto the moving mirror and to reflect *exactly* 50% onto the fixed mirror in the interferometer. The fixed mirror reflects the infrared radiation back to the beamsplitter, while the moving mirror also returns the infrared radiation, but creates a path difference between the two infrared beams.

The production of the interferogram is most comprehensive in the case of a monochromatic source. When both mirrors are at an equal distance from the beam splitter (the so-called zero path difference, ZPD), there is no path difference between the two arms of the interferometer and all the input light exits from the interferometer, half onward to the detector and half returning toward the source. As the moving mirror translates along the optical axis, the path difference steadily increases and passes through positions of constructive and destructive interference. An observer at the detector position would see an image of the source blinking as the

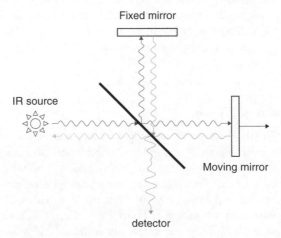

Figure 4.13
The Michelson interferometer.

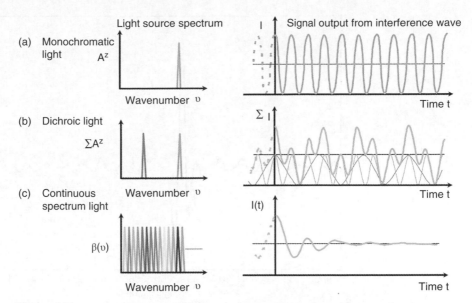

Figure 4.14

Interferograms of (a) monochromatic source, (b) dichromatic source, and (c) real polychromatic source.

mirror translates. This is called the interferogram. It is a function with light intensity as ordinate, and mirror displacement (or time) as abcis. This interferogram of a monochromatic light source can be mathematically formulated as a cosine function. This is exemplified in Figure 4.14a.

Extending to a broad-band infrared source, all wavelengths interfere constructively only at the ZPD position. This will cause the *central burst* in the interferogram. Elsewhere along the track of the moving mirror, the signal intensity is a complex sum of in-phase and out-of-phase contributions. This is exemplified in Figure 4.14b (two wavelengths) and c (full spectrum). A typical interferogram of an actual infrared source is shown in Figure 4.15.

The knowledge that a monochromatic source yields a cosine function as an interferogram is used in all modern FT–IR instruments to establish a precise tracking of the movable mirror. The interference pattern of the monochromatic light of a He-Ne laser is used to monitor the change in optical path difference. The IR interferogram is digitized precisely at the zero crossings of the laser interferogram. The accuracy of the sample spacing is solely determined by the precision of the laser wavelength. This built-in calibration of high precision is known as the Connes advantage.

As the interferogram passes through the sample, selective frequencies are absorbed and the resulting interferogram is transformed into a normal spectrum by means of a discrete Fourier transformation.

A mathematical appreciation of the interferogram can be obtained by Fourier transforming the step function, shown in Figure 4.16 (representing an ideal infrared source). This ideal source gives the same intensity of light between the "frequencies" $-a$ and $+a$ and produces no light at all outside these boundaries.

The general formula for a Fourier transformation is:

$$F(\alpha) = \frac{1}{\sqrt{2\pi}} \int_{-\infty}^{+\infty} f(u)e^{-i\alpha u}du \qquad (4.10)$$

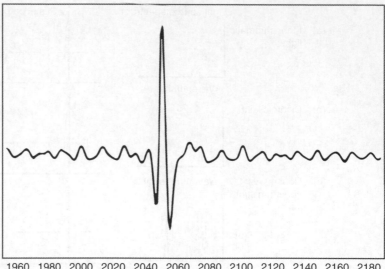

Figure 4.15
The interferogram.

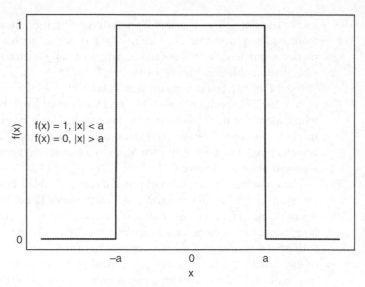

Figure 4.16
The ideal step function.

For an even function, this equation always reduces to

$$F(\alpha) = \frac{2}{\sqrt{2\pi}} \int_0^{+\infty} f(u)\cos(\alpha u)du \qquad (4.11)$$

Applying (4.11) to the step function, this further reduces to

$$F(\alpha) = \frac{2}{\sqrt{2\pi}} \int_0^a \cos(\alpha u)du \qquad (4.12)$$

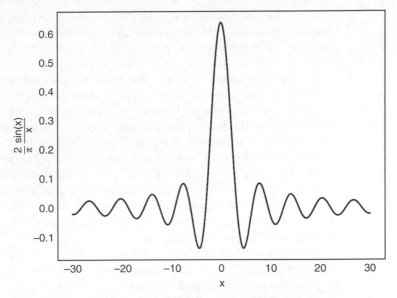

Figure 4.17

The ideal interferogram. Fourier transform of the step function of the previous figure for $\alpha = 1$.

which yields

$$F(\alpha) = \frac{2}{\pi} \frac{\sin(\alpha a)}{\alpha} \tag{4.13}$$

Equation (4.13) is graphically plotted in Figure 4.17, showing the ideal interferogram.

Jean-Baptiste Joseph Fourier (1768–1830) was a French mathematician and physicist. He acted as the scientific advisor to Napoleon. He is best known for the development of the Fourier Series, claiming that any function of a variable can be expanded into a series of sines of multiples of that variable. Lagrange and Dirichlet were able to set the correct boundary conditions for what is now known as the Fourier transform. Some people consider him as the father of the greenhouse effect, as he noted for the first time that the Earth is much warmer that it should be based on the incoming

energy of the Sun. In that way, he was the first one to present the possibility that the Earth's atmosphere might act as an insulator of some kind. He is buried at the famous graveyard in Montparnasse in Paris.

An important feature to note is that the Fourier transform algorithm consists of an integral going from $-\infty$ to $+\infty$. This would imply that the mirror moves over an infinite distance. In practice, however, the mirror movement is restricted in the order of centimeters. This forced termination is known as *boxcar truncation* and the effects on the spectrum are known as ringing. This effect is illustrated in Figure 4.18.

In order to compensate for ringing, a mathematical process known as *apodization* is used. This multiplies the collected data by a function that gradually approaches zero and has a value of 1 at the ZPD. Using apodization effectively reduces the ringing in the spectrum, but, unfortunately, also the resolution decreases. Several algorithms have been established to eliminate the effects of ringing while maintaining as much resolution in the spectrum as possible.

Another procedure commonly used is zero filling. This manipulation expands the *x*-axis of the interferogram by adding zeros at both ends, thus artificially expanding the integration interval. Due to a larger number of data points, this results in a smoother spectrum after the Fourier transformation.

4.3.3 DRIFTS – Diffuse Reflectance Infrared Fourier Transform Spectroscopy

When infrared radiation is directed onto the surface of a solid sample, two types of reflected energy can occur. One is specular reflectance and the other is diffuse reflectance. The specular component is the radiation that reflects directly off the sample surface. Diffuse reflectance is the radiation that penetrates the sample and then emerges (see Figure 4.19). A diffuse reflectance accessory is designed so the diffusely reflected energy is optimized and the specular component is minimized.

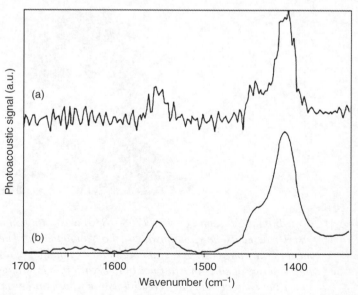

Figure 4.18
The effects of ringing, a consequence of boxcar truncation.

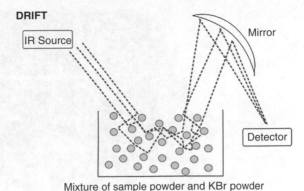

Figure 4.19
Diffuse reflection. Source: Taken from http://www.wikiwand.com/en/Geology_applications_of_Fourier_transform_infrared_spectroscopy.

The optics collect the scattered radiation and direct it to the infrared detector. The sample is usually ground and mixed with a material such as KBr, that acts as a non-absorbing matrix. By diluting the sample in a non-absorbing matrix, the proportion of the infrared beam that is diffusely reflected by the sample is increased.

A theory of diffuse reflectance at scattering surfaces is the Kubelka–Munk theory, developed in 1931. The Kubelka–Munk model relates sample concentration to the intensity of the measured infrared spectrum. The Kubelka–Munk equation is generally expressed as:

$$f(R) = \frac{(1-R)^2}{2R} = \frac{k}{s} \tag{4.14}$$

where R is the absolute reflectance of the layer, k is the molar absorption coefficient, and s is the scattering coefficient. The Kubelka–Munk theory predicts a linear relationship between the spectral intensity and sample concentration under conditions of a constant scattering coefficient and infinite sample dilution in a non-absorbing matrix. Hence, the relationship can only be applied to highly diluted samples. In addition, the scattering coefficient is a function of the particle size. The samples must be ground to a uniform, fine size as quantitatively valid measurements are desired. Finally, the equation only applies to an "infinitely thick" sample layer, which in infrared spectroscopy occurs at a sample thickness of approximately 3 mm. These restrictions reduce the number of spectra presentable in Kubelka–Munk units, but do not restrict "non-ideal" samples from quantitative diffuse reflection analysis.

DRIFT collectors can be equipped with high temperature/low pressure chambers or high temperature/high pressure chambers (Figure 4.20). This allows infrared measurements of, for instance, supported transition metal ion in "on stream" conditions during actual catalytic or thermal experiments. These chambers can also be purged with a customized gas mixture.

As an example, let us have a look at the dihydroxylation process that we discussed and showed in Figure 4.7.

In Figure 4.21, a silica gel is shown that was heated at three different temperatures. In spectrum (a), we see the silica gel after heating at 60 °C. The sharp peak at 3747 cm^{-1} is typical for isolated silanols. The sharpness of the peak is explained by the fact that all isolated (free) silanols have the same k-value or spring constant. They

Figure 4.20
A catalytic chamber, a high temperature and vacuum DRIFT cell.

are isolated, so they not perturbed by any other group. This results in a very narrow peak. Note also that this peak is at the highest frequency or energy.

Next to the sharp peak we see a broad shape that spans several hundreds of reciprocal centimeters, this is a collection of all sorts of hydroxyls that are in hydrogen interactions. The closer to the sharp free silanol peak, the less perturbed the O—H stretch vibration is by the "pulling" of a neighboring oxygen in the hydrogen bridge. Further away, around $3200 \, cm^{-1}$, we find the physically adsorbed water that is in very strong hydrogen bond interaction, weakening the O—H stretch vibrations the most.

Further down, below the $2000 \, cm^{-1}$, we find water bending vibrations and siloxane vibrations that we will not discuss further here.

Upon heating to $400 \, °C$ (b) and $700°$ (c), first the adsorbed water disappears and then also the vicinal silanols to leave only free (combination of isolated and geminal silanols) silanols at the surface, showing only this sharp peak. Its intensity will first increase (the fewer vicinal silanols, the more that are "free") and will then decrease due to continuing dihydroxylation. In fact, this spectrum can be correlated to Figure 4.7.

4.3.4 Attenuated Total Reflection

Attenuated Total Reflection (ATR) is another excellent technique to measure solids and has mainly been developed for films. It is very widely used in the polymer sciences, but recently it has expanded to almost every form of sample.

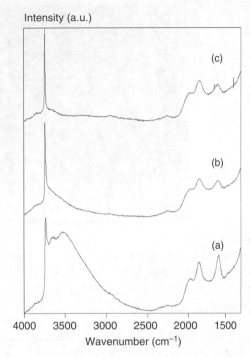

Figure 4.21
DRIFTS of silica gel at different temperatures, measured in situ. Source: Reproduced with permission of
Elsevier [11].

Its principles are founded on the well-known physics of total internal reflection.
This occurs when a wave of light strikes the boundary of its medium at an angle
that is higher than the critical angle. If the refractive index in lower in the second
medium, the wave is internally reflected. This phenomenon is mostly known as an
optical effect, but the principles apply to all sorts of (electromagnetic) waves. The
critical angle is derived from Snell's law. An example is shown in Figure 4.22.

An important side effect is the appearance of an evanescent wave beyond the
boundary surface; in other words, the wave travels "a little" into the medium with
lower density and gets adsorbed. That is why the technique is called "attenuated"
total reflectance; the light beam will fade with the number of reflections. In physics,
this phenomenon is called the Goos–Hänchen shift. We can summarize it here by the

Figure 4.22
Total internal reflection of a He-Ne laser (512 nm) in a block of PMMA. Source: Taken from https://
commons.wikimedia.org/wiki/File:TIR_in_PMMA.jpg (public domain).

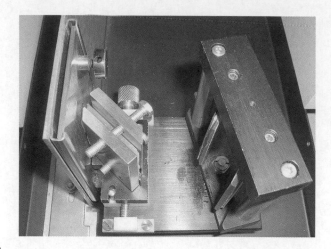

Figure 4.23
ATR attachment for infrared spectroscopy. The sample film is pressed against the pink crystal. Source: https://commons.wikimedia.org/w/index.php?curid=9519325.

following equation:

$$d = \frac{\lambda}{2\pi n_2 \sqrt{\sin^2\theta - \frac{n_1^2}{n_2^2}}} \tag{4.15}$$

with d being the penetration in the optical low density medium, λ the wavelength of the radiation, n_1 the optical index of the "sample," the optical low density medium, n_2 the optical refraction index of the dense medium (the "crystal"), and θ the angle of the light with the boundary. A typical ATR setup is shown in Figure 4.23. The sample (film or powder) is pressed against a crystal with a very high optical index. Typical crystals are ZnSe ($n = 2.4$) or Ge-crystals ($n = 4$). Advanced cells allow the operator to vary the angle of the incident beam (θ), typically between 30° and 60°.

By changing the incident angle and/or changing the refraction index of the single crystals, it is very easy to tune the penetration depth d. This is advantageous, for instance, for very opaque samples that would otherwise give completely saturated spectra. ATR cells also exist for liquids, the operator just needs to pour liquid in a little container in which the ATR is fixed.

References

1. Mackenzie, J.D. (1988). *J. Non-Cryst. Solids* 100: 162–168.
2. Iler, R.K. (1979). *The Chemistry of Silica: Solubility, Polymerization, Colloid and Surface Properties and Biochemistry of Silica*. Wiley.
3. Unger, K.K. (1979). *Porous Silica, its Properties and Use as a Support in Column Liquid Chromatography*. Amsterdam, The Netherlands: Elsevier.
4. Barby, D. *Characterization of Powder Surfaces* (ed. G.D. Parfitt and G.S.W. Sing), London, UK: Academic Press.
5. Stober, W., Fink, A., and Bohn, E. (1968). *J. Colloid Interface Sci.* 26: 62–69.
6. (a) Yoldas, B.E. (1975). *J. Mater. Sci.* 10: 1856–1860; (b) Yoldas, B.E. (1977). *J. Mater. Sci.* 12: 1203–1208.
7. Yamane, M., Aso, S., and Sakaino, T. (1978). *J. Mater. Sci.* 13: 865–870.
8. Brinker, C.J. and Scherer, G.W. (2013). *Sol-Gel Science*. Elsevier.

9. Hench, L.L. and West, J.K. (1990). *Chem. Rev.* 90: 33–72.
10. Osterholz, F.D. and Pohl, E.R. *Silanes and Other Coupling Agents* (ed. K.L. Mittal), 1992. Utrecht, The Netherlands: VSP.
11. Vansant, E.F., Van Der Voort, P., and Vrancken, K.C. (1995). *Characterization and Chemical Modification of the Silica Surface*, Studies in Surface Science and Catalysis, vol. 93. Elsevier.
12. Ying, J.Y., Benziger, J.B., and Navrotsky, A. (1993). *J. Am. Ceram. Soc.* 76: 2571–2582.
13. Rao, K.S., El-Hami, K., Kodaki, T. et al. (2005). *J. Colloid Interface Sci.* 289: 125–131.
14. (a) Vanblaaderen, A. and Kentgens, A.P.M. (1992). *J. Non-Cryst. Solids* 149: 161–178; (b) Vanblaaderen, A., Vangeest, J., and Vrij, A. (1992). *J. Colloid Interface Sci.* 154: 481–501.

5 Ordered Mesoporous Silica

5.1 MCM-41 and MCM-48 – Revolution by the Mobil Oil Company

5.1.1 The Original Papers and Patents

Zeolites and zeotypes, combined with the mesoporous amorphous silica and alumina were, and are still, the materials of choice in bulk chemistry and the petrochemical industry. They are cheap and robust and the entire industry has built large production facilities based on these porous catalysts and supports, so it is not obvious to introduce alternatives. But the small pore sizes of zeolites stimulated researchers to keep on searching for materials with larger pores.

So, it was an oil company, the then-called Mobil Oil Company (currently merged into Exxon Mobil) that patented and published the very first report on a "templated mesoporous silica" with uniform mesoporous pores in 1992 [1].

Actually, ordered mesoporous materials were reported for the first time in 1990 by Kuroda and coworkers [2]. A hydrothermal synthesis was described where sodium ions in the interlayer space of Kanemite are exchanged with alkyltrimethylammonium chloride ions. By increasing the alkyl chain in the alkyltrimethylammonium ions, mesopores between 2 and 4 nm and surface areas of about $900 \, m^2 \, g^{-1}$ were attained. In a subsequent report by Inagaki et al. [3], optimization of the reaction conditions led to a highly ordered mesoporous material with a hexagonal unit cell. The hexagonal honeycomb structure was clearly visible by Transmission Electron Microscopy (TEM). However, in the meantime the Mobil Research and Development Corporation reported their breakthrough research on a new family of ordered mesoporous materials, designated M41S [1b, c]. Most likely inspired by the development of large pore crystalline materials like zeotype VPI-5, Kresge et al. prepared mesoporous silicas with both hexagonal (MCM-41) and cubic (MCM-48) symmetry with pore sizes ranging between 2 and 10 nm, by employing surfactants. After the first patents on these materials appeared [1b, 4], a publication in *Nature* [1c] and one in the *Journal of the American Chemical Society* (*JACS*) [1b] followed. With over 12 000 citations (1992–2008), these two papers form the foundation of the field of ordered mesoporous materials.

Introduction to Porous Materials, First Edition.
Pascal Van Der Voort, Karen Leus and Els De Canck.
© 2019 John Wiley & Sons Ltd. Published 2019 by John Wiley & Sons Ltd.

Wieslaw J. Roth (left) and Charles T. Kresge (right) were two of the many authors of the Nature paper in 1992. Roth retired from ExxonMobil in 2009 and is now a professor in Krakow (Poland). Charles (Charlie) Kresge became R&D Vice President of the Dow Chemical Company after he left the Mobil Oil Company in 1999. These authors and many others of the Nature *paper received several prizes and medals for their groundbreaking work*

(Photograph taken from ref. [5]).

The idea was simply genius, briefly: dissolve a surfactant in an acid or basic solution at such conditions that it form micelles, "liquid crystals." This is obtained at the so-called Critical Micelle Concentrations (CMC). The first CMC (CMC-1) is the concentration at which spherical micelles are observed. The second CMC (CMC-2) is the concentration at which spherical micelles start to transform into rodlike micelles (Figure 5.1). The values are largely dependent on the type of surfactant, but also on the synthesis conditions. For instance, with increasing temperature, the required concentration of surfactants for a sphere-to-rod transformation to occur increases.

Now, introduce a hydrolysable silica source (typically tetraethoxysilane [TEOS], $Si(OEt)_4$) to this surfactant solution and the silica will grow and form uniform pores around the surfactant (see Chapter 4). During synthesis, long-chained organic cationic surfactants are used to assemble silicate anions from solution until the formation of a dense Si-network around the micelles. A final calcination removes the organics and leads to highly porous solids with pores ≥ 2 nm and surface areas reaching $1000\,m^2\,g^{-1}$. This is shown as Mechanism 1 in Figure 5.2.

The general class of materials was named M41S and more than 20 examples were presented, using different synthesis conditions, yielding a hexagonal, one-dimensional phase (MCM-41), a cubic, three-dimensional phase (MCM-48) or a two-dimensional lamellar phase (MCM-50). There is some confusion about the nature of the abbreviation MCM; some say it means Mobil Catalytic Materials, others that it means Mobil Composition of Matter.

Figure 5.3 shows an artistic drawing of the different MCM structures. The three major forms are MCM-41, a hexagonally ordered silica material; it has the space

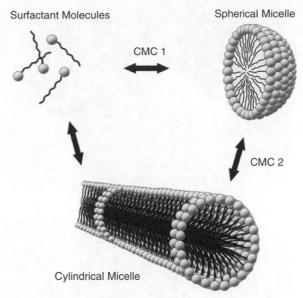

Figure 5.1

Dynamic equilibria in a surfactant-water system.

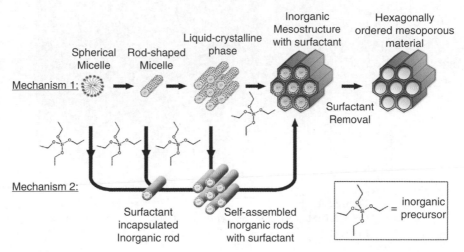

Figure 5.2

Possible mechanisms involved in mesostructure formation. Path 1 is the true liquid-crystal templating mechanism, Path 2 is designated as the cooperative liquid-crystal mechanism. Source: Reproduced with permission of ACS. Adapted from [1b].

group P6*mm*. MCM-48 is a cubically ordered silica: the light material is the silica, the two darker rods are the surfactants, they are intertwined, forming a cubic structure in the space group I$a\bar{3}d$. A third form is a lamellar silica structure, with layers of silica, a bit like a phyllosilicate clay. The typical wall thickness of these original MCM materials was around 1 nm, the pore size varied depending on the surfactant and synthesis conditions around a 2–4 nm diameter.

The hexagonal ordering of the MCM-41 materials was clearly visible in electron microscopy. In Figure 5.4, we show the original TEM image of the MCM-41

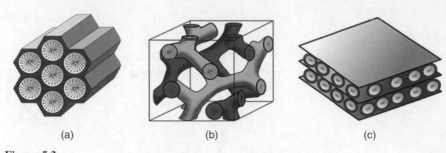

Figure 5.3

The three phases: (a) hexagonal MCM-41, (b) cubic MCM-48, (c) lamellar MCM-50. Source: Reproduced with permission of John Wiley & Sons, Ltd. Adapted from [6].

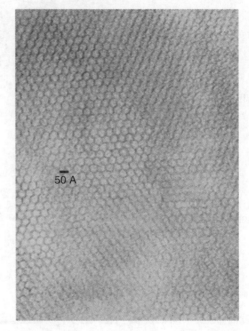

Figure 5.4

Original TEM image of the first report on MCM-41. Source: Reproduced with permission of Springer Nature [1c].

as published by the inventors. The hexagonal ordering (honeycomb ordering) of the materials is clearly visible and the uniformity of the pores is also evident.

Another indispensable technique in the characterization of these materials is powder X-Ray Diffraction (XRD). The simple hexagonal honeycomb structure of MCM-41 is a typical example. The XRD patterns of these materials (see Figure 5.5), show only the reflections of the honeycomb ordering of the pores, such long-distance reflection peaks are therefore typically at very low angles. The XRD patterns do not show any sign of atomic ordering, the silica walls themselves are amorphous, in contrast to all zeolites and zeotypes discussed earlier. Sometimes these materials are thus referred to as "semi-crystalline": the pores are ordered, but the atoms creating the walls are not ordered. The ordering of the pores is referred to as mesoscopic ordering: ordering at the meso-scale.

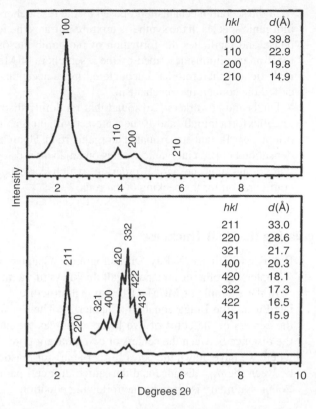

Figure 5.5
XRD patterns of MCM-41 and MCM-48. Source: Reproduced with permission of ACS [1b].

Similar observations are seen for the (more complex) $I\bar{a}3d$ structure of MCM-48.

The Mobil Oil researchers discovered a third phase, a lamellar phase, denoted MCM-50. This phase received less attention as it has little potential for applications, but it is an important phase to understand and rationalize the formation mechanism of, as we will do in subsequent sections.

Another matter that still remains a topic of interest is the manner in which the micelles aggregate into a liquid-crystal. At the time, the researchers of Mobil proposed two synthesis mechanisms to explain the formation of M41S type materials [1b, c]. These two mechanisms are illustrated in Figure 5.2. In the first mechanism, the surfactant liquid-crystal phase is formed prior to the addition of the inorganic species and directs the growth of the inorganic mesostructures. However, this mechanism did not meet much support in the literature [7]. In 1995, Cheng et al. pointed out that the liquid-crystal phase in a CTAC-water (cetyltrimethylammonium chloride) system only forms when the concentration of CTAC is higher than 40 wt% [7b]. CTAC is cetyl trimethyl ammonium chloride, the surfactant used by the Mobil researchers in their first reports; the bromide form is also often used (cetyltrimethylammonium bromide [CTAB]). In the conventional synthesis procedure of MCM-41, the CTAC concentration is much lower than 40 wt%, and only micelles can exist in solution. Because MCM-41 could be formed at surfactant concentrations as low as 1 wt%, it was very doubtful that this first mechanism occurred [7a].

In the second mechanism proposed by the researchers at Mobil, the presence of an inorganic species in the synthesis mixture initiates the formation of the liquid-crystal phase and facilitates the formation of inorganic mesostructures [1b, c]. Under the reaction conditions described by the researchers of Mobil, this mechanism is more realistic and therefore has encountered more acceptance in the literature [7b]. It is called the cooperative mechanism.

Davis and coworkers [7a] found that randomly distributed rod-shaped surfactant micelles form initially and as such interact with inorganic oligomers to form randomly oriented surfactant encapsulated inorganic rods. Upon heating, a base-catalyzed condensation between inorganic species on adjacent rods occurs. This condensation initiates long-range hexagonal ordering, which corresponds to the minimum energy configuration for the packing of the rods.

5.1.2   Calculating the Wall Thickness

Based on both the X-Ray Diffractograms (Chapter 3) and the nitrogen sorption (Chapter 2) isotherms, the pore wall thickness of the materials can be calculated.

In the example of MCM-41 hexagonal structure, the d_{100} spacing can be calculated using the Bragg equation (Chapter 3). The d-spacing is the distance between the centers of the pores of two layers, whereas the lattice parameter a_0 represents the distance between the centers of two adjacent pores. This is shown in Figure 5.6, where we have drawn the pores further apart for reasons of clarity. The relationship between the d_{100} distance and the lattice unit cell parameter a_0 for the honeycomb P6mm symmetry is given by the following equation

$$d = \frac{1}{\sqrt{\frac{4(h^2+hk+k^2)2}{3a_0^2} + \frac{l^2}{c^2}}} \tag{5.1}$$

This equation reduces for a (hkl) = (100) and c = ∞ to

$$a_0 = \frac{2}{\sqrt{3}}d_{100} \tag{5.2}$$

The pore diameter, D_p, derived directly from the pore analysis by nitrogen or argon sorption can then be used to calculate the wall thickness, t, according to Kruk, Sayari, and Jaroniec [8]:

$$t = a_0 - 0.95D_p \tag{5.3}$$

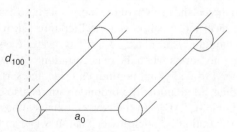

Figure 5.6
Relationship between unit cell and d-spacing in MCM-41.

Similar calculations can be made for other morphologies, for all cubic symmetries, the relation between a_0 and d is

$$a_0 = d_{hkl} \cdot \sqrt{h^2 + k^2 + l^2} \tag{5.4}$$

For the $Ia\bar{3}d$ cubic structure, the wall thickness can be calculated, by the formula derived by Ravikovitch and Neimark [9]:

$$t = \frac{a_0}{3.0919} - \frac{D_p}{2} \tag{5.5}$$

For another cubic geometry, with a $Im\bar{3}m$ space group, seen in for instance the SBA-16 (see later), the relationship is written as

$$t = \frac{\sqrt{3}}{2} a_0 - D_p \tag{5.6}$$

5.1.3 Interaction Between Surfactant and Inorganic Precursor

The first templates used were ionic surfactants with alkyl chains that are around 16 carbons in length. It was suggested that, in a basic medium, the silicate/silica species that arise from the hydrolysis of TEOS are negatively charged and are then attracted to the positively charged ammonium groups of the cationic surfactant. CTAB is the most often-used surfactant in this type of synthesis. This interaction is visualized in Figure 5.7.

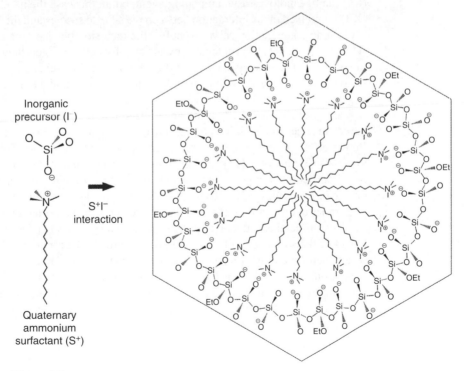

Figure 5.7
S^+I^- interaction during the synthesis of MCM-41 in basic conditions.

Strangely, the synthesis works just as well in acidic media. In this case, the chloride ions from the HCl form an ion bridge between the now positively charged silicate/silica species and the positively charged surfactant.

These types of interactions were described as S^+I^- and $S^+X^-I^+$, respectively, with S being the surfactant, I the inorganic species and X the bridging ion. Huo et al. [10] were the first to draw a general mechanism for the interaction between the inorganic precursor (I) and the surfactant (S).

The inorganic precursor should be capable of forming flexible polyionic species and should undergo extensive polymerization. Furthermore, charge density matching between the surfactant and the inorganic species should be possible. Based on these concepts, four different categories of surfactant-precursor interactions were proposed, as illustrated in Figure 5.8(a–d). The first category (a) involves the charge density matching between cationic surfactants and anionic inorganic species (S^+I^-). Considering the conventional basic (pH > 10) synthesis procedure described by the researchers of Mobil, the inorganic precursor is anionic (I^-), while the surfactant is a cationic quaternary ammonium ion (S^+).

The second category (b) involves the charge density matching between anionic surfactants and cationic inorganic species (S^-I^+) Huo et al. reported both the synthesis of iron and lead oxide mesoporous materials using anionic sulfonate surfactants [10]. The third (c) and fourth (d) categories are counterion-mediated interactions that allow the assembly of cationic or anionic inorganic species via halide ($S^+X^-I^+$) or alkali metal ($S^-M^+I^-$) ions, respectively.

This way, the synthesis of M41S type materials is feasible both under basic and acidic conditions. By operating under acidic conditions below the isoelectric point of silica (pH = 2), the silicate species are cationic (I^+). The same ammonium surfactant (S^+) can be employed as a templating agent, but in this case the halide counteranion (X^-) is involved in the interaction between the silicate species and the surfactant [10]. The halide counteranion serves to buffer the repulsion between the cationic silicate (I^+) and surfactant (S^+) molecules by means of weak hydrogen-bonding forces. On the other hand, negatively charged surfactants such as long-chain alkyl phosphates or sulfonates (S^-), can be used as templates in basic media if the interaction with the negatively charged silica species (I^-) involves a metal counterion (M^+) [10].

Soon after Huo reported on the generalized liquid-crystal templating mechanism based on electrostatic interactions between inorganic precursors and surfactants, Pinnavaia et al. proposed a fifth category to synthesize inorganic mesoporous materials (Figure 5.8e) [11]. This synthesis involves a neutral templating mechanism based on hydrogen bonding between neutral primary amines and neutral inorganic precursor molecules ($S°I°$). These materials will be discussed in Section 5.1.5.

Another hydrogen-bonding synthesis method (f), also reported by Pinnavaia et al., involves surfactants with poly(ethylene oxide) head groups [12]. Due to the adjustable length of the surfactant tail and head group, pores in the range of 2.0–5.8 nm could be attained. The poly(ethylene oxide) head group is non-ionic ($N°$), and the amine head group ($S°$) is also uncharged. The non-ionic route ($N°I°$) seemed to provide greater pore ordering than the neutral route ($S°I°$), but still lacked long-range hexagonal ordering of the pores. However, this synthesis procedure presents the advantage of using low-cost, nontoxic and biodegradable surfactants.

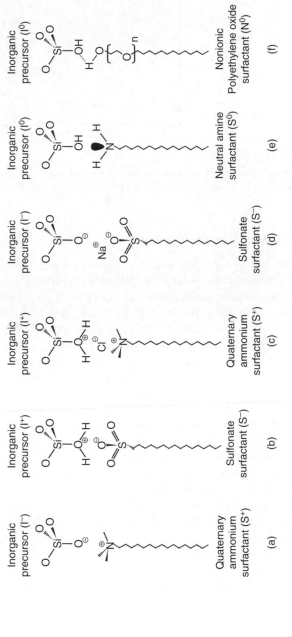

Figure 5.8

Interactions between surfactants and silica.

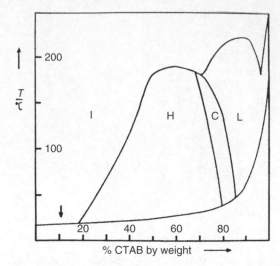

Figure 5.9
Phase diagram of CTAB. Source: Adapted from Researchgate (https://www.researchgate.net/post/What_
is_the_phase_composition_of_CTAB_water_and_hexanol_mixture_to_get_lyotropic_liquid_crystals).

5.1.4   The Surfactant Packing Parameter

But how is it possible that in some cases honeycomb structures are formed, and in other cases cubic or lamellar structures are formed?

As earlier explained, in a first approximation, it was believed that silica species are simply the negative template of the surfactant. The *soft template*, the micelle, in a hexagonal or cubic form forms the template around which the silica forms. As the template is then burned away, the remaining product is the negative copy of the surfactant. The conditions to form such a cubic phase were very critical, as can be inferred from the phase diagram of CTAB, and this has made it extremely difficult to synthesize MCM-48 in a reproducible way (Figure 5.9).

Several research groups have rationalized the synthesis methods in the past few years and have largely expanded the range of materials that can be prepared. It was soon established that the geometry of the liquid-crystal in pure water does not necessarily reflect the structure of the final inorganic mesophase, and that the mechanism of formation should be regarded as a "cooperative organization" of inorganic and organic molecular species into a three-dimensional array. Jean-Pierre Boilot and coworkers [13] published in 2003 a revised phase diagram that included both the concentration of the surfactant (CTAB) and the concentration of the silica, see Figure 5.10.

The introduction of the *surfactant ion pair packing parameter*, as a rough "molecular index" to predict the geometry of the mesophase products, can be considered as the first important step in the rationalization of the synthesis routes and the "molecular design" of new mesoporous materials. Whereas, in earlier literature, the ratio surfactant/silica source was claimed to be the major structure-directing parameter, the surfactant packing parameter explains this dependence on a more fundamental level and correlated the shape of the surfactant micelles to the silicate mesophase that is most likely formed.

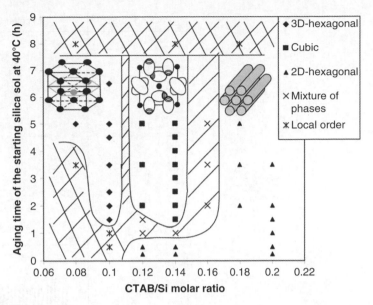

Figure 5.10

Example of a phase diagram for the cooperative mechanism, taking into account both surfactant as silica concentration. Source: Reproduced with permission of the RSC [13].

The surfactant packing parameter is defined as

$$g = \frac{V}{a_0.l} \tag{5.7}$$

with V the actual volume of the surfactant, a_0 the area of the headgroup of the micelle, and l the length of the tail (see Figure 5.11). The surfactant packing parameter is a measure for the shape of the surfactant. Stucky and coworkers [14] discussed in *Science* that, for g-values smaller than one-third, cone-like shapes are created that will pack together to spherical micelles, although a cubic and a 3D hexagonal form can also be formed. Between one-third and half wedge-like shapes are created that will aggregate to the cylindrical hexagonal phase, and surfactants with a g-value above half have a cylindrical shape that will pack together to a lamellar bilayered structure. In a narrow range between g-values of half and two-thirds, the formation of the cubic phase is possible (Table 5.1).

An obvious consequence of this was the synthesis of new surfactants that are likely to produce one mesophase in a broad range of synthesis conditions. One example was the development of the so-called *gemini surfactants*, with the general formula

$$C_nH_{2n+1}N^+(CH_3)_2 - (CH_2)_s - N^+(CH_3)_2C_mH_{2m+1}\ 2Br^-$$

abbreviated to Gem n-s-m. The $(CH_2)_s$ chain is called the *spacer*, which mainly determines the mesophase that is formed; the C_n and C_m *chains* (usually n = m) determine the average pore size of the obtained material. In this way, the gemini surfactant GEM-16-12-16 has become very important as an easy and reproducible surfactant to create MCM-48 materials [15].

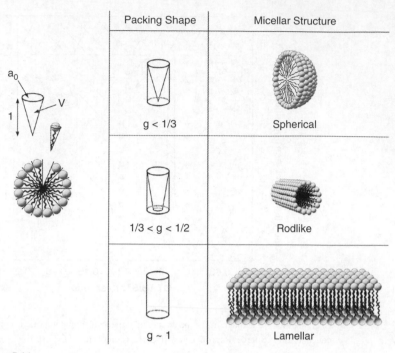

Packing Shape	Micellar Structure
g < 1/3	Spherical
1/3 < g < 1/2	Rodlike
g ~ 1	Lamellar

Figure 5.11
The surfactant packing parameter, and its effects of the geometry of the porous material.

Table 5.1 The g-value and phase relationship.

g-value	Preferred phase	Example
<1/3	Spherical (cubic) (3D-hexagonal)	
1/3 < g < 1/2	2D hexagonal	MCM-41, SBA-15
1/2 < g < 2/3	Cubic	MCM-48, SBA-16, KIT-6
2/3 < g < 1	Lamellar	MCM-50

5.1.5   Hexagonal Mesoporous Silica

In 1995, Peter Tanev and Thomas (Tom) Pinnavaia at Michigan State University published an easier route to obtain hexagonal mesoporous silica (HMS) in *Science* [11]. This route was based on a hydrogen-bonding interaction between a neutral amine and a non-ionic silica precursor. These materials were simply called HMS: Hexagonal Mesoporous Silica. Synthetically, rather than having an anionic-cationic interaction as in the case of the MCM materials (see Figure 5.8, the S^+I^- and $S^+X^-I^+$ interactions), we now have a simple $S°I°$ interaction. A typical synthesis consisted of the use of primary amine, typically C_{16}-NH_2, dissolved in a mixture of water and ethanol. The length of the chains determines largely the pore size. To this solution, TEOS is added, and the mixture is stirred and aged at room temperature. After 18 hours, the mixture is poured on a glass plate and air dried. The amine is then removed by washing with hot ethanol. The amine can be recovered and reused.

The HMS materials had a larger stability due to thicker pore walls, as the authors described in their *Science* paper, and a subsequent *Chemistry of Materials* contribution [16]. The HMS materials ($S°I°$) consistently have a large wall thickness than the

MCM-41 materials (S^+I^-), whereas the pore size are more or less the same for a similar chain length of the surfactant. In 2000, Cassiers and Van Der Voort [17] reported a method in which the amine template could be simply removed by acidified water. This procedure leads to an immediately usable material, there is no need for a subsequent calcination step. So, a simple synthesis procedure at room temperature, followed by an extraction with acidified water without any other further treatment yield a HMS material with a surface area of $1050 \, m^2 \, g^{-1}$, a pore volume of $0.83 \, ml \, g^{-1}$, and a pore size of 3.7 nm.

Thomas (Tom) J. Pinnavaia is currently a University Distinguished Professor at Michigan State University. He was an important pioneer in the development of mesoporous silicas, including the MSU-series and HMS. Before that, he was already active in this field, creating mesoporous materials based on Clays. One of these materials were the so-called PCH materials, Porous Clay Heterostructures

(*Photo:* http://cit.msu.edu/faculty/pinnavaia.html).

5.1.6   Stable Ordered Mesoporous Silica – SBA

Although the MCM-41 and MCM-48 were an enormous success, especially in research labs all around the world, the people working on catalysis were still not completely satisfied. The thin and amorphous walls of the MCM-48 and MCM-41 (about 1 nm thick) were relatively unstable. Especially in moist conditions or water, the siloxane bonds are attacked by water (see Chapter 4) and broken into silanols.

Dongyuan Zhao, currently at Fudan University, but then a post doc with Galen Stucky at Santa Barbara, delivered in 2000 a series of materials that were called the SBA materials (SBA is the acronym for Santa Barbara) [18]. They had considerably larger pores and considerably thicker walls, the hexagonal variant was called SBA-15 and the cubic variants was called SBA-16. It had a different geometry though to the MCM-48, it was a centrosymmetric cage type structure with the space group $Im\overline{3}m$ (see Figure 5.12).

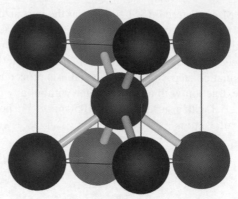

Figure 5.12
Pore structure of SBA-16.

Galen D. Stucky is a Professor at the University of California at Santa Barbara and worldwide respected for his contributions to the field of (meso)porous structured materials. He found many "other" applications for these materials and won in 2008 the ATACCC Award for developing a revolutionary blood-clotting gauze for the military.

Photo: https://labs.chem.ucsb.edu/stucky/galen/stuckygroup/biography.html.

The surfactants used were very cheap, non-ionic, and biodegradable "Pluronic®" surfactants, a brand name of BASF, for a series of polyethylenglycol – polypropyleneglycol – polyethyleneglycol (PEG-PPG-PEG) block-copolymers (Figure 5.13). Just as in the case of the MCM-41 and MCM-48 materials, the *g-value* of the surfactant and the surfactant/silica ratio will determine the final pore geometry.

The Pluronic P123 stands for $(PEG)_{20}$-$(PPG)_{70}$-$(PEG)_{20}$ block copolymer. The nomenclature needs some explanation: the "P" stands for "paste," the aggregation form of the surfactant. (Other letters are "F" for "flakes" and "L" for "liquid.") If

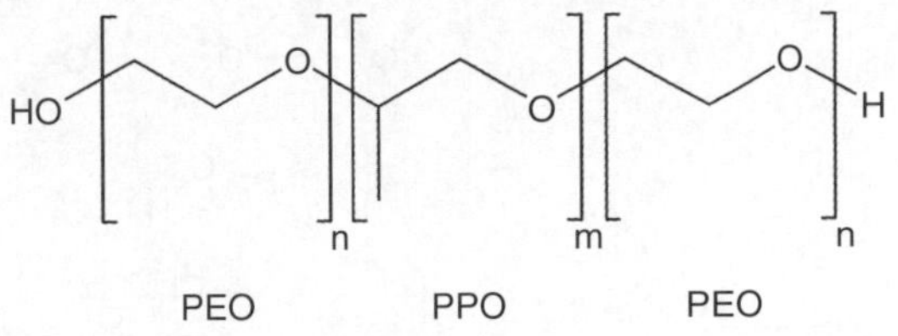

Figure 5.13
Non-ionic triblock copolymer (Pluronic) with large poly(ethylene oxide) (PEO) and poly(propylene oxide) (PPO) blocks.

Table 5.2   Basic physical properties of mesoporous MCM-41 and SBA-15 silicas.

Mesoporous silicas	Surface area $(m^2\,g^{-1})$	d_{100} (nm)	Unit cell (a_0) (nm)	Pore size (nm)	Wall thickness (nm)
MCM-41	1028	3.83	4.42	2.65	1.77
SBA-15	680	8.16	9.42	5.90	3.52

you multiply the first two numbers by a factor of 300, you will get the average molar mass, so in this case P123 has a molar mass of $12 \times 300 = 3600\,g\,mol^{-1}$; and the last number, multiplied by 10 gives the weight percentage of the hydrophilic groups (so the PEG groups), this would be in this case 30%.

Another famous one is the F127 that has a formula of $(PEG)_{100} - (PPG)_{65} - (PEG)_{100}$. This much more hydrophilic surfactant is typically used for the creation of the cubic SBA-16. This can be rationalized again by the surfactant packing parameter. The much smaller hydrophobic core compare to the hydrophilic tails yields a packing parameter much smaller than 1 and favorable for cubic structure. Note that the P123 had on the contrary a very large hydrophobic core in comparison to the hydrophilic tails leading to large g-values and thus a hexagonal packing. F127 comes in the form of flakes, has an average molar weight of $3600\,g\,mol^{-1}$ and has a weight percentage of hydrophilic groups of 70%.

As shown in Table 5.2, the pore size and wall thickness of the SBA-15 is the double of these of the MCM-41, rendering the SBA variants more stable in water and air [19]. They can host much larger molecules, which is very important in the field of adsorption and catalysis.

Next to these differences, a very important difference is the fact that for the SBA-type materials, the walls themselves are microporous, because they are perforated. This is due to the specific interaction of the silica precursor with these long tail surfactants.

Let us take the case of SBA-15. The P123 surfactant forms a hydrophobic core of the (PPG)-block, and the more hydrophilic (PEG) blocks stick out and interacts with the silica precursors and the formed silicates. In the typical acid conditions of the synthesis (there are also recipes for basic conditions), these hydrophilic PEG blocks interact with the silica according to a $S^0H^+X^-I^+$ interaction as shown in Figure 5.14. The surfactant should be regarded as a hydrophobic core with hydrophilic micelles sticking out and penetrating through the silica that is formed around them. This renders the SBA-type materials highly microporous.

We show this in Figure 5.15.

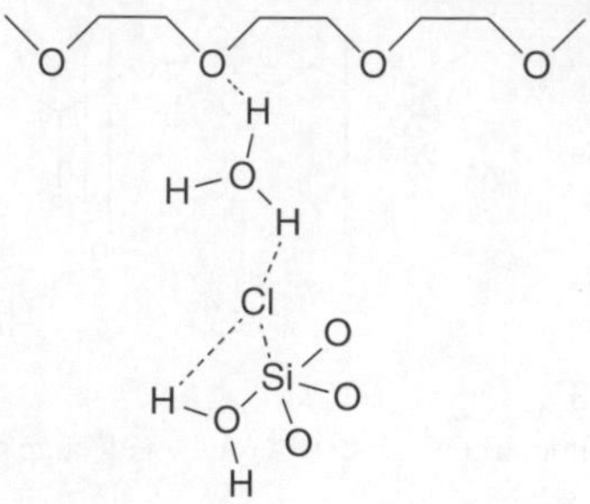

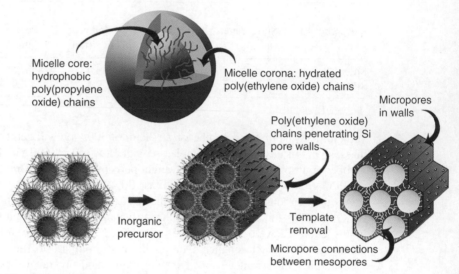

Figure 5.14
Interaction between PEG and silica in acid media.

Micelle core: hydrophobic poly(propylene oxide) chains

Micelle corona: hydrated poly(ethylene oxide) chains

Micropores in walls

Poly(ethylene oxide) chains penetrating Si pore walls

Inorganic precursor

Template removal

Micropore connections between mesopores

Figure 5.15
Micelle structure of PEO-PPO-PEO triblock copolymers. Source: Courtesy of Carl Vercaemst.

Anne Galarneau et al. [20] explained in a very visual paper in 2007 how researchers can tune the parameters of the synthesis or post-treatment to tune the pore size of the SBA-15, but also the amount and pore size of the micropores and the wall thickness. The SBA-15 and SBA-16 are therefore not only very stable, but they have a tunable combination of micropores and mesopores and can thus be referred to as *hierarchical porous materials*, a term reserved for materials that have at least two distinct different pore sizes. See Figure 5.16 showing three SBA-15 materials, synthesized at different conditions. The mesopores can then act as the "highways," guaranteeing fast transport of molecules without any diffusional limitations and the micropores (functionalized or not) can then act as "catalytic pockets" or "adsorption nests."

On a side note, usually SBA-15 materials are drawn in the literature as short and rigid and straight honeycomb ordered pores. This is not entirely the case. In collaboration with Krijn De Jong at Utrecht University, we published TEM pictures showing the curvatures in real SBA-15 (and other) samples [21]. Some of these TEM pictures are shown in Figure 5.17.

Other, but less studied, SBA materials are SBA-11 (cubic, $P m \overline{3} m$) and SBA-12 (3D-hexagonal, $P6_3/mmc$) [22]. These are usually synthesized using the commercially available surfactants Brij 56 ($C_{16}EO_{10}$) and Brij 76 ($C_{18}EO_{10}$), respectively.

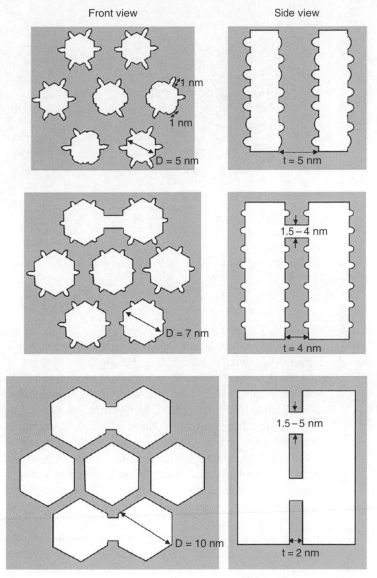

Figure 5.16

Schematic representation of three SBA-15s synthesized in different conditions (Temperature = 60, 100, and 130 °C), revealing different pore diameters, wall thicknesses, microporosities, and interconnections between the main channels. Source: Reproduced with permission of ACS [20].

5.1.7 Plugged Hexagonal Templated Silica

In 2002, Van Der Voort et al. [23, 24] reported a remarkable isotherm that was never published earlier. The isotherm is shown in Figure 5.18 as type (C).

Type (A) isotherm in Figure 5.18 represents the classic isotherm of the open cylindrical pore case in the SBA-15. According the IUPAC classification [25], this is an isotherm type IV(a), with a hysteresis loop type H1.

Type (B) isotherm in Figure 5.18 represents the typical case of "cavitation" (see Chapter 2), the hysteresis loops is forced to close at $P/P_0 = 0.42$, the cavitation

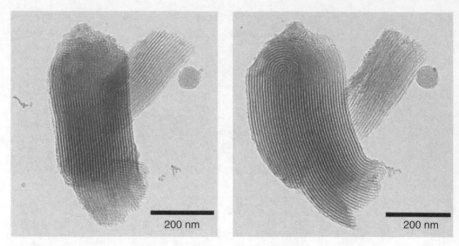

Figure 5.17

TEM pictures of SBA-15, showing the curvature in the pores [21]. Source: Reproduced with permission of the RSC.

pressure of liquid nitrogen at 77 K, below which no stable meniscus of liquid nitrogen is possible. It represents the "inkbottle" pore and is classified by IUPAC as a type IV(a) with a hysteresis loop H2(a).

At the time that Van Der Voort published the isotherm (C), this type of isotherm was not known and not indexed by IUPAC, who used a previous classification of isotherms in the period 1985–2015 [26]. The material was prepared as a regular SBA-15, but with a largely increased silica concentration. The isotherm (C) is remarkable. Combined with the typical P6*mm* XRD pattern for a hexagonal ordered structure, the following characteristic features can be observed: (i) adsorption in intrawall micropores at low relative pressures; (ii) multilayer adsorption in regular mesopores and capillary condensation in narrow intrawall mesopores; (iii) a one-step capillary condensation, indicating uniform mesopores; (iv) a two-step desorption branch indicating the pore blocking effects (sub-step at the relative pressure of around 0.42). The adsorption–desorption behavior is consistent with a structure comprising both open and blocked cylindrical mesopores. The high-pressure desorption step corresponds to nitrogen desorption from open pores. The blocked pores will remain filled until the vapor pressure is lowered below the "magical" point $p/p_0 = 0.42–0.45$, after which a spinodal decomposition of the condensed nitrogen will occur and these sections will spontaneously empty.

The authors called this material PHTS; Plugged Hexagonal Templated Silica. The capillary condensation process and the cavitation is visualized in Figure 5.19.

The open versus corrugated and blocked material is also clearly visible from the TEM recordings in Figure 5.20.

In the same paper, the authors made a comparison of the hydrothermal stability of a wide range of mesoporous silica materials, as visualized in Figure 5.21. The description of the treatments is as follows: "Mild" samples are treated in a nitrogen flow (25% water) at 1013 hPa and 673 K for 50 hours; "Medium" samples are treated in a nitrogen flow (25% water) at 1013 hPa and 673 K for 120 hours; "Hard" samples are steamed on a grid above the water in an autoclave at 393 K and autogenous pressure for 24 hours.

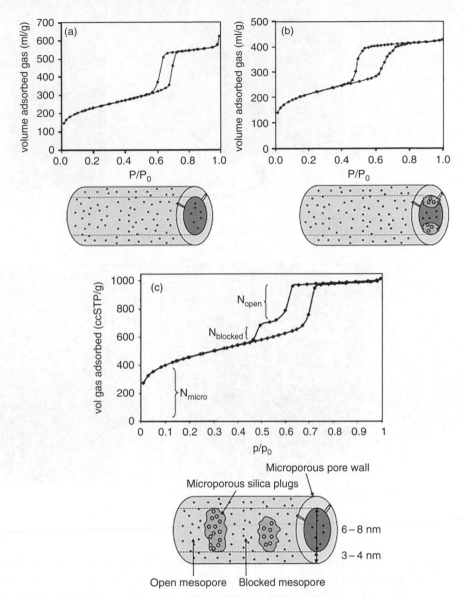

Figure 5.18

Isotherms of (a) open pores; (b) blocked or inkbottle pores and (c) "plugged" pores. Source: Reproduced with permission of ACS [23].

It is remarkable that only the SBA-15 and the PHTS materials survive this harsh treatment. All other materials collapse.

In 2015, IUPAC classified this new hysteresis profile as hysteresis loop H5 [25].

5.1.8   The New MCM-48: KIT-6

As argued before, the synthesis of MCM-48 is not too easy, it requires either a very narrow set of conditions using the regular ammonium-based surfactants, or the use of non-commercially available gemini surfactants.

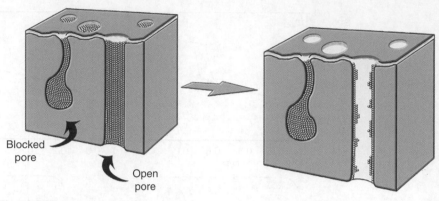

Figure 5.19
Open versus blocked pores in PHTS.

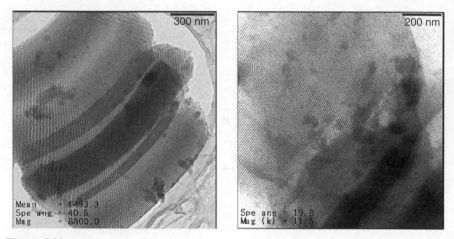

Figure 5.20
TEM images of (left) regular SBA-15 and (right) the PHTS. Source: Reproduced with permission of ACS [23].

Much later, in 2003, Freddy Kleitz, Ryong Ryoo, and coworkers [27] (Korea Advanced Institute of Science and Technology) published a novel pathway to make the I$a\bar{3}d$ cubic mesoporous silica more easily, in *Chemical Communications*; they called it KIT-6 (KIT standing for the Korea Institute of Technology). They created a large pore high quality I$a\bar{3}m$ cubic material by the simple addition of butanol to an acidified solution of P123. Remember that the molecular structure of P123 typically yielded hexagonal SBA-15 because of its high g-value. The authors reasoned that the addition of butanol is responsible for the preferred swelling of the hydrophobic volume of the block-copolymer micelles, leading first to the formation of micellar aggregates with a decreased curvature (lamellar mesophase). The lamellar form was made previously (so-called MCM-50), but has not been followed up much in the literature, as these structures have no applications; the lamellar phase is formed for g-values in between the hexagonal and the cubic phase.

Upon further reaction or increased temperature, the further condensation of the silicate region provokes folding and regular modulation of the silica surface, inducing significant changes in the micelle curvature. The interplay of the silicate formation

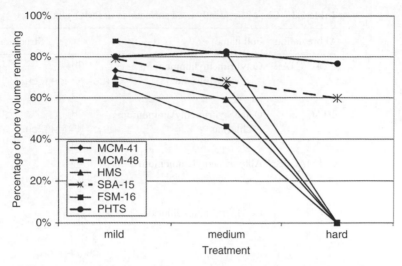

Figure 5.21
Comparison of the hydrothermal stability of several mesoporous silicas. The terms hard, medium, and soft are defined in the text. Source: Reproduced with permission of ACS [23].

and the micelle curvature was known previously (the cooperative mechanism), but this was the first time it was observed for a non-ionic surfactant.

5.1.9   Further Developments of Mesoporous Silica

Since the publication of the M41S type materials, thousands of papers have appeared on alternative preparations of porous silicas, using different surfactants, different ingredients, other structures, and so on.

It is impossible to cover all these papers, even in such a comprehensive book as this one. We can give an overview of the most important applications though. Table 5.3 shows an overview of the surfactants that have been often used in the synthesis of mesoporous silicas. The CTAB was the original surfactant for the synthesis of MCM-41, but several other cationic surfactants have been used as well. Among the non-ionic surfactants, we have already discussed the amphiphilic triblock copolymers such as P123 and F127. Another popular class of surfactants for mesoporous silica synthesis are the Brij® surfactants. Brij surfactants are also amphiphilic, and non-ionic, they are polyethylene glycol alkyl/aryl ethers. They act very similar to the triblock copolymers.

When using non-ionic structure direction agents (SDAs) such as triblock copolymers, inorganic salts will exhibit a special effect during the Periodic Mesoporous Organosilica (PMO) assembly by influencing the interaction between several parts of the polymer [28]. The salt causes a dehydration of the hydrated ethylene oxide units of the polyethylene oxide (PEO) chain, which is located next to the polypropylene oxide (PPO) chain. This results in an increased hydrophobicity of the PPO chain and significantly decreases the hydrophilicity of PEO. Many salts have shown to improve the hydrothermal stability of during the crystallization process (e.g. NaF, NaCl, KF, KCl, Na_2SO_4, …) [18, 29]. It also allows us to prepare highly ordered materials in a wide range of acidic concentrations.

Table 5.3 List of frequently used surfactants in PMO synthesis.

Abbreviation	Full name	Structural formula
CTAC/CTAB	Cetyltrimethylammonium chloride/bromide	$H_3C-\overset{\overset{\displaystyle CH_3}{\mid}}{\underset{\underset{\displaystyle CH_3}{\mid}}{N^+}}-(CH_2)_{15}CH_3 \quad Cl^-/Br^-$
OTAC	Octadecyltrimethylammonium chloride	$H_3C-\overset{\overset{\displaystyle CH_3}{\mid}}{\underset{\underset{\displaystyle CH_3}{\mid}}{N^+}}-(CH_2)_{17}CH_3 \quad Cl^-$
C_nTMACl/Br	Alkyltrimethylammonium chloride/bromide	$H_3C-\overset{\overset{\displaystyle CH_3}{\mid}}{\underset{\underset{\displaystyle CH_3}{\mid}}{N^+}}-(CH_2)_{n-1}CH_3 \quad Cl^-/Br^-$ (n = 8,10,12,14,16,18)
CPCl	Cetylpyridinium chloride	pyridinium ring $-CH_2(CH_2)_{14}CH_3 \quad Cl^-$
FC4	Fluorocarbon surfactant	$C_3F_7O(CFCF_3CF_2O)_2CFCF_3CONH(CH_2)_3-\overset{\overset{\displaystyle C_2H_5}{\mid}}{\underset{\underset{\displaystyle C_2H_5}{\mid}}{N^+}}-C_2H_5 \quad I^-$
Brij-30	Polyoxyethylene (4) lauryl ether	$HO{-}[{-}CH_2CH_2O{-}]_4 CH_2(CH_2)_{10}CH_3$
Brij-56	Polyoxyethylene (10) cetyl ether	$HO{-}[{-}CH_2CH_2O{-}]_{10} CH_2(CH_2)_{14}CH_3$
Brij-76	Polyoxyethylene (10) stearyl ether	$HO{-}[{-}CH_2CH_2O{-}]_{10} CH_2(CH_2)_{16}CH_3$
Triton-X100	Polyoxyethylene (10) octylphenyl ether	octylphenyl $-O[-CH_2CH_2O-]_{9.5}H$
P123	Pluronic P123 Poly(ethylene glycol)-poly(propylene glycol)-poly(ethylene glycol)	$HO[]_{20}[]_{70}[]_{20}H$ (EO$_{20}$PO$_{70}$EO$_{20}$)
F127	Pluronic F127 Poly(ethylene glycol)-poly(propylene glycol)-poly(ethylene glycol)	$HO[]_{100}[]_{65}[]_{100}H$ (EO$_{100}$PO$_{65}$EO$_{100}$)
B50-6600	Poly(ethylene oxide)–poly(butylene oxide)–poly(ethylene oxide)	$HO[]_{39}[]_{47}[]_{39}H$
PEO–PLGA–PEO	Poly(ethylene oxide)–poly(lactic acid-*co*-glycolic acid)–poly(ethylene oxide)	$HO[]_x[]_y[]_z[]_x H$
$C_{n\text{-}s\text{-}m}$	Divalent and gemini surfactants	$C_nH_{2n+1}-\overset{\overset{\displaystyle CH_3}{\mid}}{\underset{\underset{\displaystyle CH_3}{\mid}}{N^+}}-(CH_2)_s-\overset{\overset{\displaystyle CH_3}{\mid}}{\underset{\underset{\displaystyle CH_3}{\mid}}{N^+}}-C_mH_{2m+1} \quad 2Br^-$

The addition of KCl significantly increases the interaction between the SDA and the polysilsesquioxane and weakens the disordering of the organic units. It especially improves the interaction between the non-ionic block copolymer and the oligomers of the precursor that are formed during hydrolysis and condensation.

5.1.10 Pore Size Engineering

An important aspect of tailoring a mesoporous silica is the ability to enlarge the pore size. This is called *pore size engineering*. Particularly for applications such as the immobilization or adsorption of proteins, enzymes or drugs, this is of major importance. The employment of block-copolymers and poly(alkylene oxides) already resulted in larger pores, however values above 10 nm could not be achieved. As we discussed earlier, tuning the synthesis conditions is an important tool to engineer the pore sizes. But still the limit seems to be about 10 nm.

The addition of swelling agents to the reaction mixture is the solution for this issue. These additives are typically hydrophobic compounds that will interact with the surfactant by settling in the hydrophobic part of the polymer. An expansion of the hydrophobic core of the surfactant will occur and this results in larger pores. Typical swelling agents are 1,3,5-trimethylbenzene (TMB), 1,3,5-triisopropylbenzene (TPB), cyclohexane, dodecane, and poly(propylene glycol), but xylene, toluene, and benzene have also been reported. With the aid of TMB, the pore sizes can be enlarged to about 40 nm in acidic triblock copolymer systems (say the SBA-type materials) and to 10 nm in basic CTAB (MCM-type materials) conditions. There is a price to pay: the materials lose order and become more disordered, although the pore size remains uniform.

Table 5.4 gives an overview of the strategies to obtain a certain pore size [30].

5.1.11 Making Thin Films – The EISA Principle

The true liquid-crystal templating and the cooperative self-assembly are not suitable for the deposition of highly uniform thin films. Therefore, the evaporation-induced self-assembly (EISA) has been developed by Brinker's group [31] (Figure 5.22). In

Table 5.4 Strategies to obtain materials with a certain pore size.

Pore size (nm)	Method
2–5	Surfactants with different chain lengths including long-chain quaternary cationic salts and neutral organoamines
4–7	Long-chain quaternary cationic salts as surfactants High-temperature hydrothermal treatment
5–8	Charged surfactants with the addition of organic swelling agents such as TMB and midchain amines
2–8	Non-ionic surfactants
4–20	Tiblock-copolymer surfactants
4–11	Secondary synthesis, for example water-amine postsynthesis
10–30	High molecular weight block-copolymers, such as PI-*b*-PEO, PIB-*b*-PEO, and PS-*b*-PEO Triblock copolymers with the addition of swelling agents TMB and inorganic salts Low-temperature synthesis

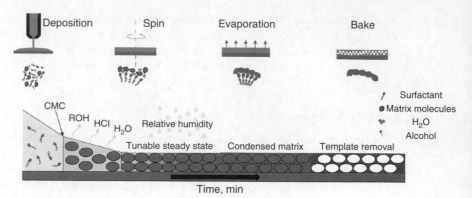

Figure 5.22
Formation of mesoporous films via the EISA.

the EISA approach, an excess of a volatile solvent is used to ensure that the surfactant concentration in the solution remains below the CMC. Additionally, the solution must be consisted of solvents and reactants which are highly volatile. Since less condensed entities (small and mobile) are preferred during self-assembly, it is important to choose conditions that favor hydrolysis but hinder condensation of the inorganic species.

Through variation of the initial alcohol/water/surfactant/matrix precursor mole ratio it is possible to follow different trajectories in composition space and to arrive at different final mesostructures [31a]. Upon addition of the solution to the substrate, the preferential evaporation of the volatile solvent during dip- or spin-coating concentrates the depositing film in nonvolatile surfactant and silica species. The evaporation of the volatile species is one of the main parameters that governs the entire film-formation process. When the surfactant concentration reaches the equivalent of the CMC for the system, micelles start to form. Eventually, an organized mesostructure film is obtained. The main challenge of the EISA compared to the previously described precipitation methods is that it is a mainly kinetically governed mechanism that requires high control of the processing conditions, and it can lead to metastable hybrid materials that require additional steps for stabilization. The quality of the final mesostructure is thus highly dependent on the processing conditions and more especially on the atmosphere composition (water and solvent relative pressures), since the latter defines the evaporation rate and the system content in water and solvent at equilibrium.

5.2   Applications of Mesoporous Silica

5.2.1   In Heterogeneous Catalysis – Functionalization of Mesoporous Silica

One of the most explored applications of the mesoporous silicas is the field of heterogeneous catalysis. It was one of the primary reasons to develop these types of materials in the early 1990s. Therefore, the first efforts in the field of heterogeneous catalysis wanted to exploit the larger pore sizes of the catalysts to convert larger substrates that could not enter the pores of zeolites and zeotypes. As the pores of the ordered mesoporous silicas are relatively uniform, researchers also investigated the *shape selectivity* in the catalytic process.

5.2.1.1   *Functionalization of the Mesoporous Silica*

A plethora of methods for preparing catalysts is available.

5.2.1.1.1   *Metal Doping*

Very similar to zeolite synthesis, mesoporous silicas can be doped with hetero-elements. The most often used dopants are Al^{3+} and Ti^{4+}, but many other dopants have been used in the past decades. The most used method is to add a hydrolysable metal salt to the synthesis mixture; alkoxides or chloride salts are often used. One has to choose the metal compound in such a way that the hydrolysis rate of the metal compound is in the same range as the silicon source. A rule of thumb is that the larger the alkoxy group is, the slower the hydrolysis occurs. For example, as Ti-alkoxides are much more reactive than the silica counterparts, a typical match is found by mixing TEOS (ethoxy ligands) with $Ti(^{i}OPr)_4$ (isopropoxy ligands, to slow down the hydrolysis rate).

Al-ions are typically introduced using $Al(^{i}OPr)_3$, aluminum chloride, hydroxide, or the sulfate. The introduction of aluminum created both Lewis and Brønsted acid sides in the material.

As an early example, Avelino Corma et al. [32] systematically investigated the catalytic cracking performance of gas oil of an Al-MCM-41 with varying Si/Al ratios (Si/Al = 14, 100, 143). Just as in zeolites, acid catalysts are created this way. The authors compared these catalysts with the more classical ones, being Ultra Stable Y (USY)-zeolite (Si/Al = 100) and amorphous silica-alumina (Si/Al = 2.5). Two main conclusions were drawn from this work: (i) the less Al^{3+} is added, the stronger these sites are (also observed in zeolites) and (ii) while the USY is 139 times more active than the Al-MCM-41 for small molecules (*n*-heptane), the USY was 11 time *less* active than the Al-MCM-41 for large molecules (gas oil). These values confirmed the size-exclusion effect in zeolites: the large gas oil molecules cannot penetrate the pore system of the USY. Still the catalytic activity of the semi-amorphous Al-MCM-41 is much lower in an equal-level playing field.

Avelino Corma is a Spanish chemist, who is internationally recognized for his lead-ing research on heterogeneous catalysis. He is Professor at the Institute of Chemical Technology (ITQ-CSIC-Polytechnical University of Valencia). He has published more than 900 research papers and is inventor on more than 100 patents. Over 12 of those patents have been applied industrially in commercial processes of cracking, desulfuration, isomerization, epoxidation, chemo selective oxidation of alcohols, and chemoselective hydrogenations.

Photo: https://commons.wikimedia.org/wiki/File:Avelino-Corma.jpg *(public domain).*

It is also possible to create basic catalysts, typically by ion-exchange – that is, exchanging the acid mobile protons, just as in zeolites – with Na or Cs ions. Van Bekkum and Kloetstra [33] made such catalysts and prepared Na-MCM-41 and Cs-MCM-41. In particular, the Cs-exchanged catalyst showed some superbase properties and was not only active in the typical Knoevenagel condensation, but also in the more demanding Michael addition.

Herman van Bekkum is a Dutch (organic) Chemist who became one of the pioneers of heterogeneous catalysis in zeolites and later also on heterogeneous catalysis using mesoporous materials. He has been the Rector Magnificus at the Technical University of Delft and has been a researcher at the Royal Dutch Shell Oil Company. He has been knighted in the Netherlands for his contributions.

Photo: http://www.delta.tudelft.nl/article/herman-van-bekkum.

Doping with transition metals then typically yields oxidation catalysts. The bench-mark in oxidation catalysis on industrial scale is the TS-1 (titania silicate) zeolite. It also suffers from small pores. Pinnavaia was one of the first to demonstrate the use of Ti-HMS and Ti-MCM-41 [34]. Just like Corma for the acid catalysis, he corrobo-rated that the TS-1 is the catalyst of choice for small molecules and becomes inactive for molecules exceeding its pore size, and that the Ti-HMS and Ti-MCM-41 are still active for larger molecules.

5.2.1.1.2   One-Pot Synthesis – Co-Condensation

The co-condensation method is a one-pot synthesis procedure. Here, the functionalized silica is prepared through co-condensation of TEOS with functional silanes in the presence structure-directing agent. This is visualized in Figure 5.23. The involvement of the organosilane in the mesostructured formation implies that the chemistry of the functional group has to be taken into account. If not, the organosilane can interfere in the formation of micellar aggregates, leading to disordered amorphous materials. For instance, if the organosilane consists of an amine group, working under acidic conditions implies protonation of this functional group that may interact with silanol groups and prevent direct interaction of the surfactant with the condensating silicate species.

As the organic functionalities of these materials are incorporated into the framework during the formation of the mesostructure, they are usually nicely homogeneously distributed throughout the network. However, the homogeneous distribution of the organic units is strongly dependent on the hydrolysis and condensation rates of the silica and organosilica precursors. Rate differences for the hydrolysis and/or condensation of mixed precursors can lead to self-condensation and phase separation.

The co-condensed materials can be used as such, for example, an amine group in a base-catalyzed reaction (Knoevenagel, Henry, aldol-condensation, Michael addition), can be further processed (e.g. the oxidation of a thiol group by a mild oxidant to a sulfonic acid group for acid catalysis), or can be the anchoring point to attach organocatalysts or nanoparticles (NPs) (e.g. gold nanoparticles that attach on thiol

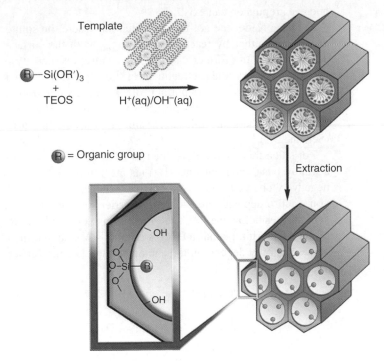

Figure 5.23

Synthesis of functionalized ordered mesoporous silica by means of the co-condensation method. Source: Figure is redrawn from Fröba's excellent review [35], Reproduced with permission of John Wiley & Sons, Ltd.

functionalities). We refer to an older, but excellent review by M. Davis for a phethora of examples [36].

The main disadvantage of this method, however, is the effect of organic content on the degree of mesoscopic order of the obtained hybrid materials. As the concentration of organosilane increases, the structural ordering of the material decreases and will ultimately lead to a completely disordered solid.

5.2.1.1.3 Dry and Wet Impregnation

Dry impregnation or "pore volume impregnation" or "incipient wetness impregnation" is the impregnation method in which the amount of liquid (solution of the precursors) used is just enough to fill the pore volume of the support. In *wet impregnation* the support is dipped into an excess quantity of solution containing the precursor(s) of the active phase. In practice, these operations are carried out in various ways. In dry impregnation the solubility of the catalyst precursors (usually soluble salts) and the pore volume of the support determine the maximum loading available each time of impregnation. If a high loading is needed, successive impregnations (and heat treatments) may be necessary. When several precursors are present in the impregnating solution simultaneously, the impregnation is called "co-impregnation."

In the first step of impregnation three processes occur: (i) transport of solute into the pore system; (ii) diffusion of solute within the pore system and (iii) uptake of solute by the pore wall. In the case of wet impregnation, a fourth process is operative, namely transport of solute to the outer particle surface. Dependent on the process conditions, different profiles of the active phase over the particle are obtained. For instance, dependent on the pH, the interaction with the support can be strong or weak, and even repulsion can exist.

Let us consider the not-unusual situation where the solute or its ions are fixed to the support either by reaction or exchange with the surface OH groups and/or by adsorption. In the former case, the concentration (density) of surface OH groups, which depends on the pretreatment of the support, is crucial. In the latter case, the surface charge plays an important role. At a pH value of the so-called *Point of Zero Charge* (PZC) the surface is electrically neutral. At pH values above PZC, the surface is negatively charged, while at pH values below PZC the surface is positively charged (Scheme 5.1).

For silica this can be illustrated as follows. At pH = 3 the surface is neutral. In a mildly basic environment, H^+ ions are removed, and, as a result, the surface is negatively charged. In an acid environment, the surface will become protonated. If you want to deposit anions onto the carrier surface, the preparation should proceed at pH values below the PZC, and for cations you would prefer a pH value above that of the PZC. Alumina has a PZC of around 8. We should mention that the exact PZC values not only depend on the chemical nature of the carrier, but also on its history

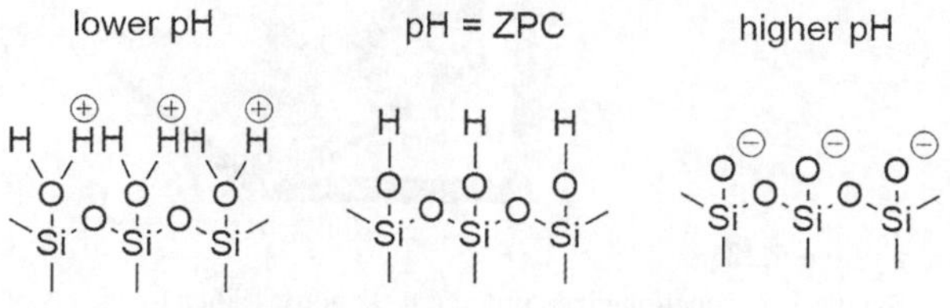

lower pH   pH = ZPC   higher pH

Scheme 5.1
Surface of silica at basic, neutral, and acidic pH.

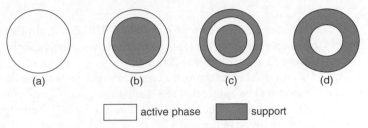

Figure 5.24
Four types of active-phase distribution. (a) uniform, (b) eggshell, (c) egg-white, and (d) egg-yolk.

and the method by which it was prepared. Of course, for the solid support a window of stability exists.

Impregnation Profiles

For impregnated catalysts a completely uniform profile of the active material over the particle is not always the optimal profile. It is possible to generate profiles on purpose, and in this way to improve the catalyst performance. Figure 5.24 shows four major types of active-phase distribution in catalyst spheres. The gray regions represent the areas impregnated with the active phase. Type *(a)* is a uniform catalyst while the others have a non-uniform active-phase distribution. They are called "eggshell," "egg-white," and "egg-yolk" catalysts. The optimal profile is determined by the reaction kinetics and the mode of catalyst poisoning. For example, an eggshell catalyst is favorable in the case of a reaction with a positive reaction order, whereas an egg-yolk catalyst is the best choice for reactions with negative orders. When pore mouth poisoning is dominant it might be attractive to locate the active sites in the interior of the catalyst particles. Another factor is attrition. If attrition is important and if the active phase is expensive (e.g. precious metals), it might be preferable to place the active phase in the interior of the catalyst particles.

The addition of a second component to the impregnating solution allows fine-tuning of the catalyst. Figure 5.25 illustrates this. Impregnation of H_2PtCl_6 is carried out in the presence of citric acid, which adsorbs more strongly than H_2PtCl_6 (and HCl). Without the presence of citric acid an eggshell type of profile for Pt is obtained. When some citric acid is present, this will adsorb first and block the outer sphere of the catalyst for the Pt-species. The Pt-species will pass the citric acid outside part and settle in the middle, creating an egg-yolk catalyst.

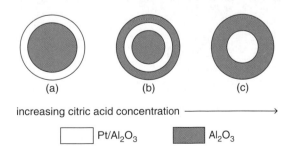

increasing citric acid concentration $\longrightarrow$

Pt/Al_2O_3 Al_2O_3

Figure 5.25
The influence of coadsorbing ions (citrate) on the Pt concentration profile.

5.2.1.1.4   Grafting

According to the recommendations of IUPAC, "deposition involving the formation of a strong (e.g. covalent) bond between the support and the active element is usually described as grafting or anchoring."

"This is achieved through a chemical reaction between the functional groups on the support and an appropriately selected inorganic or organometallic compound of the active element." Other terms like "immobilized," "heterogenized," "attached," and so on are found in literature.

Three classes of coordination metals are used very frequently for direct grafting/anchoring:

(1) metal halides and oxyhalides (e.g. $TiCl_4$, $MoCl_5$, CrO_2Cl_2)
(2) metal alkoxides ($VO(iPr)_3$) en diketonate complexes ($Cu(acac)_2$, $VO(acac)_2$)
(3) organometallics, especially metal allyls and metal carbonyls ($Cr(\eta^3\text{-}C_3H_5)_4$; $Ru_3(CO)_{12}$).

These types of organometallic complexes typically react with the silica surface, according to

$$Si - OH + TiCl_4 \rightarrow Si - O - TiCl_3 + HCl$$

Of course, several side reactions are possible (e.g. reaction with two silanols by the same complex) and the presence of water is detrimental to the reaction, as it will react directly with the very reactive metal complexes.

One of the most important problems is the poor stability of most grafted metal complexes on the silica surface in water or polar media. Also, at higher temperatures, the grafted complexes show a high mobility and tend to cluster (coalescence). This is sometimes solved by silylating the surface first.

Silylation of the Silica Surface

Silane coupling agents have the ability to form a durable bond between organic and inorganic materials. The general formula for a silane coupling agent typically shows the two classes of functionality. X is a hydrolysable group typically alkoxy, acyloxy, halogen, or amine. Following hydrolysis, a reactive silanol group is formed, which can condense with other silanol groups – for example, those on the surface of silica – to form siloxane linkages.

The R group is a non-hydrolysable organic function. The result of reacting an organosilane with a substrate is shown in Scheme 5.2.

Most of the widely used organosilanes have one organic substituent and three hydrolysable substituents. In the majority of surface treatment applications, the alkoxy groups of the trialkoxysilanes are hydrolyzed to form silanol-containing

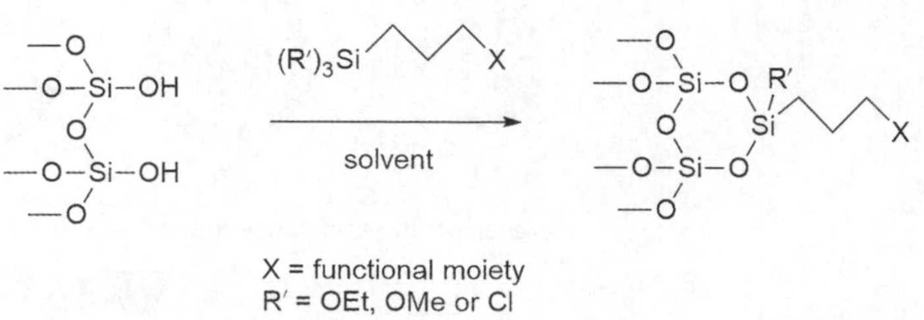

X = functional moiety
R' = OEt, OMe or Cl

Scheme 5.2
Example of grafting organosilanes onto a silanol-containing surface.

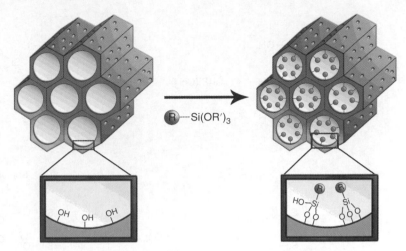

Figure 5.26
Silylation of SBA-15. Source: Figure is redrawn from Fröba's excellent review [35], Reproduced with permission of John Wiley & Sons, Ltd.

species. Reaction of these silanes involves four steps. Initially, hydrolysis of the three labile groups occurs. Condensation to oligomers follows. The oligomers then hydrogen bond with OH groups of the substrate. Finally, during drying or curing, a covalent linkage is formed with the substrate with loss of water. Although described sequentially, these reactions can occur simultaneously after the initial hydrolysis step. At the interface, there is usually only one bond from each silicon of the organosilane to the substrate surface. The two remaining silanol groups are present either in condensed or free form (Figure 5.26).

Van Der Voort gave a nice example of the combination of both techniques back in 1999 [37]. First, a silica surface was functionalized with a chlorosilane (dimethyl dichloro silane) to remove the surface silanols and to render the surface much more hydrophobic by the methylsilyl groups (species A in Figure 5.27). As new Si—Cl functionalities are created, these groups can be easily hydrolyzed (just by ambient air) and this "second generation" isolated silanols are then reacted with an active organometallic complex, vanadyl acetylacetonate, or $VO(acac)_2$ (species B). Finally, a calcination step removes the organic ligands, and a dihydroxyvanadyl species is grafted on the silica surface (species C). This species is surrounded by a hydrophobic environment and cannot "surf" on the surface. Vanadium oxide species on silica are very mobile and move rapidly on the silica surface, until they coalescence into large clusters and crystals of V_2O_5. This mechanism is prohibited here, and in fact, one of the first "single site catalysts" was created, long before the term gained widespread popularity.

5.2.1.1.5 *Gas Phase Coating Techniques – Atomic Layer Deposition (ALD)*

Coating techniques can be defined as all procedures that share the final aim of creating a thin layer on a foreign substrate. The thickness of such a coating varies from a monomolecular layer to several millimeters. The most conventional technique is Chemical Vapor Deposition (CVD) and its variants (Metal-Organic CVD [MO-CVD], Plasma Enhanced CVD [PE-CVD], and Laser CVD [L-CVD]). However, these techniques are not often used to coat mesoporous silica powders,

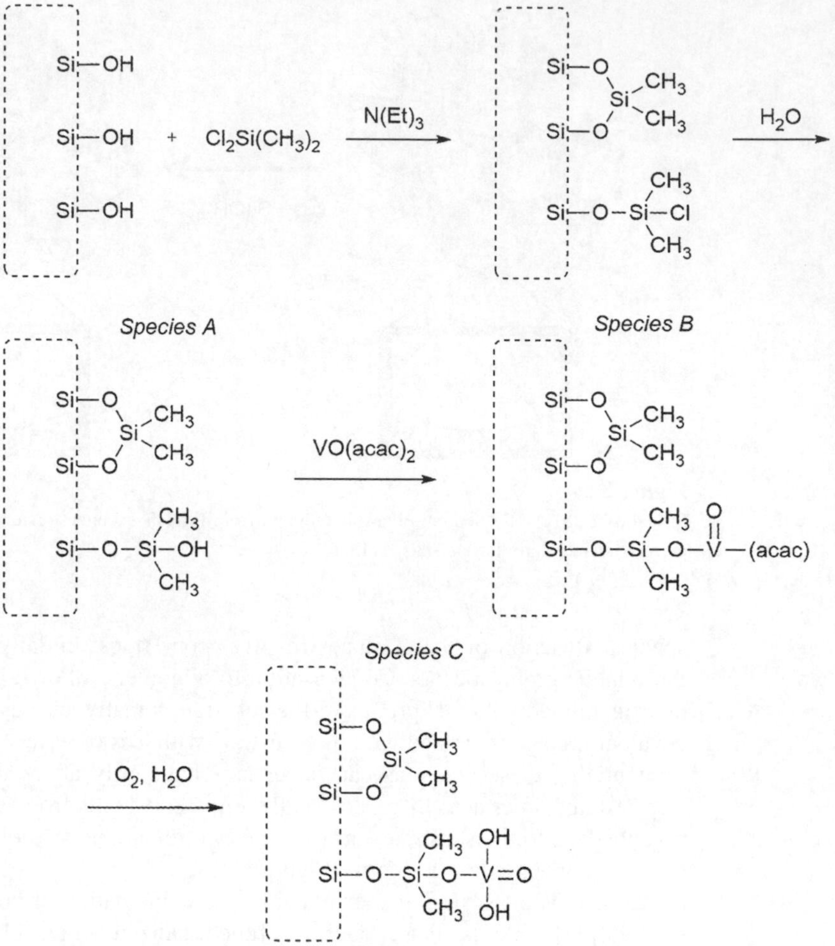

Figure 5.27
Creating isolated and stable vanadyl groups on silica by silylation. Source: Redrawn and reproduced with permission of ACS [37].

they are typically used in the fields of microelectronics and photonics, mostly on well-defined flat surfaces (silicon wafers).

We will therefore only discuss the technique of ALD. Although also originally developed to create lighting panels (information boards in airports, etc.), the technique has evolved to an advanced technique to create catalysts.

CVD involves a very complex mixture of gases and/or ions, causing numerous uncontrollable side reactions. From a chemical point of view, these reactions are extremely difficult to monitor.

This was recognized by Suntola in the 1980s when he developed a new coating technique: Atomic Layer Epitaxy (ALE) [38]. ALE is a method for producing thin films and layers of single crystals one atomic layer at a time, utilizing a self-control obtained through saturating surface reactions. ALE is based on separate surface reactions between the growing surface and each of the components of the compound, one at a time. These components are supplied in the vapor phase, either as elemental vapors or as volatile compounds of the elements.

Note

On the other side of the "iron curtain" in those days, A.A. Malygin was building a very similar approach, it was called Molecular Layering. Due to the political situation in those days, the term Molecular Layering never really broke through in the West, as most publications were in Russian. An English publication [39] has appeared in the framework of NATO's Science for Peace Program.

ALE was originally developed to meet the needs of improved ZnS thin films and dielectric thin films for electroluminescent thin film display devices. However, soon it became clear that any combination of a very reactive metal complex, followed by pulse of a second gas (usually water, ammonia, H_2S), can create any sort of thin layer. So, it was picked up again later under the name ALD. Epitaxy means crystalline overgrowth, for catalysis the layers do not have to be epitaxial, so a more general "deposition" was used. The main advantage of ALD is therefore that the layers a built up "layer by layer" (LbL) allowing fine control over the thickness of the layers. The general principle of ALD is visualized in Figure 5.28 and an example on how to make an alumina thin layer is shown in Scheme 5.3.

Research on ALD is still emerging in the field of catalysis. In a recent study, Van Der Voort has shown the amount of fine-tuning that is possible by ALD [40] (see Figure 5.29). By deposition of HfO_2 on an inkbottle type titania (compared to a planar substrate), one can clearly observe the narrowing of the pores that occurs linearly until the pore mouth becomes too narrow to allow the metal complexes, after which the pore is sealed, but is still not completely filled. This was confirmed by sorption experiments, we created "closed but unfilled" pores.

5.2.1.2 *Functionalized Mesoporous Silica in Heterogeneous Catalysis*

Silica (as gel or aerogels) are often used supports for heterogeneous catalysts. So, all the reactions that apply to silica gels obviously also apply to ordered mesoporous

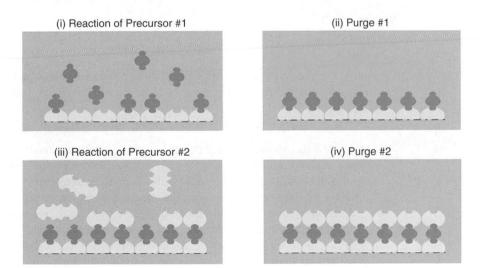

Figure 5.28

ALD Process, (i) a pulse of the first reactant is introduced; (ii) reactant 1 reacts until the surface is covered with exactly one monolayer after which residual molecules are pumped off; (iii) a pulse of reactant 2 is introduced, reactant 2 must react with reactant 1; (iv) reactant 2 forms a monolayer on top of reactant 1.

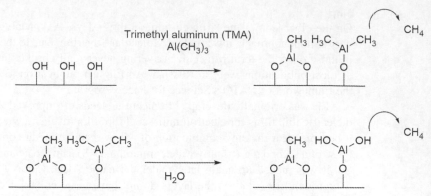

Scheme 5.3
ALD process for one monolayer of Al_2O_3: Trimethyl aluminum, reacts with the silanols with release of CH_4; as a second reactant water is introduced creating Al-OH sites, again with release of CH_4. A second cycle would start again with TMA, reacting again with the Al-OH sites. The layer forms "atom per atom" on the surface.

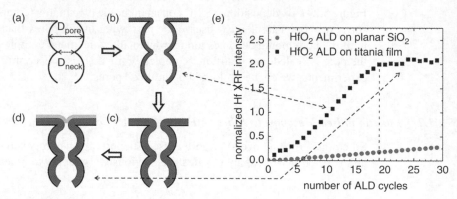

Figure 5.29
ALD of HfO_2 on an inkbottle titania pore.

silicas. The ordered materials offer the advantage of a much higher surface area (more active sites possible per gram material) and the ordered mesopores (offering improved diffusion and possible shape selectivity).

Silica as a support has a number of important disadvantages:

(1) The stability of the anchored groups on silica in humid air and water is relatively low. As we discussed in Chapter 4, the siloxane bond is prone to hydrolysis; one siloxane bond (Si—O—Si) splits with water into two hydroxyl groups (Si—OH HO—Si). This not only reduces the structure of the entire silica structure over a period of time; even more importantly, the functional groups that are usually anchored with reaction with the silanols are equally prone to hydrolysis. This results in a complete loss of functional groups when the catalyst is used in moist conditions. Only water-free processes are suitable for long-term catalysis life. (Of course, as in Fluid Catalytic Cracking [FCC] where the catalyst is continuously refreshed, this might still be worthwhile in certain applications.)

(2) Typical "acid oxides," with a low PZC (e.g. vanadium oxides, titanium oxides, tungsten oxides, chromium oxides, etc.) are highly mobile on the acid surface

Table 5.5 Predicted surface VO_x species as a function of the acidity of the support.

Oxide support	pH of support (at PZC)	Predicted VO_x species
MgO	11	$VO_3(OH)$
Al_2O_3	8.9	$VO_3(OH)$
TiO_2	6.0–6.4	$VO_2(OH)_2(VO_3)_n$
ZrO_2	5.9–6.1	$VO_2(OH)_2(VO_3)_n$
Nb_2O_5	4.3	$V_{10}O_{27}(OH)(VO_3)_n$
SiO_2	3.9	$V_2O_5V_{10}O_{26}(OH)_2$

of silica. This means that even carefully grafted metal oxides as single sites are highly mobile. As soon as the reaction temperature increases, these species become much more mobile and will "skate" or "surf" on the silica surface and will cluster together toward first nanosized clusters that eventually grow into large metal oxide clusters. A typical example is shown in Table 5.5: Predicted surface VO_x species as a function of the acidity of the support. The very acid VO_x species (V_2O_5 has a PZC of 2) bond very strongly to the basic hydroxyl sites on the MgO surface and only the single site V species are observed. MgO is not always a good catalytic support, it has a very low surface area, is very soft (attrition), and dissolves in many solvents including water. When the PZC of the supports increases, the isolated V-sites do not longer form, and on the very acidic silica support mainly large clusters of V species are formed.

Strategies have been developed to avoid this, see for example earlier in this chapter on the hydrophobized silica surrounded VO_x species (Figure 5.27).

Let us see if the promised advantages (higher catalytic activity due to higher surface area and shape selectivity) have found some practical applications. Also, the larger pores can accommodate much larger catalytic groups, so larger organometallics can be loaded in the mesoporous silica pores as well.

5.2.1.2.1 *Bifunctional Catalysts*

We start with the anchoring of two functions at the same time. This would not work in the homogeneous phase because the active groups would react or interact with each other and get neutralized.

Huang et al. [41] published the following procedure in 2011 to synthesize bifunctional acid/base catalysts (Figure 5.30). Please note that this represents just one example out of hundreds of papers that have appeared on the synthesis of bifunctional mesoporous silica catalysts. By combining two silanes, being the regular TEOS and a mercaptopropyl trimethoxysilane (called STMOS by the authors) a mesoporous silica (in the form of nanoparticles) was synthesized using the classic synthesis with CTAB in basic media. This way, both silanes co-condense to form a mercapto-groups containing silica nanoparticles. These functional groups are inside the pore system of the nanoparticles. Subsequently, without removing the surfactant, these particles are treated with an amine containing silane, aminopropyl trimethoxysilane (called APTMOS by the authors) in dry toluene. As the pores are still completely filled by the CTAB surfactant, the APTMOS can only react on the outer surface of the nanoparticles, creating aminopropyl functional groups on the particles. In the final step, several things occur simultaneously: the thiol (mercapto) groups are easily oxidized by H_2O_2, the surfactant is washed out, and the material

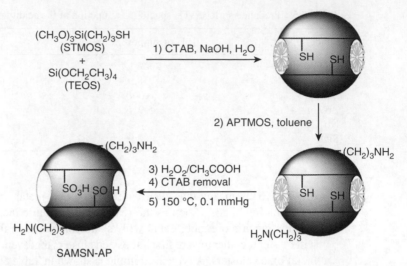

Figure 5.30
Synthesis of a bifunctional acid/base catalyst according to Huang. Source: Reproduced with permission of John Wiley & Sons, Ltd [41].

Table 5.6 One-pot reaction cascades composed of acid-catalyzed hydrolysis and base-catalyzed Henry reaction.

Entry	Catalyst	B (%)	C (%)	Conv. of A (%)
1	SAMSN-AP	2.3	97.7	100
2	SAMSN/APMSN	4.5	95.5	100
3	SAMSN	100	0	100
4	APMSN	0	0	0

is dried under vacuum at 150 °C. The final product is a bifunctional catalyst with strong acidic groups in the pores and mild basic groups on the outside of the particles (Figure 5.30).

Authors subsequently tested this bifunctional catalyst in a *cascade reaction*. A cascade reaction is a reaction that normally requires two different catalysts after each other. The authors chose a cascade of an acid-catalyzed hydrolysis of an acetal (product A in Table 5.6) toward an aldehyde (product B). This step is acid catalyzed only. The following reaction, the Henry reaction, is strictly base catalyzed and yields the end product (C). The entire reaction is shown in Scheme 5.4.

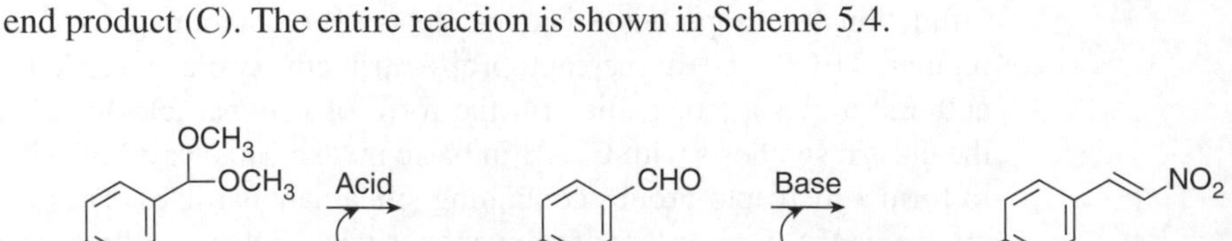

Scheme 5.4
Reaction conditions: Catalyst A: (100.0 mg, 1.5 mmol), H_2O (1.5 mmol) CH_3NO_2 (1.0 ml), 80 °C, 48 h. Conversion and yields were determined using GC data. AP: 1-aminopropane, PTSA: *p*-toluenesulfonic acid.

When the authors used the bifunctional catalyst (see Table 5.6), they called it SAMSN-AP, it is clear that the cascade reaction goes to completion, yielding 100% conversion, and 97.7% yield of the end product. Some intermediate product is still left, but this should be no problem for the synthesis of C.

The authors also tested – and this is commendable – a 50/50 mixture of nanoparticles only functionalized with amino groups and only functionalized with sulfonic acid groups. This resulted in almost the same product distribution (entry 2 in Table 5.6).

Using only the acid catalyst, 100% of A is converted to 100% B, but B does not react further, as could be expected (entry 3).

Finally, using only the base catalyst, nothing happens at all, as the starting product A cannot be converted by a base catalyst (entry 4).

5.2.1.2.2 Accommodating Large Organometallic Complexes

The large pore diameter of mesoporous silicas is able to accommodate much larger catalytic functions than that the zeolites or zeotypes. Organometallic catalysts are known to be very reactive and selective and won the Nobel Prize for Chemistry not so long ago. They are usually very expensive, not as much compared to the costs of the expensive noble metals (Pd, Pt, Au, Ag, Ru, Ir, …) but more importantly by the costs of the sometimes very expensive ligands.

Photo: U. Montan
Yves Chauvin
Prize share: 1/3

Photo: R. Paz
Robert H. Grubbs
Prize share: 1/3

Photo: L.B. Hetherington
Richard R. Schrock
Prize share: 1/3

The Nobel Prize for Chemistry was in 2005 awarded to three scientists: Chauvin, Grubbs, and Schrock. All three scientists had independently developed important catalysts for the metathesis of olefins, a very important chemical reaction. Metathesis is ancient Greek for "changing position."

(Photo: https://www.nobelprize.org/prizes/chemistry/).

The metal catalyzed olefin metathesis roughly follows Scheme 5.5:

Scheme 5.5
Metal catalyzed olefin metathesis.

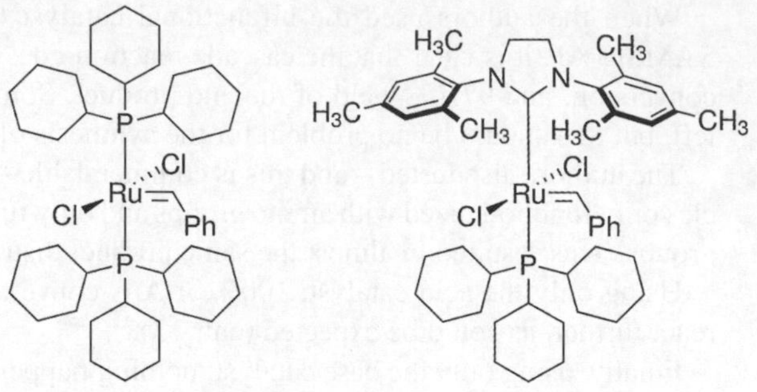

Scheme 5.6
Grubb's first- and second-generation catalyst.

However, the ligands that are coordinated to the metal site are crucially important. As an example, we show the Grubbs first and second generation catalysts that are commercially available now. They come at a price though. Checked on January 4, 2018 at the site of a very large commercial supplier, 10 g of Grubbs second generation catalyst costs €1805 or about USD\$ 2175. Normally, these catalysts get lost during the reaction. So, next to the loss of expensive catalyst, and especially in the pharmaceutical industry, an even more expensive purification step is required to remove the toxic catalyst from the end product. For this reason, the heterogenization of such a catalyst on a porous support would be highly interesting and beneficial. The big questions are however: (i) Is the heterogeneous catalyst still active enough? (ii) Is there no leaching of toxic materials out of the heterogeneous catalyst? (iii) Is the catalyst easily recyclable? (iv) Is the recycled catalyst still active for another run? Finally, (v) What is the lifetime of the catalyst? (Scheme 5.6).

In a very recent and lengthy review in *Chemical Reviews*, Kühn and coworkers [42] discussed all possible strategies and applications. We just show one of the many methods that uses the convenient *click chemistry*. It is not about the Grubbs catalysts, but a very similar metal complex. We will discuss the advantages of click chemistry further in the next chapter on PMO materials.

Although Cai and He [43] reported this procedure for a Merryfield resin, the procedure would also be valid for a mesoporous silica. In that case, a chlorosilylated mesoporous silica is exchanged with N_3^- functions. The group performs a simple click reaction by use of a Cu(I) catalyst, known as the CuAAC click reaction, with the general Scheme 5.7.

So, adding 3-methyl 1-propargylimidazolium bromide to the azide functionalize silica yields easily, by the aforementioned click reaction compound (3). Finally, adding the [Rh(COD)Cl$_2$] yields the final heterogeneous catalysts. The catalyst had excellent properties and was fully recyclable. Using click chemistry, these catalysts

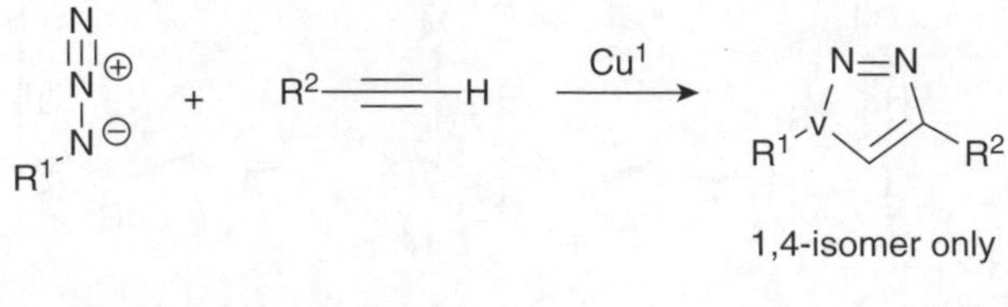

1,4-isomer only

Scheme 5.7
Cupper mediated click reaction.

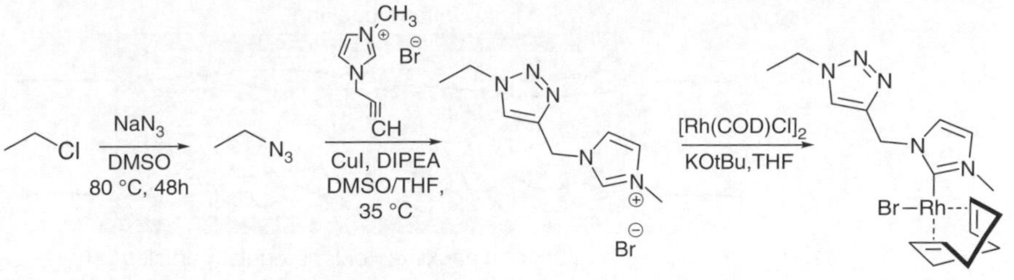

Scheme 5.8
Synthesis of a heterogeneous organometallic catalyst. Source: Redrawn from ref. [43] with permission.

can be easily prepared. Hundreds of other examples in the synthesis of heterogeneous catalysts based on mesoporous silicas can be found in the review that we mentioned earlier [42]. In this book, we will explore another type of click reaction (thiol-ene click reaction) in great detail in the chapter on PMOs (Scheme 5.8).

5.2.2 In Adsorption

5.2.2.1 Sorption of Metal Species

One of the earliest applications of mesoporous silicas was the adsorption of heavy metal ion (toxic ions) in water. Due to the large surface area and mesopores, the silica can be functionalized with metal attracting functional groups. The Soft and Hard Acid and Bases theory by Pearson is an easy starting point. The heavy metals, such as mercury and lead, bond preferably to soft ligands. Two very good soft ligands are amines and thiols.

Pinnavaia and coworkers [44] showed how thiol modified mesoporous silica easily captures Hg-ions, with little competition from typical concurring ions, such as Cd-ions, Pb-ions, and Zn-ions. The selectivity of the adsorbent is just as important as the capacity of the adsorbent. Usually, the targeted ions are present in a much lower concentration than the competing ions (think about Na^+, K^+, NH_4^+ ions ...) or even other soft ions. It is of paramount importance that only the targeted ions are captured. The materials were easily synthesized by reaction the mesoporous silica with mercaptopropyltriethoxysilane.

We will return more extensively on the sorption of metal ions in aqueous media in the following chapters.

5.2.2.2 Chromatography

Out of all applications domains for mesoporous silica particles, we believe the chromatography is the highest commercialized at the moment, as many columns can be purchased that contain porous silica particles. You will notice that – contrary to most other sections – most references originate in this section from patent literature.

The performance of chromatographic columns is strongly influenced by particle design. Uniform particles allow for a more homogeneous packing of a column and reduce the *Eddy diffusion* through the column, which leads to reduced peak dispersion. Since chromatography is in essence a diffusion-controlled process, the architecture of the pores in the material plays an important role in both the efficiency and the retention experienced with a particular type of packing material in a column. Current

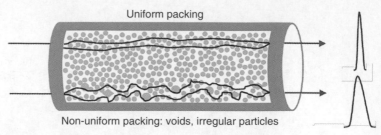

Figure 5.31
Graphical representation of the Eddy diffusion term throughout a column. Uniform pathways result in narrow peaks (top), poor quality of the column causes peak broadening (bottom).

silica based stationary phases exhibit a broad pore size distribution. This means that no two pathways throughout the particle are identical and that not every functional group is equally accessible (see Figure 5.31).

To reduce diffusion times through the particle, superficially porous particles were introduced. Superficially porous particles successfully improve the chromatographic performance. However, inherent to this particle design a reduced phase ratio leads to a lower sample loadability and a reduced retention. To deal with this drawback an ordered pore system with a strongly increased surface area could offer a solution.

We will not discuss the fundamentals of chromatography here. The interested reader is referred to some excellent books on this topic [45].

The Achilles heel of HPLC still is the hydrolysis of silica. Due to this susceptibility of silica materials to hydrolysis, conventional column performance is easily affected when using harsh conditions such as elevated pH or higher temperatures in combination with a highly aqueous mobile phase. This instability of silica packing materials is reflected in a reduced column efficiency and a reduced retention as a function of the number of chromatographic runs.

Endcapping of the silanols was the first solution and more recently researchers started to add more carbon to the stationary phase with semi and full hybrid types of silica, even fully carbonated stationary phases were tried.

Ordered mesoporous silica particles offer the advantage of uniform pore sizes and enhanced surface areas. The famous group of Klaus Unger [46], well known silica specialist and chromatographist, in collaboration with Ferdi Schüth, compared already back in 1996 the then "novel" MCM-41 particles with commercial materials, including the LiCrosphere Si 100 (actually a fumed silica or an aerosol, see Chapter 4). As no protocols to produce spherical mesoporous silicas were in place yet, these particles were just grinded to proper dimensions and irregular in shape. In this experiment, the LiChrosphere particles outperformed the MCM-41, proving again the importance of monodispersed spheres for chromatography.

It became clear that uniform and spherical particles are extremely important for a good separation. The characteristics of an "ideal" chromatographic particles are summarized as follows:

- Particles should be spherical.
- Particles should be monodisperse, $D_{90/10} < 1.6$.[1]
- Particles must be between 2 and 5 μm diameter for analytical purposes and between 5 and 15 μm for preparative purposes.

[1] $D_{90/10}$ is a statistic describing the dispersion on the particle size. D_{10} is the particle diameter where the cumulative volume of the particles reaches 10%. D_{90} is the corresponding value at 90 v%.

- Surface area should be high, at least higher than $50\,m^2\,g^{-1}$, but much higher is better.
- Pore size should be large in the mesopore range, diameter should be larger than 6 nm.
- There should be as little micropores as possible, $S_\mu/S_{BET} < 0.1$.
- Mechanical stability should resist the harsh packing and analysis pressures.
- Particles should be hydrolytically stable between below pH = 2 and above pH = 12.
- Particles should be stable at temperatures above $100\,^\circ C$.
- There should be no (metal) impurities.

5.2.2.3   Methods to Synthesize Spherical Mesoporous Particles

5.2.2.3.1   Fully Porous Spherical Particles

Monodisperse silica particles were first described by Stöber in 1968, who performed the hydrolysis and condensation of TEOS in water using ethanol as a dispersing co-solvent and ammonia as the catalyst [47]. This delivered, depending on the reaction conditions, solid particles with a controllable size between 50 and 2000 nm. Based on this groundbreaking invention, multiple pathways toward porous particles have been developed. In the first approach, these solid nanoparticles are fused in a controlled way to form a spherical particle with mesoporous voids between the original nanoparticles. This is done by *coacervation*, a technique where silica nanoparticles are brought together during a polymerization reaction. As described by Destefano and Kirkland [48], the silica nanoparticles are mixed with monomers of melamine or urea and formaldehyde. As polymerization takes place, the nanoparticles co-precipitate and after removal of the polymer by calcination, uniform micron-sized spheres are found (see Figure 5.32).

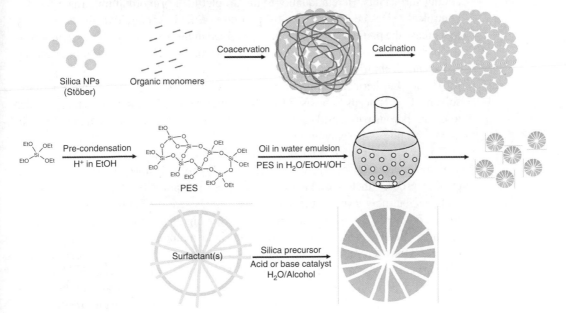

Figure 5.32

Graphical representation of the coacervation method (top), the Unger method (middle), and the adapted Stöber method (bottom) to obtain fully porous silica spheres.

Secondly, Unger and Schick–Kalb developed a Stöber-like method in 1971, where pre-condensed TEOS, that is, poly(ethoxysiloxane) or PES, is used as starting point instead of a silica sol [49]. This PES is subsequently emulsified in a water/ethanol mixture in which it forms microdroplets. Addition of a base catalyst starts hydrolysis of the PES, which results in the formation of porous beads. The condensation degree, measured by the viscosity of the PES, is claimed to control porosity, while the particle size is controlled by the stirring speed.

A third, similar option is to take the original Stöber method and add pore generating surfactants as described by Unger's group in 1997 [50]. Depending on the surfactant, pore size and ordering can be controlled, however, it seems more practical to expand smaller pores with a post-synthetic hydrothermal treatment. This approach has further been investigated by other groups, who managed to produce spheres in an acidic medium by using hydrochloric acid as a catalyst. The properties of the particles remained roughly the same except these could reach diameters up to 1 mm [51]. As a main benefit for this templated method, highly monodisperse particles are obtained that do not require physical separation by means of classification.

In 1998, another synthesis pathway for mesoporous silica spheres was introduced by Qi et al. [52]. A mixture of CTAB and the non-ionic surfactant Brij-56 combined with TEOS as silica source was used for liquid-crystal templating under acidic conditions but without the addition of a co-solvent. The resulting particles reveal a particle diameter of 2–6 μm and all characteristic properties of the ordered mesoporous materials.

Next to this, spherical particles of porous silica can also be obtained by employing water-in-oil emulsions [53] or by spray drying [54]. Both processes essentially create droplets in which xerogel particles form. Inside these spherical microdroplets, the porosity can be controlled, either by the sol-gel process occurring inside or by adding surfactants. Herein, quality of the droplet is of primordial importance as this controls the size and dispersion of the particles. With these colloid methods, it is easy to increase the particle size above those obtained in Stöber-type syntheses, while control of particle size and pore size is effectively decoupled. However, with spray drying, it does not seem trivial to obtain monodisperse particles in the size range applicable for HPLC. Furthermore, considering particles with ordered porosity, an evaporating solvent (EtOH) needs to be used together with a pore generating surfactant in order to obtain porous particles. Due to these constraints, it remains challenging to obtain large mesopores via spray drying.

Finally, another interesting procedure is called *pseudomorphic transformation*. This method, described by Anne Galarneau, highly resembles a redeposition/pore etching process that is often used to enhance the porosity of mesoporous silica particles, but now involves the addition of a surfactant, or SDA. The process starts from commercial porous silica particles. Stirring these particles in an alkaline solution (NaOH), water, and CTAB at elevated temperatures partly dissolves the particle. However, redeposition of this dissolved silica is believed to occur around the surfactant. As a result, ordered MCM-41 type pores are obtained (CTAB inducing a pore size of approximately 4 nm), while the spherical morphology is maintained. Later MCM-48 type materials were developed via this method (see Figure 5.33) [55].

Commercially, many columns packed with fully porous silica particles, both pure and modified, are available.

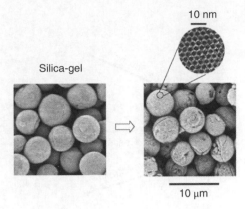

Figure 5.33

SEM images of Nucleosil 100-5 before and after its pseudomorphic transformation into as material with ordered pores (MCM-41). Source: Reproduced with permission of ACS [55].

5.2.2.3.2   *Core-Shell Particles*

Core-shell or superficially porous particles were envisaged by Horvath and Lipsky [56] as early as 1969 and subsequently developed by Kirkland [57] (Figure 6.22). These particles were prepared by coating a glass bead with poly(diethylaminoethylmethacrylate)acetate. Thereafter, 200 nm silica nanoparticles, as prepared by Stöber, are added at pH 3.6. Coulombic attraction between silica and polymer causes a layer of nanoparticles to stick on the surface of the glass bead. After four growth rounds and a sintering step, this resulted in a stable core-shell particle. These materials were commercialized as Zipax® and investigated as supports for the now abandoned technique of liquid–liquid chromatography.

Bad results in the application, combined with a large particle size and ongoing advantages in fully porous particles, caused reduced interest in core-shell particles for many years.

Core-shell particles were revisited in 1992, again by Kirkland, who developed a technique using simultaneously spray drying of a mixture of large solid silica particles, obtained after sintering of monodisperse fully porous Zorbax particles, together with a sol of silica NPs [58]. The average size of the latter was 44 nm and resulted in a wide pore size distribution from 10 to 60 nm. Both this and the presence of fully porous aggregates, which were hard to separate, limited the use of these pioneering Poroshell® particles.

A major improvement was found in using a coacervation method as described earlier for fully porous particles, but using a large core particle or using a polyelectrolyte during the coating process [59]. Optimization of these methods has at least led to several commercial materials,

The Unger group, followed by others, developed another popular LbL growing approach that employs a template surfactant to grow a porous shell on top of a silica core [60]. After one synthesis round generally, a very thin 60–75 nm layer is grown with small pore sizes. However, when suitable templates are used and pore swellers are employed one is able to boost the pore size.

Wei introduced another possible method that starts from a solid silica particle with a mean size of 3.1 µm. The selective etching of the particle's surface now generates superficially porous particles [61]. An overview of the different methods is presented in Figure 5.34.

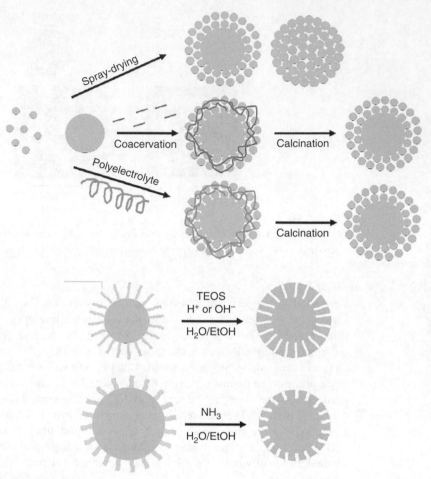

Figure 5.34
Graphic representation of synthesis methods to obtain core-shell type particles. Top: The methods of Kirk-land using mixtures of large and small solid silica particles. Middle: The Unger and Eiroshell method applying surfactants to grow porous layers on top of a solid particle. Bottom: The Wei method taking advantage of selective etching in the presence of surfactants.

5.2.3   As a Drug Carrier

Mesoporous ordered silica, due to its large and uniform pores and its biocompatibility is ideally suited to adsorb biomolecules. A next step would be the controlled release of the biomolecules, this is usually called *controlled drug release*.

Especially for cancer chemotherapy, very toxic drugs are introduced in the body that are very damaging to the good cells as well. Researchers have been studying for decades ways to introduce the drugs at the location of the cancer cells only. The ideal transport vehicle would be a *smart material*, that responds to an *external trigger* (light, pH, heat, …). If one would be able to introduce the drug in a mesoporous silica and close the pores, this would travel innocently in the body until a trigger "opens" the pores and the drug are slowly released at the desired location.

In a very recent review, Zhu et al. describe the recent progress in this field extensively [62]. We will limit our contribution here to a few illustrative examples.

Single stimulus responsive drug delivery systems are materials that release their encapsulated drugs upon one stimulus only. The stimuli are either pH, redox potential, enzyme interaction, light, temperature, magnetism, or ultrasound. For cancer treatment, the pH and the redox responsive systems are especially important, as cancer cells typically have a lower pH than healthy cells and contain a high concentration of GSH (glutathione, L-γ-glutamyl-L-cysteinyl-glycine, an active redox component).

The extracellular pH of most tumor issues is more acidic (pH = 6.5–6.8) than normal tissues (pH = 7.4) due to the so-called Warburg effect. Moreover, the pH in the endocytic visicle drops to 5.5–6.0 in the endosomes and to 4.5–5.0 in the lysosomes. So, pH-sensitive molecules can be used as "gate keeper" to keep the pores with the drugs inside closed until a low pH is encountered. Typical chemical gatekeepers, stable in neutral and basic media, but unstable in acid media are acetals, amines, boronates, and hydrozones (Figure 5.35a).

Other pH-sensitive gatekeepers would be acids with a pKa close to the tumor pH. A small change in the pH would protonate multiple sites at the gatekeeper, changing its solubility (Figure 5.35b). A nice example of this strategy was provided by Bilalis et al. [63]. They capped the drug loaded mesoporous silica particles with poly-L-histidine, with a pKa around 7.0. This closed the pores, until they reach the tumor cells with acidic pH, became protonated, unfolded from the surface and released the drugs.

5.2.3.1 Light Responsive Drug Delivery Systems

Light responsive systems have the advantage that they be controlled from outside, providing that the penetration depth of the laser in the human tissue in deep enough. So near-infrared light is the best option, it has the deepest penetration depth and is less energetic, and thus less damaging to normal tissue.

The classical example was provided by Abe et al. [64]. They used a coumarin modified mesoporous silica (Figure 5.35g) as a gate keeper. When irradiated at 300–350 nm dimerization of the coumarin occurs. When it is irradiated with light below 260 nm, however, the dimer will dissociate and open the gate. The Figure 5.35g also reveals that there is still a lot of work is ahead of us. The ideal light responsive materials operation at the NIR wavelengths is still not available at this moment.

Figure 5.35 shows many more examples that we cannot discuss within the scope of this book. We refer the reader to the very recent review of Zhu et al. [62] or to one of the many other reviews on this topic. We also refer to the works of Maria Vallet-Regi and coworkers [65].

5.2.4 Low-*k* Dielectrics

Porous silica and organosilicas are very important dielectrical barriers (low-*k* materials) in microelectronic chips. In the continuing miniaturization of electronic devices, manufacturers increase transistor speed, reduce its size, and pack more transistors on a single chip. Nowadays, it is possible to put more than 2 billion transistors on a chip. However, when the interconnecting wires also reduce in size, the resistance of these wires is increased.

$$R = \rho \frac{l}{a} \tag{5.8}$$

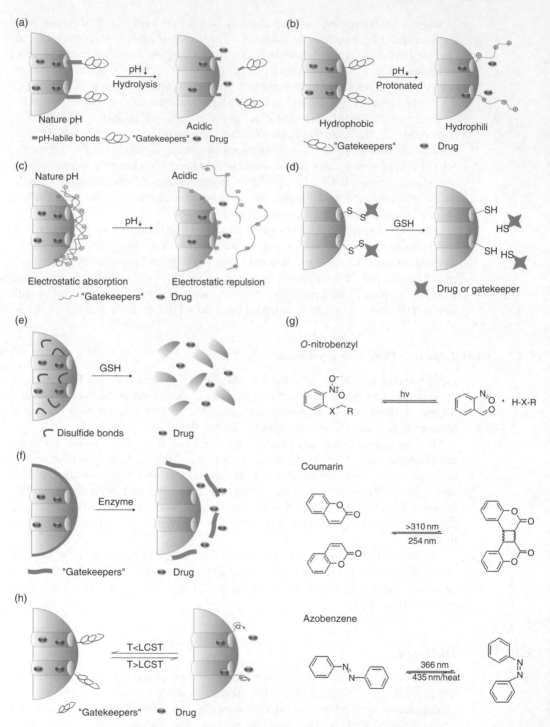

Figure 5.35

Schematic diagram of different stimuli response mechanisms: (a) the first approach for synthesis of a pH-responsive drug delivery system, in which the linker would cleaved under acidic conditions; (b) another approach for the synthesis of pH-responsive nanocarriers, in which the pKa value of the gatekeeper is near the tumor interstitial pH; (c) a type of charge switching pH-responsive drug delivery system; (d) disulfide-linked drugs or gatekeepers; (e) degradable MSNs; (f) enzyme-sensitive drug delivery vehicles; (g) the light-response mechanism of o-nitrobenzyl, coumarin, and azobenzene; (h) a thermoresponsive drug delivery system. Source: Reproduced with permission of the RSC [62].

This phenomenon is described by the relationship in Eq. (5.8), which is known as Pouillet's law and sometimes, incorrectly, referred to as Ohm's law from which it is derived. R is the total resistance, ρ is the material's resistivity, l is the length of the specimen, and a its cross section.

These interconnects come closer together and electrical interference ("cross-talk") occurs, which is highly undesirable. A very good insulator is required to isolate the different interconnects.

For a long time, this has been normal silica, it has a good mechanical and thermal stability, low leakage current and very high electrical breakdown. However, still RC-delay occurs (resistance rapacitance), and better insulators are required.

One pathway is to change the silica insulator ($k = 3.9$) by an insulator with a k-value of 2.2 or below. As dry air is the perfect insulator ($k = 1$) and water is the worst insulator ($k = 80$), a hydrophobic highly porous systems seems ideal. Mesoporous silica films, prepared by the EISA method has gained a lot of attention in this field.

We will discuss the low-k materials in more detail in Chapter 6 on PMOs, as these materials are more promising to reach the targets. We refer you to an excellent review [66], and to the pioneering papers on this matter by Brinker, Ozin, Landskron, and others.

This was not an exhaustive overview of the applications of mesoporous silica. Applications are in biomedicine, sensing, luminescence, and many others. For a quick overview of some of the most important research in the past decade, we refer you to the reviews in *Chemical Reviews* or *Chemical Society Reviews* [67].

References

1. (a) J. S. Beck, C.-W. Chu, I. Johnson, C. Kresge, M. Leonowicz, W. Roth, J. Vartuli, US Patent, No. US 5098684 1992; (b) Beck, J.S., Vartuli, J.C., Roth, W.J. et al. (1992). *J. Am. Chem. Soc.* 114: 10834–10843; (c) Kresge, C.T., Leonowicz, M.E., Roth, W.J. et al. (1992). *Nature* 359: 710–712.
2. Yanagisawa, T., Shimizu, T., Kuroda, K., and Kato, C. (1990). *Bull. Chem. Soc. Jpn.* 63: 988–992.
3. Inagaki, S., Fukushima, Y., and Kuroda, K. (1993). *J. Chem. Soc., Chem. Commun.* 680–682.
4. Landskron, H., Schmidt, G., Heinz, K. et al. (1991). *Surf. Sci.* 256: 115–122.
5. Kresge, C.T. and Roth, W.J. (2013). *Chem. Soc. Rev.* 42: 3663–3670.
6. Descalzo, A.B., Martinez-Manez, R., Sancenon, R. et al. (2006). *Angew. Chem. Int. Ed.* 45: 5924–5948.
7. (a) Chen, C.-Y., Burkett, S.L., Li, H.-X., and Davis, M.E. (1993). *Microporous Mater.* 2: 27–34; (b) Cheng, C.F., He, H.Y., Zhou, W.Z., and Klinowski, J. (1995). *Chem. Phys. Lett.* 244: 117–120.
8. Kruk, M., Jaroniec, M., and Sayari, A. (1997). *Langmuir* 13: 6267–6273.
9. Ravikovitch, P.I. and Neimark, A.V. (2000). *Langmuir* 16: 2419–2423.
10. Huo, Q.S., Margolese, D.I., Ciesla, U. et al. (1994). *Nature* 368: 317–321.
11. Tanev, P.T. and Pinnavaia, T.J. (1995). *Science* 267: 865–867.
12. Bagshaw, S.A., Prouzet, E., and Pinnavaia, T.J. (1995). *Science* 269: 1242–1244.
13. Besson, S., Gacoin, T., Ricolleau, C. et al. (2003). *J. Mater. Chem.* 13: 404–409.
14. Huo, Q.S., Leon, R., Petroff, P.M., and Stucky, G.D. (1995). *Science* 268: 1324–1327.
15. Van der Voort, P., Mathieu, M., Mees, F., and Vansant, E.F. (1998). *J. Phys. Chem. B* 102: 8847–8851.
16. Tanev, P.T. and Pinnavaia, T.J. (1996). *Chem. Mater.* 8: 2068–2079.
17. Cassiers, K., Van Der Voort, P., and Vansant, E.F. (2000). *Chem. Commun.* 2489–2490.
18. Zhao, D.Y., Sun, J.Y., Li, Q.Z., and Stucky, G.D. (2000). *Chem. Mater.* 12: 275–280.
19. Lin, H.P., Tang, C.Y., and Lin, C.Y. (2002). *J. Chin. Chem. Soc.* 49: 981–988.
20. Galarneau, A., Nader, M., Guenneau, F. et al. (2007). *J. Phys. Chem. C* 111: 8268–8277.

21. Janssen, A.H., Van Der Voort, P., Koster, A.J., and de Jong, K.P. (2002). *Chem. Commun.* 1632–1633.
22. Zhao, D.Y., Huo, Q.S., Feng, J.L. et al. (1998). *J. Am. Chem. Soc.* 120: 6024–6036.
23. Van Der Voort, P., Ravikovitch, P.I., De Jong, K.P. et al. (2002). *J. Phys. Chem. B* 106: 5873–5877.
24. Van Der Voort, P., Ravikovitch, P.I., De Jong, K.P. et al. (2002). *Chem. Commun.* 1010–1011.
25. Thommes, M., Kaneko, K., Neimark, A.V. et al. (2015). *Pure Appl. Chem.* 87: 1051–1069.
26. Sing, K.S.W., Everett, D.H., Haul, R.A.W. et al. (1985). *Pure Appl. Chem.* 57: 603–619.
27. Choi, M., Heo, W., Kleitz, F., and Ryoo, R. (2003). *Chem. Commun.* 1340–1341.
28. (a) Zhai, S.R., Park, S.S., Park, M. et al. (2008). *Microporous Mesoporous Mater.* 113: 47–55; (b) Zhai, S.R., Kim, I., and Ha, C.S. (2008). *J. Solid State Chem.* 181: 67–74.
29. Guo, W.P., Park, J.Y., Oh, M.O. et al. (2003). *Chem. Mater.* 15: 2295–2298.
30. Chen, L.H., Zhu, G.S., Zhang, D.L. et al. (2009). *J. Mater. Chem.* 19: 2013–2017.
31. (a) Brinker, C.J., Lu, Y.F., Sellinger, A., and Fan, H.Y. (1999). *Adv. Mater.* 11: 579–585; (b) Lu, Y.F., Ganguli, R., Drewien, C.A. et al. (1997). *Nature* 389: 364–368.
32. Corma, A., Grande, M.S., GonzalezAlfaro, V., and Orchilles, A.V. (1996). *J. Catal.* 159: 375–382.
33. van Bekkum, H. and Kloetstra, K.R. (1998). *Stud. Surf. Sci. Catal.* 117: 171–182.
34. Tanev, P.T., Chibwe, M., and Pinnavaia, T.J. (1994). *Nature* 368: 321–323.
35. Hoffmann, F., Cornelius, M., Morell, J., and Froba, M. (2006). *Angew. Chem. Int. Ed.* 45: 3216–3251.
36. Wight, A.P. and Davis, M.E. (2002). *Chem. Rev.* 102: 3589–3613.
37. Van Der Voort, P., Baltes, M., and Vansant, E.F. (1999). *J. Phys. Chem. B* 103: 10102–10108.
38. (a) Ahonen, M., Pessa, M., and Suntola, T. (1980). *Thin Solid Films* 65: 301–307; (b) Suntola, T. and Hyvarinen, J. (1985). *Annu. Rev. Mater. Sci.* 15: 177–195.
39. Malygin, A.A. (1999). *NATO ASI Ser., Ser. E* 362: 487–495.
40. Dendooven, J., Goris, B., Devloo-Casier, K. et al. (2012). *Chem. Mater.* 24: 1992–1994.
41. Huang, Y.L., Xu, S., and Lin, V.S.Y. (2011). *Angew. Chem. Int. Ed.* 50: 661–664.
42. Zhong, R., Lindhorst, A.C., Groche, F.J., and Kühn, F.E. (2017). *Chem. Rev.* 117: 1970–2058.
43. He, Y. and Cai, C. (2011). *Chem. Commun.* 47: 12319–12321.
44. Brown, J., Mercier, L., and Pinnavaia, T.J. (1999). *Chem. Commun.* 69–70.
45. (a) Snyder, L.R. and Kirkland, J.J. (1979). *Introduction to Modern Liquid Chromatography*, 2e. Wiley; (b) Lembke, P., Henze, G., Cabrera, K. et al. (2008). Liquid chromatography. In: *Handbook of Analytical Techniques* (ed. H. Günzler and A. Williams), 261–326. Wiley-VCH Verlag GmbH.
46. Grun, M., Kurganov, A.A., Schacht, S. et al. (1996). *J. Chromatogr. A* 740: 1–9.
47. Stober, W., Fink, A., and Bohn, E. (1968). *J. Colloid Interface Sci.* 26: 62–69.
48. Destefano, J.J. and Kirkland, J.J. (1974). *J. Chromatogr. Sci.* 12: 337–343.
49. Berg, K. and Unger, K. (1971). *Kolloid Z. Z. Polym.* 246: 682.
50. Grun, M., Lauer, I., and Unger, K.K. (1997). *Adv. Mater.* 9: 254–257.
51. Yang, H., Coombs, N., Dag, O. et al. (1997). *J. Mater. Chem.* 7: 1755–1761.
52. Qi, L.M., Ma, J.M., Cheng, H.M., and Zhao, Z.G. (1998). *Chem. Mater.* 10: 1623–1626.
53. M. Nyrstrom, W. Herrmann, B. Larsson, Patent, Silica particles, a method for preparation of silica particles and use of the particles, No. EP0298062, 1991.
54. Ide, M., Wallaert, E., Van Driessche, I. et al. (2011). *Microporous Mesoporous Mater.* 142: 282–291.
55. Martin, T., Galarneau, A., Di Renzo, F. et al. (2004). *Chem. Mater.* 16: 1725–1731.
56. Horvath, C. and Lipsky, S.R. (1969). *J. Chromatogr. Sci.* 7: 109.
57. Kirkland, J.J. (1969). *Anal. Chem.* 41: 218.
58. Kirkland, J.J. (1992). *Anal. Chem.* 64: 1239–1245.
59. J. J. Kirkland, T. J. Langlois, US Patent, Process for preparing substrates with a porous surface, No. US2009297853 2009.
60. (a) Buchel, G., Unger, K.K., Matsumoto, A., and Tsutsumi, K. (1998). *Adv. Mater.* 10: 1036–1038; (b) J. Glennon, J. Omamogho, International Patent, A process for preparing silica particles, No. WO2010061367 2010.
61. T. C. Wei, W. Chen, W. E. Barber, Polym. Chem. 2016, 7, 1475–1485.

62. Zhu, J.H., Niu, Y.M., Li, Y. et al. (2017). *J. Mater. Chem. B* 5: 1339–1352.
63. Bilalis, P., Tziveleka, L.A., Varlas, S., and Iatrou, H. (2016). *Polym. Chem.* 7: 1475–1485.
64. Abe, E., Pennycook, S.J., and Tsai, A.P. (2003). *Nature* 421: 347–350.
65. (a) Baeza, A., Ruiz-Molina, D., and Vallet-Regi, M. (2017). *Expert Opin. Drug Delivery* 14: 783–796; (b) Castillo, R.R., Baeza, A., and Vallet-Regi, M. (2017). *Biomater. Sci.* 5: 353–377; (c) Castillo, R.R., Colilla, M., and Vallet-Regi, M. (2017). *Expert Opin. Drug Delivery* 14: 229–243; (d) Doadrio, A.L., Salinas, A.J., Sanchez-Montero, J.M., and Vallet-Regi, M. (2015). *Curr. Pharm. Des.* 21: 6189–6213; (e) Martinez-Carmona, M., Colilla, M., and Vallet-Regi, M. (2015). *Nanomaterials* 5: 1906–1937; (f) Vallet-Regi, M. (2006). *Chem. Eur. J.* 12: 5934–5943; (g) Vallet-Regi, M. (2010). *J. Intern. Med.* 267: 22–43; (h) Vallet-Regi, M., Balas, F., Colilla, M., and Manzano, M. (2007). *Solid State Sci.* 9: 768–776; (i) Vallet-Regi, M., Colilla, M., and Gonzalez, B. (2011). *Chem. Soc. Rev.* 40: 596–607; (j) Vallet-Regi, M.A., Ruiz-Gonzalez, L., Izquierdo-Barba, I., and Gonzalez-Calbet, J.M. (2006). *J. Mater. Chem.* 16: 26–31.
66. Lionti, K., Volksen, W., Magbitang, T. et al. (2015). *ECS J. Solid State Sci. Technol.* 4: N3071–N3083.
67. (a) Chaudhuri, R.G. and Paria, S. (2012). *Chem. Rev.* 112: 2373–2433; (b) Ciriminna, R., Fidalgo, A., Pandarus, V. et al. (2013). *Chem. Rev.* 113: 6592–6620; (c) Valtchev, V. and Tosheva, L. (2013). *Chem. Rev.* 113: 6734–6760; (d) Deng, Y.H., Wei, J., Sun, Z.K., and Zhao, D.Y. (2013). *Chem. Soc. Rev.* 42: 4054–4070; (e) Han, W.S., Lee, H.Y., Jung, S.H. et al. (2009). *Chem. Soc. Rev.* 38: 1904–1915; (f) Innocenzi, P. and Malfatti, L. (2013). *Chem. Soc. Rev.* 42: 4198–4216; (g) Li, Z.X., Barnes, J.C., Bosoy, A. et al. (2012). *Chem. Soc. Rev.* 41: 2590–2605; (h) Parlett, C.M.A., Wilson, K., and Lee, A.F. (2013). *Chem. Soc. Rev.* 42: 3876–3893; (i) Perego, C. and Millini, R. (2013). *Chem. Soc. Rev.* 42: 3956–3976; (j) Qiu, H.B. and Che, S.N. (2011). *Chem. Soc. Rev.* 40: 1259–1268; (k) Wagner, T., Haffer, S., Weinberger, C. et al. (2013). *Chem. Soc. Rev.* 42: 4036–4053; (l) Walcarius, A. (2013). *Chem. Soc. Rev.* 42: 4098–4140; (m) Wen, J., Yang, K., Liu, F.Y. et al. (2017). *Chem. Soc. Rev.* 46: 6024–6045; (n) Wu, S.H., Mou, C.Y., and Lin, H.P. (2013). *Chem. Soc. Rev.* 42: 3862–3875; (o) Yang, P.P., Gai, S.L., and Lin, J. (2012). *Chem. Soc. Rev.* 41: 3679–3698; (p) Zaera, F. (2013). *Chem. Soc. Rev.* 42: 2746–2762.

6 Carbons

The previous chapters in this book describe inorganic materials that are mainly silicon-based solids (Chapters 3, 4, and 5). They are purely inorganic, although organic functionalities can be built in to extend their applicability. This chapter describes a chemically completely different class of materials; that is, the carbon-based materials. They entirely consist of carbon but are synthesized in a different manner to organic polymers. Structurally, they more closely resemble the (ordered) mesoporous silica materials as they also are porous and can be ordered depending on the synthesis procedure used.

6.1 Activated Carbon

Activated carbon, also known as active carbon [1] (AC), is the most widespread and commonly known carbon-based material. It is a generic term and refers to highly porous carbonaceous materials. It is also called activated charcoal, activated coal, or activated coke depending on the source it originates from. It has been known by men for centuries as it was used in Ancient Egypt. They used, as an example, charcoal for the purification of oils, but also applied it on ships to the inside of wooden barrels that contained drinking water. Since then and up to now, it is mainly used in environmental applications as an adsorbent for the removal of pollutants and the purification of water and air.

AC is a form of graphite with a random porous structure and it can contain pores over a wide variety of pore sizes, including micro-, meso-, and macropores. AC is obtained from sources that contain a high amount of carbon. The precursors are very diverse, fueled by the need for low-cost carbons. In fact, it can be made from any agricultural byproduct and everyday waste material. Precursors that are used include wood, coal, coconut shells, coffee beans, distillery waste, kelp and seaweed, palm tree nut shells, bamboo, lignin, and rubber waste [2], to name a few. Figure 6.1 shows that different carbon precursors can result in a variety of AC structures.

Many methods exist to produce AC or to "activate" the carbon in the source. Mostly, four basic steps are required: raw material preparation, pelletizing, low-temperature carbonization, and activation. Two activation methods are available that introduce the porosity in the material and expand the surface area significantly.

1. *Thermal or physical activation (PA method):* This type of activation is usually a two-step process. The carbon source is treated at elevated temperature and first involves carbonization at 400–500 °C. This removes the bulk of the volatile matter. During carbonization, the carbon atoms are rearranged and form a graphite-like

Introduction to Porous Materials, First Edition.
Pascal Van Der Voort, Karen Leus and Els De Canck.
© 2019 John Wiley & Sons Ltd. Published 2019 by John Wiley & Sons Ltd.

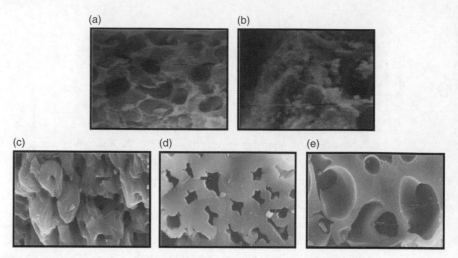

Figure 6.1

Scanning electron microscopy images of activated carbon prepared with different carbon precursors: (a) walnut shell; (b) olive pit; (c) oil palm fruit bunch; (d) coconut shell, and (e) bamboo stem. Source: Reproduced with permission of Elsevier [2].

structure. Afterward, a partial gasification is performed with a mild oxidizing agent (e.g. CO_2, steam, flue gas) at 800–1000 °C. This last step creates the final porosity.

2. *Chemical activation (CA method):* This type of activation is usually performed in one step. The carbon source is chemically treated with inorganic additives such as acids, bases or salts such as H_3PO_4, KOH, NaOH, $CaCl_2$, $ZnCl_2$, and so on (see Figure 6.2). These additives degrade during carbonization at 250–650 °C and help to dehydrate the carbon precursor. Generally, lower temperatures and shorter reaction times are needed in comparison with the physical carbonization treatment described before. The type and amount of additive will affect the resulting pore structure of the carbon significantly.

Activated carbon can be produced using either the physical or CA method. However, a combination of both methods can also be used.

The resulting chemical and physical characteristics of AC will highly depend on the carbon source used and the activation process needed as it will mainly determine the porosity (surface area, distribution of pores) and the surface functional groups present on the solid. The PA method usually uses higher temperatures than the CA method and therefore the former also results in a lower carbon yield. Also, the porosity is different. The use of PA leads to smaller surface areas, and narrow micropores and mesopores.

AC can be classified in diverse ways according to the carbon source, activation method, and behavior but a general classification is generally made based on physical properties: Powdered Activated Carbon (PAC), Granular Activated Carbon (GAC), Extruded Activated Carbon (EAC), and so on. For example, PAC or pulverized AC is an activated carbon in the form of a powder and is produced by milling or pulverizing the AC. PAC contains particles smaller than 80 US Mesh or 0.180 mm according to ASTM D5158 [3].

Activated carbon is probably the most widely used and most important commercial adsorbent due to its low price, ease of synthesis, versatility, and, moreover, many

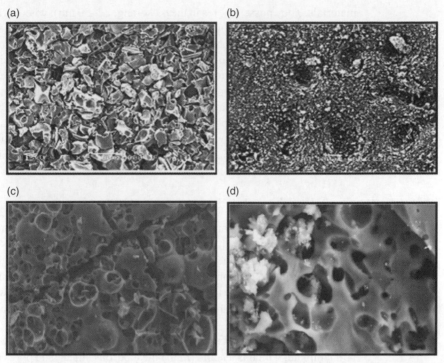

Figure 6.2
Scanning electron microscopy images of activated carbon prepared via the chemical activation method with different additives: (a) $ZnCl_2$; (b) KOH; (c) K_2CO_3; and (d) $ZnCl_2$. Source: Reproduced with permission of Elsevier [2].

precursors can be used for the production of AC. Currently, it is applied to a whole array of applications due to its porous structure combined with high surface area: gas separation, gas storage, air treatment, and also military use in, for example, gas masks, catalysis, potable and effluent water treatment, decolorizing, solvent recovery, precious metal recovery, and so on.

Although activated carbon clearly possesses many advantages, it also holds some drawbacks, especially when high-end or specialty applications are targeted. AC is less suitable if, for example, different type of metals must be separated from each other as it is a general adsorbent. Additionally, due to its inherent pore structure and large amount of micropores, bulky molecules (e.g. dyes, organic matter, enzymes, humic acids, etc.) cannot access the pores, which inevitably leads to lower adsorption capacity.

Therefore, researchers have made efforts to develop carbons that possess larger (in the meso range) and ordered pores to allow macromolecules in the pore system. It was the start for research toward mesoporous polymers and carbons.

6.2 General Introduction to Mesoporous Carbons

Ordered Mesoporous Carbons (OMCs) are a more advanced class of porous materials in comparison with activated carbon. They combine high porosity, just like mesoporous silicas, with the physico-chemical properties of organic polymers and carbon materials. They possess pores in the mesorange (2–50 nm), whereas activated carbons in most cases only exhibit microporosity. Mesoporous carbons generally combine mesoporosity and microporosity, and in some cases also have macropores. They possess unique properties concerning porosity but also hydrophobicity, which makes them especially suitable for environmental applications. OMCs can be prepared via two different synthesis procedures, that is, the hard-templated method (see Section 6.2.1) and the soft-templated method (STM) (see Section 6.2.2). Both methods first prepare an ordered mesoporous polymer (OMP), which is transformed afterward into an OMC. A general overview of the different steps in the hard-templated method or also nanocasting or exocasting strategy and the STM is shown in Figure 6.3.

6.2.1 Synthesis of Hard-Templated Mesoporous Carbons

In 1999, Ryoo et al. reported the preparation of mesoporous polymers and carbons using a hard template [5]. It was the first report of an ordered mesoporous material with a carbon-based framework. In this method, mesoporous silica materials are used as template, hence the name "hard-templated method" or indirect method, nano-, or exocasting. Basically, the template is filled with a carbon precursor with the formation of a composite or OMP and afterward the template is removed, resulting in a replica (see Figure 6.4).

In more detail, Ryoo and coworkers used in that first report an MCM-48 material as a template, which is a cubic ($Ia\overline{3}d$) mesoporous silica. The pores of this solid were impregnated (thus entirely filled) with a carbon source, that is, sucrose, dissolved in an aqueous solution containing sulfuric acid. After impregnation, the silica was heated to a range of 1073–1373 K under vacuum or in an inert atmosphere.

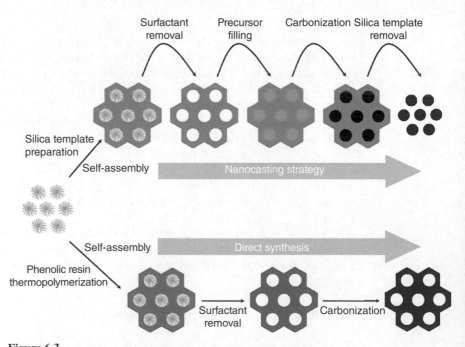

Figure 6.3
The difference between the hard-templated (nanocasting or exocasting strategy) and the soft-templated (direct synthesis) methods. Source: Reproduced with permission of the RSC [4].

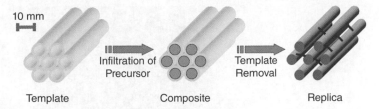

10 mm

Infiltration of
Precursor

Template
Removal

Template   Composite   Replica

Figure 6.4

Schematic illustration of the principle of the hard-templated or indirect synthesis method for ordered mesoporous carbons. Source: Reproduced with permission of John Wiley & Sons, Ltd [7].

*Ryong Ryoo is director of the Center for Nanomaterials and Chemical Reactions at
the Institute for Basic Science and Distinguished Professor in the Chemistry Department at Korea Advanced Institute of Science and Technology (KAIST) in South-Korea.
He obtained his M.S. degree at KAIST in 1979. Afterwards, he studied for a Ph.D. in
the field of heterogeneous catalysis at Stanford University and obtained his doctoral
title in 1986. After a post-doctoral research stay at the University of California in
Berkeley, Ryong Ryoo became an assistant professor at KAIST where he is now Distinguished Professor. He is famous for his pioneering work in the field of OMC, which
resulted in the well-known Carbon Material Korea (CMK) structures. As an appreciation of his work, he received in 2007 the award "Leading Scientist in Research
Front" from Thomson Scientific and the "Breck Award" in 2010 from the International Zeolite Association. Additionally, he was listed among the Top 100 Chemists
of the decade 2000–2010 by UNESCO & IUPAC, based on Thomson Reuters citation
impact data.*

Photograph: www.ibs.re.kr/eng/sub02_04_01.do.

This process converts the sucrose into carbon where sulfuric acid acts as a catalyst.
An OMP is formed inside the pores of the silica. Afterward, the silica is dissolved with
a NaOH/EtOH treatment. This results in a porous carbon material, called CMK-1,
where CMK stands for *Carbon Material Korea*. The solid showed a regular and uniform pore system as can be seen from TEM images (Figure 6.5a), whereas the SEM

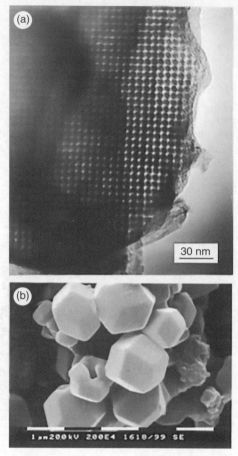

Figure 6.5
Electron microscopy images of a CMK-1 material: (a) transmission electron micrograph showing the pore system; and (b) scanning electron micrograph of CMK-1 particles. Source: Reproduced with permission of ACS [5].

image clearly shows the formation of carbon particles (Figure 6.5b) with similar morphology to silica particles.

The template synthesis did not result in a mere replication of the structure of the mesoporous silica. X-ray diffraction analysis of the different steps during the synthesis showed that the XRD pattern of the material after the impregnation step with the carbon source is identical to the pattern of the MCM-48 material (Figure 6.6). Only some lattice contraction and loss of intensity is observed. The contraction is usually observed when MCM-48 undergoes a heat treatment and is not due to the impregnation procedure. The XRD patterns clearly show a $Ia\bar{3}d$ space group. The removal of the silica framework results in a systematic transformation of the carbon into a new ordered structure, however with lower symmetry. The resulting CMK-1 structure exhibits both micropores and mesopores of 0.5–0.8 and 3.0 nm in diameter, respectively. A high specific surface area of 1380 m^2 g^{-1} was obtained and the micropore and mesopore volume resulted into 0.3 and 1.1 ml g^{-1}, respectively.

The sucrose as carbon source can be replaced by other carbon precursors, such as glucose, fructose, xylose, and so on. Moreover, with a variety of ordered mesoporous silica materials available, the applicability of this hard-templated method can

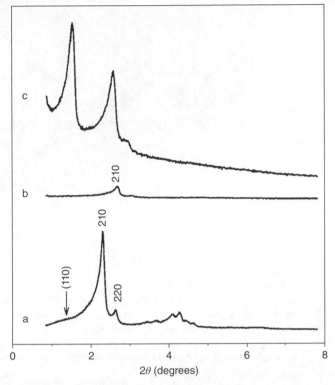

Figure 6.6

X-ray diffraction patterns of (a) MCM-48; (b) MCM-48 after impregnation and carbonization of the sucrose within the pores; and (c) CMK-1 after removal of the silica framework. Source: Reproduced with permission of ACS [5].

be extended to different pore structures and sizes. The same group headed up by Ryoo reported the synthesis of CMK-3 based on the same principle as described before, but using a SBA-15 structure as template [6]. The silica framework was removed either with a NaOH/EtOH solution or a hydrofluoric acid treatment at room temperature. In contrast to CMK-1, CMK-3 is a true inverse replica of the SBA-15 (Figure 6.7).

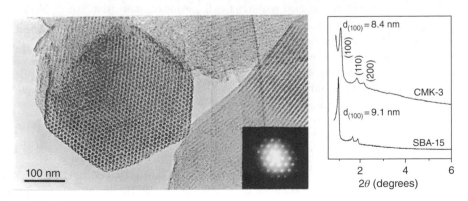

Figure 6.7

(left) Transmission electron microscopy image of CMK-3; and (right) powder XRD pattern of SBA-15 before impregnation and the resulting CMK-3. Source: Reproduced with permission of ACS [6].

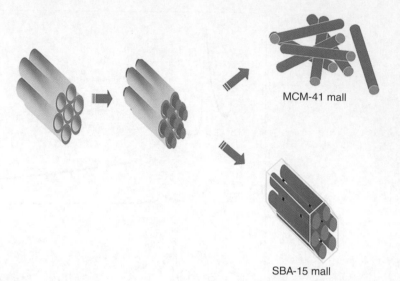

Figure 6.8

The synthesis of an ordered mesoporous carbon material in two different pores systems an isolated (MCM-41) and a continuous (SBA-15) pore system. The use of SBA-15 as template will result in the carbon CMK-3. Source: Reproduced with permission of the RSC [8a].

Both materials display a 2D hexagonal space group (*P6mm*) with the (100), (110), and (200) diffractions.

Several OMCs have been reported that start from different silica templates such as SBA-1, SBA-3, SBA-15, HMS, KIT-6, and so on [7, 8]. It must be noted that not every mesoporous ordered silica material can be used as hard template [8a]. The template SBA-15 contains both meso- and micropores. Its pore system is completely interconnected and can be entirely or partially filled with the carbon precursor. This will result in CMK-3 (Figure 6.8) or CMK-5, respectively. The latter is a carbon material made of hollow carbon tubes. The use of MCM-41 as template can, however, never result in an OMC material. This silica template consists of a non-interconnected pore system and results in non-porous carbon rods.

However, one must note that this hard-templated method also shows some clear disadvantages. The synthesis procedure is time-consuming as multiple steps are required. It also requires the hazardous or corrosive HF or NaOH solutions for the removal of the SiO_2 framework. These aspects also make the method unsuitable for large-scale production. Moreover, the exact tuning of the wall thickness and pore diameter is limited to the availability of silica templates. These drawbacks encouraged researchers to find an alternative route for the synthesis of OMPs and OMCs. Table 6.1 shows an overview of the mesocarbons, prepared by the exocasting technique.

6.2.1.1 Double Templating

Overall the use of silica as the hard template is limited, because it can only be removed under strong alkaline conditions or by HF. However, such chemical conditions are often not compatible with many different oxides, since these are also attacked by these reagents. Therefore, carbon templates provide an interesting alternative to silica, since carbon can be removed by a simple combustion at mild temperatures or by a treatment with other highly reactive gases. When an OMC is used, prepared by the nanocasting technique, it forms a "negative" mall.

Table 6.1 Summary of the reported mesoporous carbons, generated by exocasting.

Name	Template space group	Space group carbon structure	Carbon source
CMK-1	MCM-48 $Ia\bar{3}d$	$I4_1$ (or lower)	Sucrose, phenol resin
CMK-2	SBA-1 $Pm3n$	Cubic (space group unknown)	Sucrose
CMK-3	SBA-15 $P6mm$	$P6mm$	Sucrose, furfuryl alcohol
CMK-4	MCM-48 $Ia\bar{3}d$	$Ia\bar{3}d$	Ethylene
CMK-5	SBA-15 $P6mm$	$P6mm$	Furfuryl alcohol
NCC-1	SBA-15 $P6mm$	$P6mm$	Furfuryl alcohol
(No name)	SBA-16 $Im\bar{3}m$	$Im\bar{3}m$	Sucrose, furfuryl alcohol

The resulting metal oxide after removal of the carbon hard template will be a "positive" copy of the original silica ordered mesoporous material, albeit with a reduced pore size after every nanocasting procedure. This concept was proven by Schüth [9] who prepared nanocasted silica (NCS-1) by repeated exocasting: first a mesoporous "negative" template carbon replica of SBA-15 was prepared, and then the nanocasted silica was prepared by filling this CMK-3 again with a silica source and calcining the carbon hard template. Schüth showed that the morphology of the NCS-1 strongly resembles the original SBA-15 structure, although obviously the pore size and surface area were somewhat reduced.

The use of carbon malls is a viable method for the synthesis of compositions that are difficult to prepare via direct surfactant templating. This is typically the case for basic oxides, because the solubility of the oxide is much too high under typical synthesis conditions. Using CMK-3 as a mall, Roggenbuck et al. prepared mesoporous magnesium oxide with a hexagonal $P6mm$ symmetry [10]. Via the same method, even mesoporous boron nitride (BN) can be prepared with a specific surface area of $500 \, m^2 \, g^{-1}$. A range of porous metal oxides, including Al_2O_3, TiO_2, ZrO_2, and V_2O_5 has been prepared by this method [11] with high thermal stability and often a crystalline framework.

Ferdi Schüth (in English often Schueth) is a German chemist, who played an important role in the development of the exocasting technique, and on many developments in the field of mesoporous materials and heterogeneous catalysis in general. He is currently Director of the Max-Planck-Institute für Kohlenforschung and is Honorary

Professor and the University of Bochum and Vice-President of the Max Planck Society.

Photograph: https://www.kofo.mpg.de/en/research/heterogeneous-catalysis

6.2.2   Synthesis of Soft-Templated Mesoporous Carbons

To counter the disadvantages of the hard-templated method, researchers developed a direct or soft-templated synthesis procedure to prepare OMCs. The procedure is very similar to the preparation of ordered mesoporous silicas [12].

Zhao et al. developed direct STM in 2005 to synthesize highly ordered and stable mesoporous polymers [13]. Zhao was the pioneer in the development of SBA type materials (see Chapter 5) and re-used their successful surfactant-templated approach [12b].

Dongyuan Zhao (1963, Shenyang, China) received his master's degree in chemistry at Jilin University (China) in 1987 and his Ph.D. Degree from Jilin University and Dalian Institute of Chemical Physics in 1990. In 1992 he became Associate Professor at the Shenyang Institute of Chemical Technology where he was lecturer in the Chemical Engineering Department. Zhao performed several stays abroad at the University of Regina (Canada), Weizmann Institute of Science (Israel), University of Houston (USA) and the University of California Santa Barbara (USA). In 1998, he became Professor at the Department of Chemistry of Fudan University. Zhao is specialized in the development of ordered mesoporous and hierarchical materials for bio-applications, electrochemistry, catalysis, and water treatment applications. He is considered as the pioneer in the development of SBA materials (Chapter 5) and Mesoporous Polymers/Carbons.

Photograph: https://www.imms9.org/speakers

The synthesis procedure is outlined in Figure 6.9. It starts with the polymerization of a carbon precursor around a surfactant via the Evaporation-Induced Self-Assembly (EISA) principle. Zhao chose a cheap resol, a low molecular weight polymer of phenol (P) and formaldehyde (F) as carbon precursor. The resol is dissolved in a volatile solvent that evaporates during the polymerization. During the evaporation, the critical micelle concentration for the structure-directing agent is reached and the structure-directing agent (SDA) will form the required micelles to induce the porosity. The carbon precursor will interact with the SDA via hydrogen bond formation and an

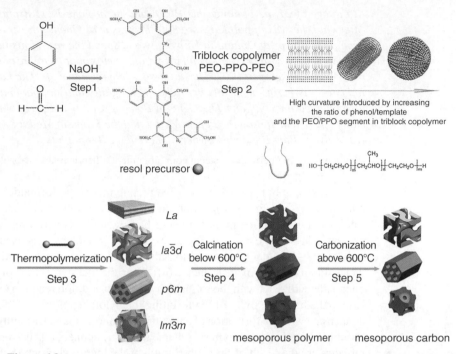

Figure 6.9
Schematic representation to prepare ordered mesoporous polymers and ordered mesoporous carbons via the soft-templated method. Source: Reproduced with permission of ACS [14].

ordered structure will be formed. Afterward, the SDA is removed via calcination or solvent extraction. After this step, a mesoporous polymer, also called a mesoporous resol, is formed. A final carbonization step at 900 °C will render the OMC material. Two different structures were synthesized, that is, FDU-15 (*Fudan University*, China) and FDU-16 with a hexagonal and cubic mesostructure, respectively.

The selection of the resol precursor is the key to success here as it contains many hydroxy groups that can strongly interact with the triblock copolymer. The use of phenol and formaldehyde as monomers was not completely new. It was discovered by the chemist Leo Baekeland in 1907 who developed the famous thermosetting phenol-formaldehyde resin also known as Bakelite [15].

Leo Henricus Arthur Baekeland (1863–1944) is a chemist with Belgian-American roots. He was born in Ghent (Belgium) and studied chemistry at Ghent University.

At the age of 21, he obtained his Ph.D. and he became First Professor at Bruges. Afterward, he was appointed Associate Professor at Ghent University. In 1917, he became professor at Columbia University and spend the rest of his time in America. He is well-known for two important inventions: Velox photographic paper in 1893 and the polymer Bakelite in 1907. The latter gave him the name of "The Father of Plastics Industry" as he invented this very cheap, versatile, and nonflammable plastic. He also founded the company The General Bakelite Co. in 1910. He was honored with many awards, including the Perkin Medal and the Franklin Medal in 1916 and 1940, respectively. Baekeland was owner of more than 100 patents.

Photograph: http://periodieksysteem.com/biografie/leo-henricus-arthur-baekeland

The first step in the formation of the prepolymer is the formaldehyde addition to phenol [16]. Two different reaction mechanisms are known and are pH dependent. The prepolymer can be prepared under basic conditions and a reaction pathway is presented in Scheme 6.1. Usually an excess of formaldehyde is used, that is, a P : F ratio of 1 : 1–3. Under these conditions the phenol is deprotonated and forms the resonance stabilized phenoxide (Scheme 6.1(1)). In the first step of the polymerization the phenoxide will react with the formaldehyde via an addition reaction (Scheme 6.1(2)). Due to the high reactivity of the resulting methylolphenol this will further react and condensate with the formation of a mixture of mono-, di- and tri-methylolphenols and eventually the final prepolymer (Scheme 6.2). The connections between the aromatic moieties are the result of S_N2 substitution of the hydroxyl group (Scheme 6.1(3)) or a Michael addition of the *o-* or *p-*chinonmethide (Scheme 6.1(4)). Other precursors can be used (see later) and the same type of reactions will take place.

The synthesis can also be performed under acid conditions with an excess of phenol to formaldehyde. In this case, Novolacs are obtained instead of resols. The different pH results in mostly linear and slightly branched oligomers with low molecular weight that are linked to each other via the methylene bridges. The reaction mechanism under acid conditions and the resulting Novolac prepolymer are presented in

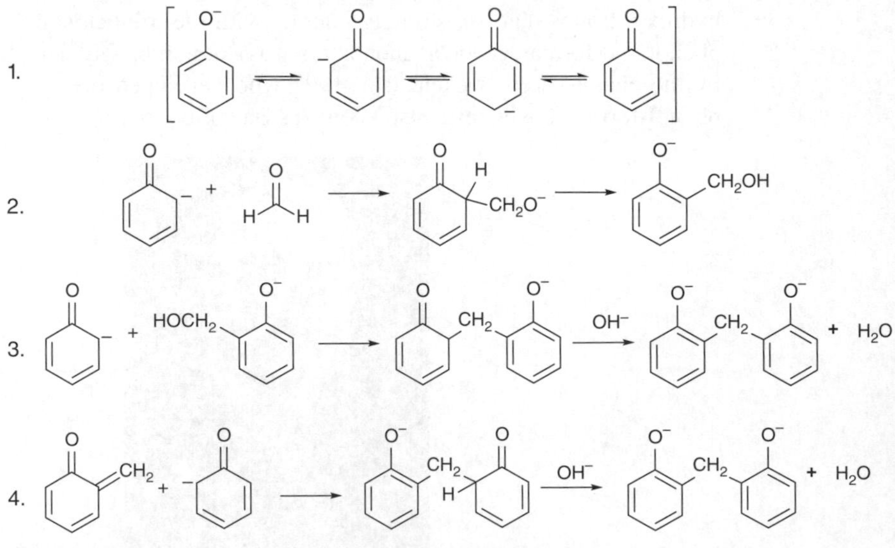

Scheme 6.1
Overview of the different reactions that take place during prepolymerization under basic conditions.

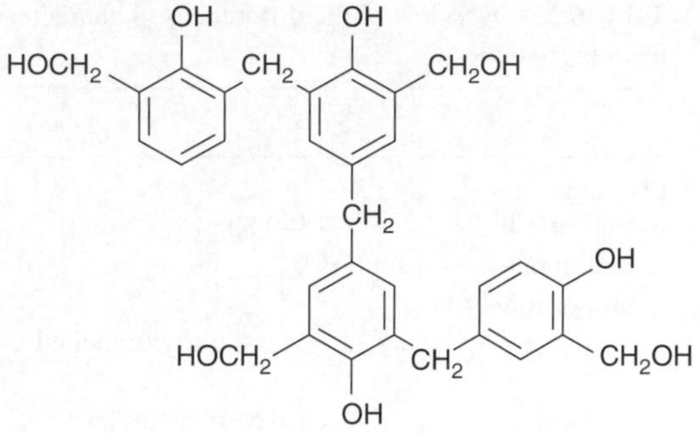

Scheme 6.2

The prepolymer "resol" with low molecular weight formed out of phenol and formaldehyde under basic conditions.

Schemes 6.3 and 6.4, respectively. Thus, depending on the pH during synthesis, a different prepolymer will be formed that will also behave differently. An overview can be found in Table 6.2. In the end, both type of prepolymers will form the same 3D interconnected polymer network in the OMP material [16].

6.2.3   Influence of Synthesis Conditions on the Soft-Templated Method

The synthesis of OMPs and OMCs via the STM is influenced by many factors: the choice of precursor, the acid or base catalyst, the surfactant, the precursor surfactant ratio, the synthesis method, the carbonization temperature, and so on.

The influence of the pH during the synthesis has already been discussed in Section 6.2.2 as this affects strongly the formation of the prepolymer. The other synthesis parameters will be briefly presented in the following subsections. The reader is referred to some excellent reviews for more information concerning the influence of the different synthesis parameters [4, 8a, 16, 17].

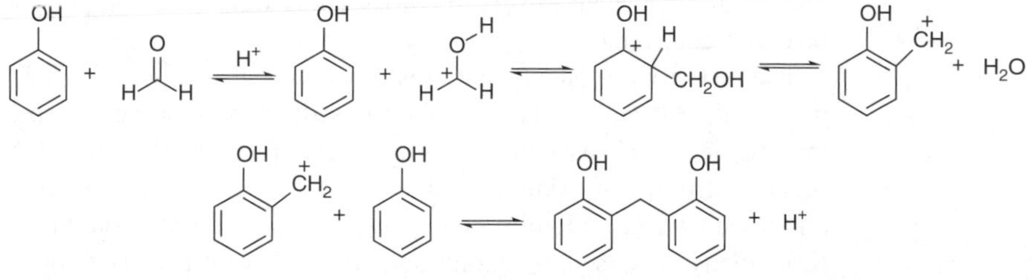

Scheme 6.3

Overview of the different reactions that take place during prepolymerization under acidic conditions.

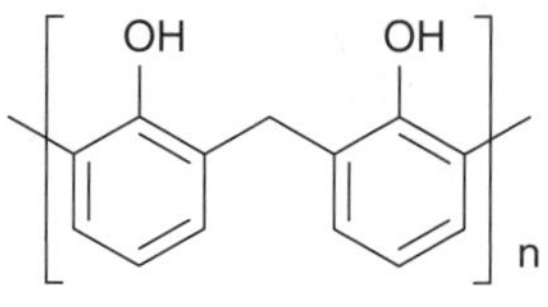

Scheme 6.4

The prepolymer "Novolac" with low molecular weight formed out of phenol and formaldehyde under acidic conditions.

Table 6.2 Overview of the differences in characteristics between the Novolac and resol prepolymers.

	Novolac	Resol
pH range	<4	>5
Molar ratio F: P[a]	0.75–0.85 : 1	1.00–3.00 : 1
Reaction rate proportional to …	[H+]	[OH−]
Structure	Linear, slightly branched	Three-dimensional, cross-linked
Before curing	Exclusively methylene bridges	Mainly methylene and ether bridges
After curing	Methylene bridged 3-D network	Methylene bridged 3-D network
Molecular weight	Lower (MW = 2000)	Higher (MW = 500–5000)
Prepolymer type	Thermoplastic	Thermoplastic
Soluble in …	Not soluble in organic solvents and water	Organic solvents and water

[a]F = formaldehyde and P = phenol.
Source: Reproduced with permission of Elsevier [16].

6.2.3.1 *Precursor*

Several precursors are commonly used in the synthesis of mesoporous carbons. These phenolic resin monomers are phenol, resorcinol (benzene-1,3-diol) and phloroglucinol (benzene-1,3,5-triol) (Scheme 6.5) and they all contain an aromatic ring with one or more hydroxyl functions. The hydroxyl functions will allow the precursor to interact via hydrogen bonding or electrostatic interactions ($I^+ X^- S^+$, see Chapter 5) with the polyethylene oxide (PEO) chains of the SDA.

The reactivity of the precursor highly depends on the amount of hydroxyl groups present. This means that the reactivity of phloroglucinol is the highest, followed by resorcinol and finally phenol. Phenol is often selected as the precursor despite its lower activity, because of its low price and the fact that the polymerization rate can be easier controlled.

The toxicity of phenol and resorcinol is an additional aspect that cannot be neglected. As an alternative, plant-based or bio-sourced carbon precursors are investigated, such as tannin from, for example, wattle barks [18], chestnut [19], or larch from, for example, sawdust [20], and lignin from hardwood [21], and all proved to be suitable as mesoporous polymer/carbon precursors. Normally, formaldehyde is used as second precursor where it acts as a sort of "cross-linking" agent. However, also this molecule is toxic. A possible substitute that has been

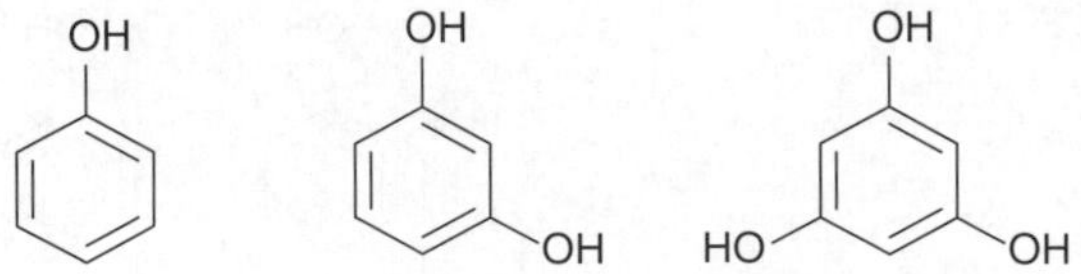

Scheme 6.5
A selection of phenolic resin monomers for the synthesis of mesoporous carbons: phenol, resorcinol, and phloroglucinol.

proposed and investigated is furfural [22]. It is also worth mentioning that in certain cases hexamethylene tetramine (HMT) is added during synthesis. This additive will produce *in situ* formaldehyde and ammonia, where the ammonia further catalyzes the polymerization reaction between the alcohol and aldehyde [22, 23].

Next to pure carbon-based mesoporous carbons, the inclusion of nitrogen has also been investigated. Here, N-containing precursors are used such as melamine [24], dicyandiamide [25] or urea [26]. They can be used as sole precursor or combined with phenol or resorcinol. An example of the resulting structure, starting from the N-rich melamine, formaldehyde, and phenol, is presented in Scheme 6.6. An overview of reported carbons and N-doped carbons, including synthesis conditions, has recently been reviewed and can be found in reference [17a].

6.2.3.2 Structure-Directing Agent

The SDA can also be varied and influences the porosity of the resulting mesoporous ordered polymers or carbons. The same type of SDAs is used as for OMS materials. A strong interaction between SDA and precursor molecules is also a condition that must be met for OMPs or OMCs. Initially, the cationic quaternary ammonium SDAs were used. However, only disordered or collapsed structures were obtained due to the weak interaction between SDA and precursors. Additionally, the SDAs turned out to

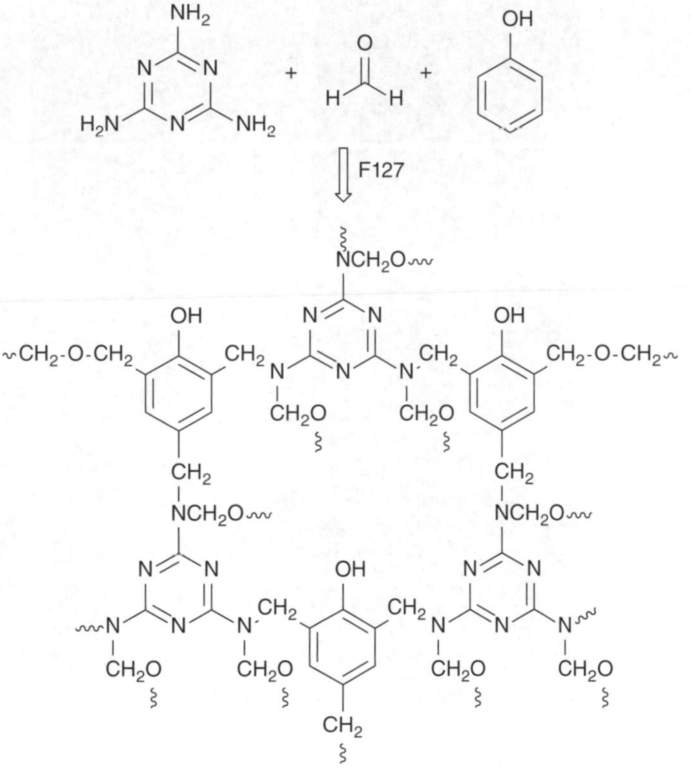

Scheme 6.6
The formation of a melamine resin with melamine as N-source, phenol as C-source, formaldehyde as cross-linking agent, and F127 as structure-directing agent.

be hard to remove. The triblock copolymers such as P123, F127, and F108 were better candidates. They possess PEO and polypropylene oxide (PPO) blocks. The interaction occurs via hydrogen bonds, which is a much stronger interaction. The alcohol precursor is generally hydroxyl rich and this further enhances the affinity between precursor and SDA. Depending on the SDA and the reaction conditions (especially precursor to surfactant ratio), different mesostructures can be obtained, analogous to OMS materials. For example, 3D cubic (Figure 6.10), 2D hexagonal (Figure 6.11), and cubic bicontinuous (Figure 6.12) structures can be obtained, denoted as FDU-16, FDU-15, and FDU-14, respectively [14]. An example is presented in Figure 6.9. Here, phenol and formaldehyde are employed in basic environment with different triblock copolymers.

Pore enlargement can also be achieved by adapting the SDA and multiple examples are readily available in literature [17a]. By using for example Vorasurf 504, pore sizes ~20 nm can be obtained [27]. This triblock copolymer contains a polybutylene oxide moiety instead of PPO and is represented by $(EO)_{38}$-$(BO)_{46}$-$(EO)_{38}$. The addition of 1,3,5-trimethyl benzene even enhanced the pore diameter to 27 nm [28]. The combination of PEO and polystyrene (PS) also resulted in materials with increased pore

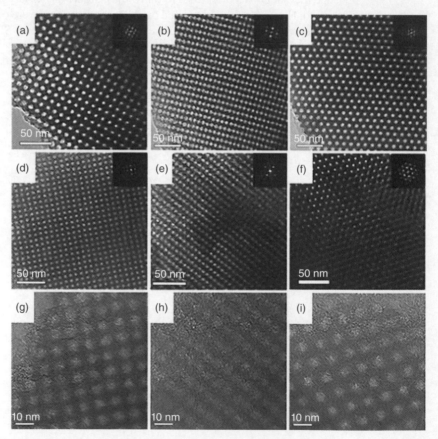

Figure 6.10

Transmission electron micrographs (a–f) and high resolution TEM images (g–i) of ordered mesoporous carbons, FDU-16, using F127 as SDA. Two calcination procedures were used: at 350 °C under O_2/N_2 flow (images a–c) and at 1200 °C under Ar (images d–i). Images a, d and g are viewed along the [100] direction; b, e, h along [110]; and c, f, i along [111]. Source: Reproduced with permission of ACS [14].

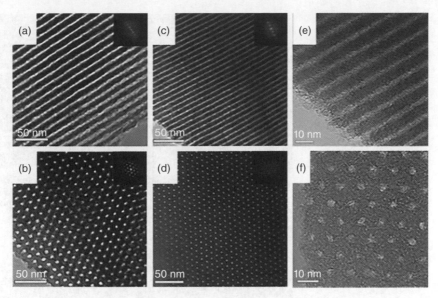

Figure 6.11
Transmission electron micrographs (a–d) and high resolution TEM images (e, f) of ordered mesoporous carbons, FDU-15, using F127 as SDA. Two calcination procedures were used: at 350 °C (images a, b) and at 1200 °C (images c–f) under Ar. Images a, c and e are viewed along the [110] direction; b, d, f along the [001] direction. Source: Reproduced with permission of ACS [14].

sizes from 23 nm to even ~33 nm [29]. In general, tuning of the pore size can be easiest performed by the proper selection of the SDA. P123, F127, and F108 will normally generate mesoporous carbons with pores between 3 and 10 nm. More specialized SDA, having a core that is more hydrophobic, will produce carbons with 10–30 nm pores. The use of pore swellers can further increase the pore size with a few nanometers.

6.2.3.3 *Precursor to Structure-Directing Agent Ratio*

The precursor to SDA ratio or P/S ratio is an important parameter that influences in the first place the mesophase structure. It should be noted that certain SDAs, under certain conditions, can only form one certain type of mesophase. For example, going back to the example with SDAs P123, F127, and F108 as introduced earlier, F108 will cause materials with $Im\bar{3}m$ mesophase (Figure 6.9) [14]. Whereas for P123 and F127, the mesophase is highly dependent on the P/S ratio. These SDAs can generate materials with completely different mesophases. Low P/S ratios resulted in 2D structures. By using higher P/S ratios, 3D symmetries are observed. This can be explained as follows: Altering the precursor (e.g. phenol) to SDA ratio, changes the hydrophilic/hydrophobic ratio of the resol-surfactant mesophase. This also changes the interfacial curvature that results in a different mesophase. The same is valid when changing the amount of PEO moieties, that is, by adding a different SDA, as this changes the hydrophilic/hydrophobic ratio as well. In the following series of SDAs, the EO:PO ratio increases, which influences the hydrophilicity/hydrophobicity of the SDA: P123 = $(EO)_{20}(PO)_{70}(EO)_{20}$; F127 = $(EO)_{105}(PO)_{70}(EO)_{105}$; F108 = $(EO)_{132}(PO)_{50}(EO)_{132}$. A proper choice of

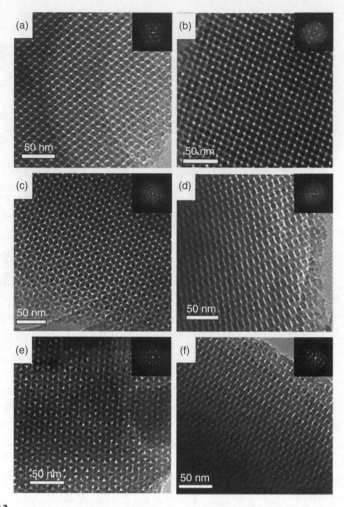

Figure 6.12

Transmission electron micrographs (a–d) and high resolution TEM images (e, f) of ordered mesoporous carbons, FDU-14, using P123 as SDA. Two calcination procedures were used: at 350 °C under N$_2$ (images a–d) and at 600 °C (images e, f) under Ar. Image (a) is viewed along the [531] direction; b [100]; c, e [111]; and d, f [311]. Source: Reproduced with permission of ACS [14].

the SDA and P/S ratio must be done in advance if one attempts to prepare materials with a certain structure (see also Chapter 5).

Phase diagrams reported in literature can help with the selection (Chapter 5). Secondly, the P/S ratio also affects the pore size. This was observed by several researchers [17a]. In many cases, the P/S ratio simultaneously influences the pore size and mesophase structure.

6.2.3.4   Synthesis Method

In general, three different synthesis methods are used to prepare OMPs and OMCs: the hydrothermal aqueous procedure, the EISA and the two-phase system-based method [16]. Figure 6.13 shows pictures of the OMPs synthesized via the various

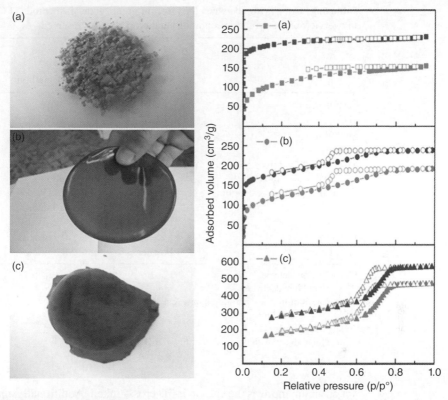

Figure 6.13

(left) Macroscopic view of the As-synthesized red-orange colored resins; (right) Nitrogen sorption isotherms of the calcined polymers (lower isotherm) and carbons (upper isotherm). Three different synthesis procedures were used: (a) hydrothermal aqueous procedure, (b) evaporation-induced self-assembly (EISA), and (c) two-phase system-based pathway. Source: Reproduced with permission of Elsevier [16].

methods that can result in a powder, a self-supporting membrane or a monolith. Clearly, the macroscopic appearance of the OMPs highly depends on the preparation method and moreover, it also influences the porosity. The isotherms in Figure 6.13 differ in shape and hysteresis loop. A summary of the synthesis conditions and eventual properties of the OMPs/OMCs is presented in Table 6.3.

The first route is the hydrothermal aqueous pathway that is a regular batch process at elevated temperatures, as reported by Van Der Voort [16, 30]. Phenol and formaldehyde are used as precursors in aqueous environment under alkaline conditions. Performing this method in neutral or weak acidic environment slows down the reaction. It is believed that this procedure works via a true liquid crystal templating mechanism and it leads to light orange colored powders. This method requires very long synthesis times of more than 100 hours, not taking the required calcination step into account, which is an immense drawback. Also, a very precise control of the pH is needed. Only when the pH ∼ 9, a stable network will be obtained. As can be seen from Table 6.3 and the isotherms in Figure 6.13, these materials are mainly microporous with a small amount of mesopores and possess a high specific surface area. It must be noted that the isotherm of the OMP in Figure 6.13 is not closed at relative low pressures (P/P_0) due to swelling of the polymer in liquid nitrogen during the nitrogen sorption measurement. After carbonization, this swelling phenomenon does

Table 6.3 Comparison between the three different synthesis methods of OMPs and OMCs: an overview of the synthesis conditions and properties is presented.

	Hydrothermal Aqueous method	EISA method	Two-phase system-based method
Carbon source	P/F[a]	P/F[a]	R/F[b]
SDA	F127	F127	F127
Medium	NaOH	NaOH	HCl
Solvent	H_2O	EtOH	H_2O/EtOH
Synthesis temperature	65–70 °C	(1) 70 °C; (2) RT	RT
Total synthesis time (h)[c]	100–125	45–50	35–40
Isotherm type/hysteresis loop	Type I/–	Type IV/H1	Type IV/H2
S_{BET} (m^2 g^{-1})	350–400[d]	400–500	600–650
dp (nm)[e]	< 2	2–7	5.5–8.5
V_p (ml g^{-1})	0.15–0.2	0.4–0.5	0.6–0.7
Morphology	Powder	Membrane	Monolith

[a]Phenol/Formaldehyde.
[b]Resorcinol/Formaldehyde.
[c]As-synthesis of the mesoporous polymer (calcination step excluded).
[d]$S_{langmuir} = 650$–$700 \, \text{m}^2 \, \text{g}^{-1}$.
[e]Total pore volume.
Source: Reproduced with permission of Elsevier [16].

not occur anymore as the carbon is 3D cross-linked. Additionally, the reproducibility of this type of synthesis can be an issue.

The second route is the EISA method and shows a lot of similarities with the previously discussed method [4, 16, 31]. It makes also use of phenol (P), formaldehyde (F), F127 and alkaline environment. The most significant differences are the reaction time (shorter), the solvent type (ethanol) and the molar ratios of P, F, and F127. The concentration of the SDA is at the beginning relatively low and below critical micelle concentration. A cooperative self-assembly of F127 and the carbon sources occurs when the ethanol evaporates during the EISA process and the critical micelle concentration is reached. Although the reaction time is significantly shorter in comparison with the hydrothermal aqueous method, the EISA method requires more individual synthesis steps. The resulting materials are mesoporous with a broad pore size distribution and cage-like pores with narrow pore entrances. This method can also be performed under acidic conditions and in fact works in a broad pH range.

The third and final synthesis method is called the two-phase system-based method or also macroscopic phase separation method that makes use of resorcinol instead of phenol [16, 32]. This precursor is more active, but is also more expensive. The polymerization process takes place in acidic environment in the presence of ethanol. In brief, a mixture is made of all the different components and allowed to stand for a certain period of time. A two-layered system will gradually appear with an ethanol-water rich phase at the top and a polymer rich yellow phase at the bottom. A monolith resin is obtained after separation of the phases and a thermopolymerization process on the polymer rich phase. This monolith can be pulverized to the required particle size. Relatively short reaction times are required and the number of different steps is very limited, which makes this method ideal for upscaling. A mesoporous material is obtained with high specific surface area and large and uniform cylindrical pores.

The carbonization does not influence the properties that indicates a very high thermal stability.

6.2.3.5 Structure-Directing Agent Removal

The SDA present in the pores must be removed to obtain the porosity. This can be performed via solvent extraction or calcination. The latter required an inert atmosphere as oxygen will enhance the oxidation of the SDA and will damage the polymer as well. Moreover, the heating rate needs to be well-controlled. Heating too fast will cause a quick release of gases during calcination that can also destroy the polymer network. Generally, a heating rate of $1–5\,°C\,min^{-1}$ is suitable and a calcination time for 4–6 hours at 350–400 °C suffices.

During this step, a shrinkage of the polymer can already occur. SDA removal via solvent extraction will avoid this as it is a milder procedure. It will result in materials with thicker pore walls and larger pores, however, the specific surface area and pore volume will be lower in comparison with materials that underwent a calcination procedure. Normally, H_2SO_4 can be used in the solvent extraction.

6.2.4 Transformation of Polymer into Carbon, the Carbonization Temperature

The ultimate step in the preparation of mesoporous carbons is the carbonization of the cross-linked polymer network. This is achieved by a relatively slow temperature treatment under inert atmosphere (N_2 or Ar) where the resin or polymer is transformed in a carbon. Oxygen must be avoided as oxidation of the carbon structure can take place instead of carbonization. The carbonization process takes place in several stages and in most cases the surfactant removal and carbonization itself is done subsequently. First the resin is heated slowly until 400 °C. A slow heating rate is preferred such as 2 °C per min. The SDA is removed and an ordered mesoporous resin or polymer is obtained. At higher temperatures, starting from 400 °C, the actual carbonization process takes place and shrinkage of the network slowly occurs until a temperature of 600 °C has been reached. In certain cases, the structure can even completely collapse if an unstable material is obtained [33]. During this step, the methylene bridged phenolic moieties are converted into coalesced carbon rings [34]. Then, above 600 °C, the heating rate can be increased significantly, normally 10 °C per min, and no shrinkage is observed anymore [14]. These materials are already called OMCs. In the temperature range of 600–800 °C, the porosity is remarkably increased as micropores are formed and the high specific surface area is generated. The carbon structure is completely formed if the temperature is increased to or beyond 800 °C. Although no shrinkage occurs, a distortion of the structure is observed in XRD measurements. Literature reports have shown that during the carbonization process different forms of pore are observed via transmission electron microscopy, that is, circular, asymmetric and slightly more asymmetric, at 400, 600, and 800–1000 °C, respectively [35]. The carbonization step mostly determines two important properties of the resulting materials: the hydrophobicity and the pore size. The former increases with increasing temperature during the final steps of the treatment. The latter is also temperature dependent as higher temperatures induces shrinkage of the network.

The resulting mesoporous carbons, treated at >800 °C, possess >90 wt% of carbon whereas the mesoporous polymers exhibit 70–85 wt% of carbon [36]. An OMC material can eventually be further graphitized at 2600 °C, while maintaining its meso-porosity; of course, only if it exhibits a high thermal stability to survive these very high temperatures [32a].

6.2.5 (Hydro)Thermal and Mechanical Stability

The thermal, hydrothermal, and mechanical stability of a material is a very important property as it will influence the applicability.

6.2.5.1 Thermal Stability

The thermal stability is determined via dynamic thermogravimetric measurements where the decomposition of the polymer or carbon is monitored as a function of the temperature. In the presence of oxygen, OMPs are generally stable up to 300–350 °C in air. Increasing the temperature above these values results in a destruction or decomposition of the polymer. Weight loss is observed in several temperature intervals and with the possible formation of the following degradation products: formaldehyde, phenol, toluene, benzene, cresols, xylenols, methane, CO, CO_2, and water [37]. The thermal stability of the OMPs increases when an inert atmosphere is used. The materials can withstand 350–400 °C. Temperatures above 400 °C will induce the carbonization process and the material is transformed from an OMP to OMC (see Section 6.2.4). The carbon material itself exhibits a very high thermal stability as it is stable up to 1400 °C in an inert atmosphere [13, 14].

6.2.5.2 Hydrothermal Stability

The remarkable high mechanical but also hydrothermal stability is due to the three-dimensional covalently bonded network, entirely made of carbons. This avoids the occurrence of hydrolysis, which is often the cause of structural collapse of silica materials. The latter consists of weak hydrolyzable siloxane bridges that are absent in OMPs and OMCs.

6.2.5.3 Mechanical Stability

The mechanical stability of the OMP was investigated by placing the material under a certain amount of lateral pressure. Figure 6.14 shows the minimal pressure that could be applied at which the structure will lose its porosity. The difference between the polymer and the silicas is significant. No collapse of the structure, or even a decrease in porosity was observed at 740 MPa. This was the highest pressure that could be experimentally applied [16].

Additionally, a comparison was also made between carbons prepared by the soft-templated (Section 6.2.2) and the hard-templated (Section 6.2.1) methods, which are represented by FDU-15 and CMK-3, respectively [14]. Zhao's group placed both materials under a pressure of 500 MPa for 10 minutes and evaluated the materials with X-ray diffraction. They showed that FDU-15 is mechanically more stable than CMK-3, probably due to the continuous framework of FDU.

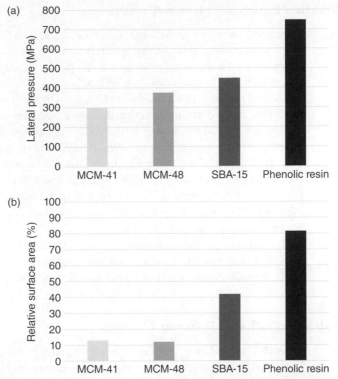

Figure 6.14
A comparison between MCM-48, MCM-41, SBA-15 and ordered mesoporous polymers. (a) Investigation of the mechanical stability: the lateral pressure is shown at which the structure collapses (>55% decrease in specific surface area); (b) Hydrothermal stability was investigated via the relative decrease in specific surface area with a treatment of steam for 24 h under autogenous pressure and 100% relative humidity. Source: Adapted from the Ph.D. thesis by Ilke Muylaert, Ugent, 2012: Development of Ordered Mesoporous Phenolic Resins as Support for Heterogeneous Catalysts, promoter: P. Van Der Voort (p. 1.9).

6.3   Surface Modification of Mesoporous Polymers and Carbons

OMPs chemically consist of a high amount of carbon-based moieties; that is, aromatic rings and methylene bridges but also some hydroxyl functions. The carbons are further pyrolyzed and mainly consist out of aromatic rings. The presence of these functions is beneficial for some applications where a hydrophobic surface is required or where the presence of aromatics is beneficial. This can be desired in, for example, adsorption processes of aromatic or hydrophobic compounds for environmental applications (see Section 6.5.1). It also means that on both materials the amount of surface functionalities is rather limited which can be used for further modifications. The introduction of more advanced functionalities (heteroatom or metal-based compounds) is sometimes required in more high-tech applications, for example, in the field of catalysis and electronics, but also in separation.

Two possibilities exist to introduce functions in the polymers and/or carbons. This can be done prior to the synthesis of the solids or via a post-modification procedure. Many examples can be found in literature [4, 16]. A few examples are highlighted next to give a general idea of the possibilities.

6.3.1 Pre-Modification of Polymers/Carbons

The introduction of heteroatoms (N, S, F, or metals) can be achieved before the start of the synthesis of the polymer/carbons. It is a sort of "pre-modification" of the polymer or carbon as the modification occurs at the precursor stage before polymerization or carbonization. The precursors can be functionalized with heteroatoms, or precursors that already include heteroatoms can be used. The most famous example is the use of melamine as N-containing or doped precursor (see Section 6.2.3.1) in the STM. Fluor can be introduced by mixing Fluor containing precursors such as *p*-fluorophenol together with phenol and formaldehyde [38]. These materials can withstand temperatures of 900 °C while keeping the C—F bonds. Also, the inclusion of metals is possible where for example nickel [39] is included in the mesoporous carbon. Here, the hard-templated method can be used, where the metal precursor $(Ni[NO_3]_2.6H_2O)$ is mixed with the carbon precursor furfuryl alcohol. Also, silicon, titanium and iridium have been included [32b, 38].

A disadvantage of this method, regardless if the hard- or soft-templated procedure is used, is that the material must undergo a carbonization treatment. This can remove some functional groups and any weak C—X bonds (based on, e.g. N, O, etc.) [4].

6.3.2 Post-Modification of Polymers/Carbons

A second possibility is the post-functionalization of the polymer or the carbon. Silica materials generally exhibit a rather high amount of silanol functions that can act as anchoring positions for further functionalization. This is the well-known grafting procedure (see Chapter 5). Unfortunately, the surface of the polymers and especially the carbons do not possess a high amount of surface functionalities. In fact, the surface of carbons is rather inert. This low reactivity is rather problematic when post-modification must be done. Possible post-synthetic treatments are: impregnation, adsorption, anchoring, and ion-exchange.

A handy trick to introduce nitrogen doping on an inert surface is treating the carbons with ammonia [40]. The carbons encounter an NH_3 flow at high temperature. The ammonia can replace the oxygen containing moieties by nitrogen-containing groups. Or the carbon is etched by radical formation, which is generated by the destruction of the ammonia. This process also takes place under elevated temperature and results in a nitrogen species that is embedded completely in a stable carbon network.

Oxygen-based functionalities such as carboxylic acids, esters, or quinones can be introduced by treating the carbons with acids, hydrogen peroxide, or ozone [8c, 29, 41]. All these treatments are based on the partial oxidation of the carbon surface. It must be noted that these oxidative methods should be used carefully as damage to the surface and pore structure easily occurs.

6.4 Nanocarbons

A different group of carbon-based materials are the nanocarbons. Just as activated carbon, they are graphitic like materials but they possess an additional important feature. At least one dimension of the material is <100 nm. The material contains sp^2 hybridized carbons and it was the beginning of an entire carbon-based materials group that now include fullerenes, graphene quantum dots, graphene nanoribbons, nanodiamonds, nano-onions, carbon nanotubes (CNTs), and so on [42]. These materials are promising in the application fields of nano-electronics, energy storage, and so on.

Graphene is normally considered as a monolayer of sp^2 bonded carbons in a two-dimensional lattice [43]. It is in fact the basic building block of all other structures such as fullerenes and CNT, which are the most famous and researched nanocarbons out of this list (Figure 6.15). These nanocarbons can be considered as large building blocks and can assemble on a macroscopic scale via intermolecular forces. Eventually, they form one-, two-, and three-dimensional structures such as fibers, films, and monoliths, respectively. The properties of these superstructures can extend the characteristics of the individual nanocarbon units.

6.4.1 Fullerenes

Fullerenes were one of the first nanocarbon allotropes discovered and were reported by Kroto, Curl, and Smalley in 1985 [44]. The first fullerene consisted of 60 carbons (C_{60}) that are organized into 12 pentagons linked with 20 hexagons (Figure 6.16). The authors called this structure buckminsterfullerene or the "buckyball" as it forms a hollow sphere in the shape of a European soccer ball with a diameter of 7.1 Å [46]. The structure was produced by the vaporization of carbon from a solid disk of graphite into a high-density helium flow. A focused pulsed laser was used for the vaporization process. Since 1990, it has been possible to produce buckyballs in larger quantities [47], which resulted in a significant increase of research performed on this topic. The

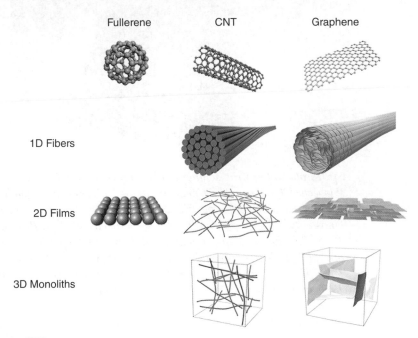

Figure 6.15
Schematic representation of nanocarbons: fullerene, carbon nanotube (CNT), and graphene as a 1D fiber, 2D film, and 3D monolith. Source: Reproduced with permission of ACS [42b].

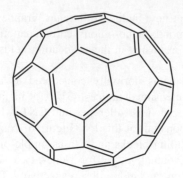

Figure 6.16

The structure of C_{60}, buckminsterfullerene. It exhibits a t-icosahedral symmetry. Source: Reproduced with permission of Springer Nature [45].

authors started from pure graphitic carbon soot, which was produced by evaporating graphite electrodes in a He atmosphere. The soot contained some weight percentages of C_{60} molecules that was subsequently extracted.

Sir Harold Walter Krotoschiner or also known as Sir Harry Kroto was born in 1939 in Wisbech (UK). He studied Chemistry at the University of Sheffield (UK) and obtained his Ph.D. in Molecular Spectroscopy in 1964. In 1967, he performed research at the University of Sussex (UK) after which he became a Full Professor in Chemistry in 1975. Together with his colleagues Robert Curl and Richard Smalley and graduate students James R. Heath, Yuan Liu, and Sean C. O'Brian, he demonstrated in 1985 that C_{60} molecules could be formed and named it Buckminsterfullerene. In 1996 he received the Nobel Prize for Chemistry together with Curl and Smalley for this outstanding work. He also received many other awards (Knight Bachelor and Michael Faraday Prize) and highly promoted science education. In 2004 Kroto joined the Florida State University (USA) as Francis Eppes Professor. Kroto died in 2016.

Photograph: www.sheffield.ac.uk/alumni/news/professor-sir-harry-kroto-1.572392

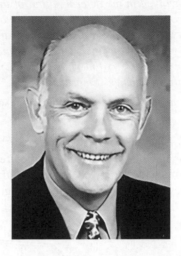

Richard Errett Smalley was born in 1943, Akron (USA) and obtained his Ph.D. at Princeton University in 1973. He then performed a post-doctoral stay at the University of Chicago where he was the pioneer concerning the development of supersonic beam laser spectroscopy. After joining Rice University in 1976, Smalley obtained the Gene and Norman Hackerman Chair in Chemistry. He became a Full Professor in Physics in 1990 and in the same year he founded the Center for Nanoscale Science and Technology of which he became director in 1996. Together with Kroto and Curl, he received the Nobel Prize for Chemistry in 1996. Next to this, he won several prizes and awards. He initiated the creation of Rice's Center for Nanoscale Science and Technology (CNST) that was renamed The Richard E. Smalley Institute for Nanoscale Science and Technology after his death in 2005. He was honored in a resolution as the "Father of Nanotechnology" by the US Senate.
Photograph: https://www.nobelprize.org/nobel_prizes/chemistry/laureates/1996/smalley-bio.html

Robert Floyd Curl was born in 1933 in Alice (US) and studied at Rice University. He obtained his Ph.D. in Chemistry from the University of California (US) where

he was involved in the determination of the bond angle of disiloxane with infrared spectroscopy. After his post-doctoral stay at Harvard University, he returned to Rice University. Curl received many awards and honors, including the Nobel Prize for Chemistry in 1996 together with Kroto and Smalley. He retired in 2008 and is now Professor of Chemistry Emeritus at Rice University and Pitzer-Schlumberger Professor of Natural Sciences Emeritus.

Photograph: https://www.bakerinstitute.org/experts/robert-curl

The C_{60} material has received the most research attention up until now as it is considered as one of the most stable structures. However, the fullerene material group consists of a wide variety of structures with various numbers of carbons and symmetries. A few examples are presented in Figure 6.17 [45].

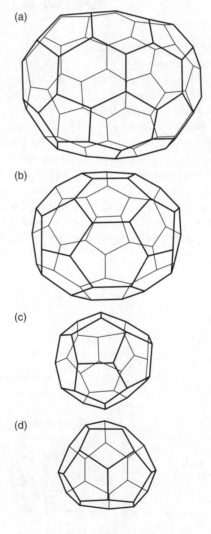

Figure 6.17
A series of fullerene structures: (a) C_{70} fullerene which resembles an American football; (b) C_{50} fullerene; (c) C_{32} fullerene; and (d) C_{28} fullerene. Source: Reproduced with permission of Springer Nature [45].

The functionalization of fullerenes has also been attempted. C_{60} showed an electrophilic and dienophilic character toward organic reagents as it is an electron-deficient molecule with a conjugated π-system and several organic reactions (Diels–Alder, 1,4-additions, nucleophilic additions, etc.) were successful. Also, it can coordinate to metals and act as a ligand [42a, 48]. An overview of some chemical reactions that can be performed on the C_{60} fullerene is presented in Figure 6.18.

Fullerenes have showed extraordinary physical and chemical properties, such as photosensitizing and electron acceptor characteristics, which make these structures promising for catalysis, medical and biological applications, and photovoltaics. More information and references on fullerenes and their applications can be found in the following reviews [42, 49].

As shown in Figure 6.15, the nanocarbons can connect with each other and form 2D and 3D structures. This assembly is a thermodynamic process, as agglomeration of the separate structures will lower the free energy of the system as the surface area

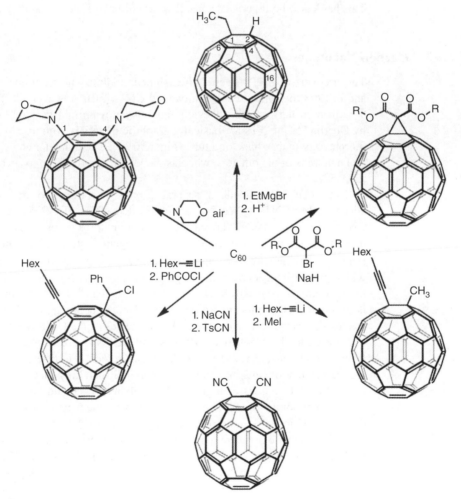

Figure 6.18
Addition reactions that can be performed on the C_{60} structure. Based on reference [48b]. Source: Reproduced with permission of ACS [42a].

will be enlarged. The interactions between the building blocks are in this regard very important. Many interactions may occur depending on the structure, for example, van der Waal's attractions, π-π, electrostatic and hydrophobic interactions, and hydrogen bonding. Although it appears simple to assemble the building blocks to superstructures, a mere stacking of the blocks will probably not result in the desired formation.

Connecting the nanocarbons must be well-controlled to avoid any random assembly and many factors play a vital role, such as the mechanical strength of the structures. Two interactions are mostly dominant, that is, the π-π interaction between the sp^2 carbons of the nanocarbons and hydrogen bonding, which is a very strong interaction, when, for example, hydroxyl groups are present on the nanocarbon. More details about nanocarbon self-assembly can be found in reference [42b].

Fullerenes typically exhibit a spherical shape with a curved surface that can only result in a point contact. This contact mode is incomplete and only leads to rather weak interactions. Unfunctionalized fullerenes do not possess any dipolar interactions and therefore will not form any superstructures in solid state. Most assemblies take place in solution-based processes where the choice of solvent is very important. This has led to liquid crystals, needles, and films of fullerene aggregations [42b].

6.4.2 Carbon Nanotubes

The discovery of CNTs is generally assigned to Iijima who reported in *Nature* in 1991 and in 1993 the synthesis of multiwalled CNTs [50]. However, CNTs were already described earlier by Radushkevich and Lukyanovich in 1952 and later on in 1976 by Oberlin [51]. CNTs are basically graphitic materials with a seamless cylindrical morphology in the shape of a tube (Figure 6.15). The diameter can vary around several hundreds of nanometers, whereas the length can reach up to several centimeters [50b, 52].

CNTs can be divided in several groups according to their wall number: single-walled CNTs (SWNTs), double-walled CNTs (DWNTs), and multiwalled CNTs (MWNTs). SWNTs are basically one-atom thick graphite sheets that are rolled up into tubes. The graphene possesses a honeycomb structure as can be observed in Figure 6.19. The way the graphene is wrapped is indicated by the chiral vector $\vec{C}$ with a pair of integers (n, m). The graphene sheet can form different constructions, that is, chiral $(n > m > 0)$, zigzag $(n, 0)$ and armchair (n, n). The chiral construction is considered as a non-standard type of CNT construction and will lead to additional electrical, mechanical, and optical characteristics of the CNT. SWNTs only contain one graphene sheet, whereas DWNTs and MWNTs are nanotubes that are made from several concentric NTs. The diameter of each NT differs from each other in such a way that one fits into another (Figure 6.20) and is generally in the nanometer range. MWNTs possess a higher stability than SWNT and have the possibility of independent doping of the inner and outer tubes. The wall number will depend upon the synthesis method and will eventually result in CNTs with different properties. Several methods exist to prepare CNTs and an overview of the different synthesis methods with some key features are presented in Table 6.4 [51]. Well-known methods are arc discharge, laser ablation, and chemical vapor deposition (CVD).

A first method uses an arc discharge between two carbon electrodes under inert atmosphere [53]. The type of inert gas and the pressure will influence the yield, purity, and quality of the end-product. If the electrodes are made from graphite mixed with nanoparticles (Ni, Fe, Co, Pt, Rh, or their alloys), SWNTs are formed. On the other

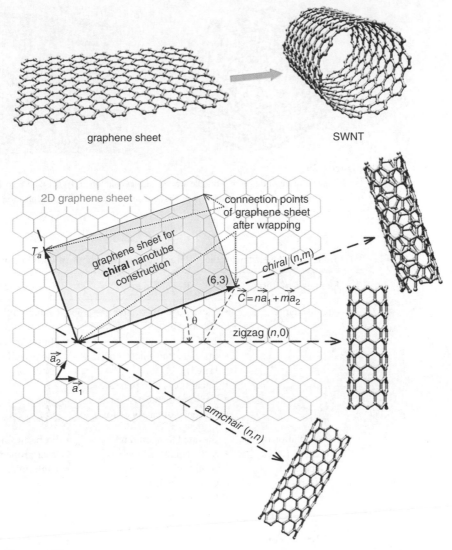

Figure 6.19

Schematic representation of the formation of a single-walled carbon nanotube (SWNT). Top: A graphene sheet is rolled up into a SWNT. Bottom: Diverse ways of folding the graphene sheet along the chiral vector $\vec{C}$ results into chiral, zigzag, or armchair construction. Source: Reproduced with permission of RSC [51].

hand, if no metal nanoparticles are added, this will result in MWNTs. DWNTs are formed under very specific conditions, for example, using H_2 or Ar gas with Ni, Co, Fe, and S as catalysts or by pulsed arc discharge with a Y/Ni alloy.

The use of the laser ablation step in a pulsed laser deposition process can also lead to CNT. The resulting properties are highly influenced by the laser characteristics and the chemical and structural properties of the target material, the flow and pressure of the buffer gas, and so on. The CNT is formed when the laser hits a graphite pellet in the presence of a metal catalyst (Ni, Co). It is an excellent method to produce high-quality SWNTs with a relatively pure composition [51].

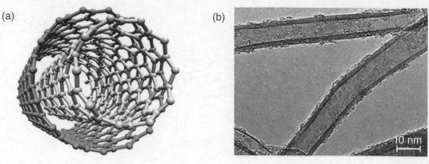

Figure 6.20

(a) Schematic representation of a double-walled carbon nanotube (DWNT); and (b) TEM image of multiwalled carbon nanotubes deposited via plasma enhanced chemical vapor deposition (PECVD). Source: Reproduced with permission of the RSC [51].

Table 6.4   Overview of the different synthesis methods of carbon nanotubes.

Synthesis Method	Requirements	Key features
Arc discharge	• Elevated temperatures above 1700 °C • Special atmosphere (inert, H_2 CH_4, …) • Subatmospheric pressure • Two carbon electrodes • Presence of catalysts for SWNTs	• Small number of structural defects • High purity and yield under well-controlled conditions
Laser ablation (Pulsed Laser Deposition Process)	• Elevated temperatures • Strict control of the conditions • Graphite pallet • Lasers mostly used: Nd:YAG and CO_2 • Catalysts (e.g. Ni, Co)	• Properties influenced by laser properties • High-quality and purity
Catalytic chemical Vapor deposition (CCVD)	• Mostly low-temperature < 800 °C • Can be: thermal, plasma enhanced, water assisted, oxygen assisted, hot-filament, microwave plasma, radiofrequency	• Well-controlled • High purity • Relatively large-scale • Economically viable

Source: Reproduced with permission of the RSC [51].

Another method uses CVD process with a metal catalyst. A vapor that consists of methane, ethane, acetylene, ethylene, H_2/CH_4, or ethanol is led over a metal catalyst. The latter usually consist of metal nanoparticles (Fe, Co, Ni, or an alloy). The vapor undergoes subsequently a thermal or plasma irradiation treatment that results in the formation of CNTs. An example of the CNTs prepared via CVD is shown in Figure 6.20 [51]. Currently, thermal, or plasma enhanced CVD processes are the standard preparation methods for the synthesis of CNTs.

The described methods mostly provide CNTs is a relatively low yield. A large quantity of other particles is generally also formed such as nanocrystalline graphite, amorphous carbon and fullerenes. Additionally, other impurities are the remaining metal catalysts which are required in some cases during the synthesis process. Although many synthesis methods can provide relatively pure CNTs, depending on the exact synthetic conditions used, a small amount of impurities is always present. This is still one of the major challenges in the production of CNTs as these impurities will limit its use in applications. Extensive research is performed on purification processes and involve chemical (oxidations) or physical (filtration, centrifugation, etc.) treatments, or the combination of both, leading sometimes to quite complex processes [54].

6.5 Application of Porous Carbon-Based Materials

Porous carbon has been a candidate for almost every possible application you can think of, as it has been available since the time of Ancient Egypt and easy to obtain. The applicability of other "younger" materials such as OMPs and OMCs, is less well researched although this has increased during recent years. The most important implementations of porous carbon-based materials probably are: environmental applications such as the removal of pollutants, in the field of heterogeneous catalysis, and energy storage such as batteries.

6.5.1 The Adsorption of Pollutants

The basic principles of adsorption are elaborately explained in Chapter 2 and it is clear from that chapter that adsorption still is a valuable technique to purify waste water or gas streams. Porous activated carbon is by far the material of choice for general purification processes as it is a cheap, all-round, effective adsorbent that exhibits a large internal surface area and pore volume with both micro- and mesopores. This adsorbent is even now used in massive quantities for all kinds of adsorption processes going from inorganic pollutants, such as metals, to small and bulky organic contaminants.

Activated carbon has many advantages, but also some clear drawbacks. The adsorbent is not target specific and can adsorb many molecules or atoms at the same time. This is highly undesired in certain high-end purification application when only one very specific pollutant must be removed. Here, the other more advanced porous carbon-based materials come into place. Alternately, additional functional groups can induce affinity for the target molecule.

Regardless of this, certain aspects or characteristics are important to consider when designing an adsorbent [1a, 55]. These are for example: the production procedure of the adsorbent, the pore size, the interactions between adsorbent and adsorbate (hydrophobicity, aromaticity, etc.).

6.5.1.1 The Production Procedure

The production or preparation procedure of the adsorbent normally dictates which functional groups will be present on the surface of the material and this will eventually

Table 6.5 Overview of the characteristics of activated carbon (AC) and AC treated with HCl and HNO$_3$.

	AC	AC-HCl	AC-HNO$_3$
S_{BET} (m^2 g^{-1})	972	1015	987
V_{tot} (ml g^{-1})	0.53	0.55	0.53
R (nm)[a]	1.09	1.08	1.09
PZC[b]	8.60	6.70	2.98

[a]Pore radius determined from the adsorption isotherm.
[b]Point of zero charge: the pH value that is required to give a zero net-charge of the surface.
Source: Reproduced with permission of Elsevier [56].

influence the adsorption capacity. For AC, two types of activation methods can be used, that is, the PA and CA (see Section 6.1). A chemical treatment with certain acids will induce more oxygen containing functionalities on the surface of the solid; for example, carboxyl functions, phenolic hydroxyl groups, carbonyl moieties, and so on.

For example, an activated carbon (AC) was treatment with two different acidic solutions, 2M HCl and 2M HNO$_3$, which resulted in AC-HCl and AC-HNO$_3$, respectively [56]. Some basic properties are presented in Table 6.5. The authors found that the acids influenced the functional groups on the surface of the AC. The AC-HNO$_3$ material possessed more active acidic groups such as carboxyl and lactone moieties. Hydrochloric acid on the other hand decreased the amount of active acidic groups significantly. This can also be easily deducted from the Point of Zero Charge (PZC) values (Table 6.5). AC has a basic surface, whereas AC-HCl is more neutral and AC-HNO$_3$ is obviously more acidic. The influence on the adsorption of three dyes was evaluated with methylene blue, crystal violet, and rhodamine B (Scheme 6.7).

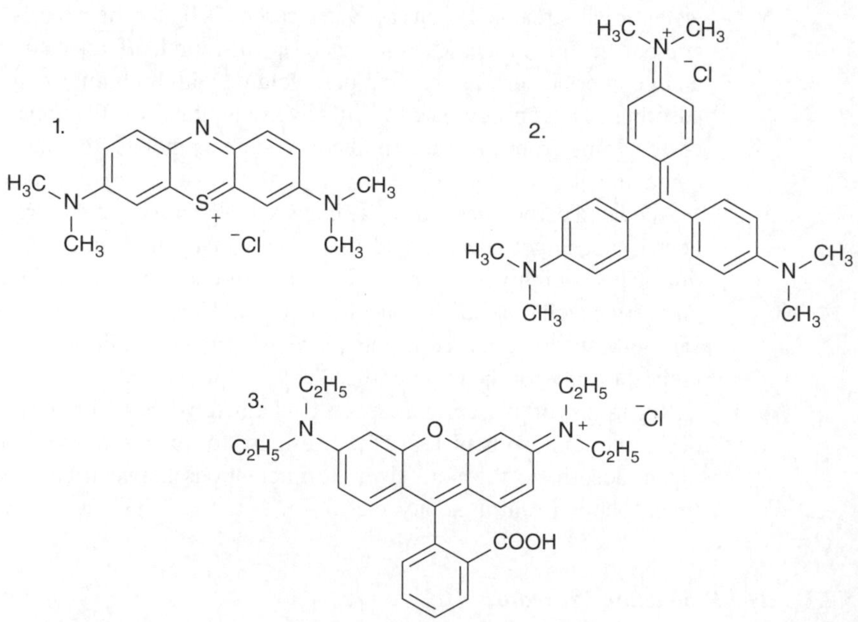

Scheme 6.7
The chemical structures of the cationic dyes: 1. Methylene blue; 2. Crystal violet; and 3. Rhodamine B.

For the large dyes, crystal violet and rhodamine B, the following sequence in adsorption capacity was found: AC-HCl > AC > AC-HNO$_3$. The basic AC-HCl showed the highest adsorption capacity of the cationic dyes. For methylene blue, a slightly different sequence is found: AC > AC-HCl > AC-HNO$_3$. This dye has a smaller molecular size (Scheme 6.7) and can penetrate the adsorbents much easier into the inner smaller pores. This is an additional aspect which can also influence the adsorption capacity.

6.5.1.2 The Pore Sizes

The example in the previous paragraph already shows a difference between the adsorption capacity of molecules with a different molecular size. The adsorption of large organic molecules, with even bigger sizes than the previously mentioned dyes, is still a major challenge. Examples are enzymes (e.g. lysozyme) and the large organic pollutant humic acid (HA). The pores of activated carbon are just not large enough to adsorb and remove these bulky molecules. This issue was especially one of the major driving forces behind the development of the mesoporous polymers and carbons with larger pore sizes (see Section 6.2).

Humic acids are especially problematic when producing drinking water [57]. These molecules can form carcinogenic byproducts when they encounter Cl$_2$ that is often used as a disinfectant. The removal with activated carbon is not ideal due to its limited pore size. In fact, humic acids are generally a mixture of macromolecules, are rather bulky with molecular weights between 350 and 50 000 g mol^{-1}.

The difference between the adsorption capacity of activated carbon and mesoporous carbons for humic acid is clearly shown in Figure 6.21 and Table 6.6 [57a]. The AC is commercially available. The mesoporous carbons were synthesized via the acid-catalyzed EISA method using resorcinol, formaldehyde, and F127 as a soft template. This has led to mesoporous carbons with large pores to almost 50 nm (not

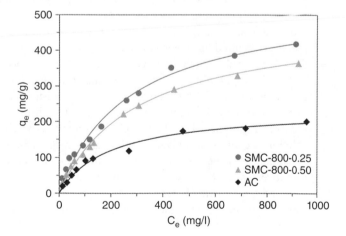

Figure 6.21

Adsorption isotherms of two mesoporous carbons prepared via the soft-templated method (SMC) and activated carbon. Each time, 10 mg of adsorbent was added to 50 mL of HA solution with various concentrations (C$_e$) (25 °C, 7 days, 200 rpm). A Langmuir fit was used. Source: Reproduced with permission of ACS [57a].

Table 6.6 Material characteristics of the adsorbents, together with the adsorption capacity.

	S_{BET} $(m^2\ g^{-1})$	V_{tot} $(ml\ g^{-1})$	D_p $(nm)^a$	q_e $(mg\ g^{-1})^b$
AC	1027	0.50	2.3	189 ± 17
SMC-800-0.50^c	663	1.26	23.7	294 ± 17
SMC-800-0.25^d	534	1.37	43.8	352 ± 20

^{a}Pore diameter determined via the BJH method using the desorption branch.
^{b}Experimental adsorption capacity for 10 mg of adsorbent, 50 ml of 500 mg l^{-1} HA, RT, 200 rpm, pH = 7.
^{c}Mesoporous carbon obtained after a carbonization temperature of 800 °C and a precursor to surfactant ratio of 0.50.
^{d}Mesoporous carbon obtained after a carbonization temperature of 800 °C and a precursor to surfactant ratio of 0.25.
Source: Reproduced with permission of ACS [57a].

shown in Table 6.6). The authors showed that an increase in pore size resulted in an increase in adsorption capacity. This is also called the size exclusion effect where large molecules cannot enter small pores. The research showed that the high molecular weight fractions of HA were not adsorbed by AC, explaining the lower adsorption capacity. The material SMC-800-0.25 with the largest average pore size and a broad pore size distribution could adsorb HA molecules of the entire molecular weight distribution of HA. The authors also mention that the adsorption in influenced by a hydrophobic π-stacking mechanism. Indeed, the interactions between adsorbent and adsorbate are of major importance for the adsorption process.

6.5.1.3 *Interactions Between Adsorbent and Adsorbate*

During the adsorption process, several interactions can play a vital role; for example, electrostatic interactions, hydrogen bonding, Van der Waals interactions, π-π stacking, and so on [55]. A high affinity between the adsorbent and the adsorbate will improve the adsorption capacity significantly. Most carbon-based materials are hydrophobic and therefore ideal for adsorbing organic molecules.

As an example, the adsorption of tannic acid is examined. Tannic acid has some remarkable antioxidant properties for all kinds of diseases including cancer [58]. The authors compared the adsorption capacity of a carbon nanocage (CNC) with CMK-3 for a dye (Alizarin Yellow) and a mixture of tea compounds including tannic acid.

The CNC material is a nanocarbon material that consists of regularly structured cage-type mesopores [59]. It is prepared via a hard-templated method using the mesoporous silica KIT-5 as template. The authors first observed a high affinity of CNC for the dye. The material could completely remove Alizarin Yellow from an aqueous solution in contrast to activated carbon and CMK-3 (Figure 6.22A). The result of the combination of the size and aromatic system of the adsorbate. Second, the authors also investigated the adsorption of bioactive tea compounds such as caffeine, catechin and tannic acid (Figure 6.22B). Especially for tannic acid, the CNC material showed a higher affinity and thus a larger adsorption capacity. It was concluded that the different adsorption behavior originated from the increased hydrophobicity and the stronger π-π interactions between the adsorbate and the surface of the CNC.

It is clear from the previous paragraphs that the adsorption process is influenced by many aspects. It is an interplay between the properties of the adsorbent, the characteristics of the adsorbate, but also the conditions that are used during adsorption. To some degree, the adsorption can be predicted by considering the known interactions

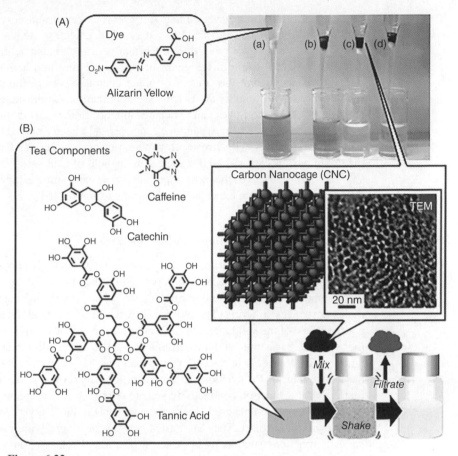

Figure 6.22

(A) The filtration of the dye Alizarin Yellow by different carbon-based materials, that is, (a) no material; (b) activated carbon; (c) carbon nanocage; and (d) CMK-3. The carbon nanocage can completely remove the dye. (B) The one-pot separation of tea components (caffeine, catechin and tannic acid) on the carbon nanocage material. Source: Reproduced with permission of ACS [58].

and affinity in certain cases. However, a full prediction of the adsorption behavior is difficult as it always is a combination of several aspects that play a role at the same time.

Next to the adsorption of dyes and humic acids, also other pollutants can be adsorbed that will not be discussed in detail. This includes heavy metal ions (Hg^{2+}, Cd^{2+}, Cu^{2+}, Fe^{3+}, etc.), anions (chromates, phosphates, selenates, etc.), organic contaminants (dyes, toxins, etc.), gases (hydrogen, carbon dioxide), small biomolecules (amino acids, vitamins, etc.), and larger biomolecules (peptides, proteins, enzymes, etc.). More information on the adsorption of these adsorbates can be found in the following reviews on carbon-based materials and adsorption processes [1a, 55, 60].

6.5.2 As Catalytic Support or Direct Heterogeneous Catalyst

Catalysis is a very important application for many types of material. As many chemicals are produced via catalytic processes, the search for catalysts to increase the rate and selectivity is a never-ending story. Porous carbon-based materials play a vital role in this quest as they can be used as a direct, heterogeneous catalyst. This is also known as carbocatalysis. More importantly, they can also act as a support and contribute actively to the heterogenization of interesting and active homogeneous catalysts. Especially in the context of green chemistry where the search for environmentally friendly materials is still on-going, carbon-based materials can be an alternative, cheap, non-toxic, and valuable support. Some carbon materials, such as activated carbon, have been around for centuries and already played their role in heterogeneous catalysis. Many research papers, reviews, and books have appeared on the use of porous carbon-based materials in the field of catalysis. Here, we can only show a brief overview of the most important aspects and some examples. We refer to the literature for a more extended overview and examples [61].

6.5.2.1   *Carbon-Based Materials Supporting Catalytic Species*

Carbon-based porous materials are very often used as catalytic support. This means that catalytically active species are deposited on the surface of the solid. This will disperse the active species and normally keep it that way during catalysis and eliminates or reduces sintering. The physical separation of the catalytic species over the entire surface of the solid increases the active surface of the catalyst significantly and therefore increases the catalytic activity. Moreover, in gas flow reactions, the reactants can diffuse better to the catalytic species and have a better contact. Contingent heat formed during the reaction can also be more easily removed. Also, porous carbon materials show electron conductivity and are relatively easily functionalized.

The porous carbon materials can support metals, organocatalysts and organometallic catalysts [61a, b, e]. They are basically used for a very wide range of different catalytic reactions such as hydrogenation reactions [62], environmentally oriented catalysis (e.g. Bromate reduction) [63], hydroprocessing [64], and many more. The latter is probably one of the most important catalytic application for carbon-based materials. The applicability of carbon-based materials recently also extends to photocatalysis, electrocatalysis, and fuel cells [61e]. Furthermore, the reduction of NO, N_2O, and in extension CO_2, is an emerging and important application for metal(oxide) supported on carbons [65]. We have selected a few examples to show the versatility of porous carbon-based materials as a catalytic support, using tree type of catalysts (metal, organo-, and organometallic) on three different supports for three important but different catalytic reactions.

A first example shows the simplest system and describes a very common metal supported catalyst for hydrogenation reactions. Hydrogenation is the reaction between molecular hydrogen gas (H_2) and a substrate, very often an alkene, alkyne, aldehyde, ketone, and so on with the formation of alkanes or (primary or secondary) alcohols, and so on depending on the substrate. This important chemical reaction takes place in the presence of a homogeneous or heterogeneous catalyst. The latter frequently consists of Pd, Ni, Pt, and suchlike, supported on a solid. Isomerization of alkenes can also sometimes occur under the same reaction conditions.

In this example, the hydrogenation of acetylene to ethylene with a palladium catalyst supported on porous carbon has been discussed [62b]. Pd acetate was first prepared in situ by dissolving anhydrous $Pd(OAc)_2$ in methanol. Afterward, the solid carbon support ($S_{BET} = 274\,m^2\,g^{-1}$) was added to this solution. This method is also known as the alcohol reduction method. A 1 wt% Pd/C catalyst was obtained.

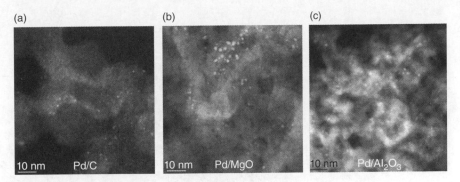

Figure 6.23

STEM images of the as-prepared 1 wt% Pd containing catalysts on (a) carbon; (b) magnesia; and (c) alumina. Source: Reproduced with permission of Elsevier [62b].

For comparison, other supports were also used such as MgO ($S_{BET} = 32 \, m^2 \, g^{-1}$) and Al_2O_3 ($S_{BET} = 160 \, m^2 \, g^{-1}$). A visualization of the catalysts can be found in Figure 6.23, where STEM images are shown. The Pd particles have particle sizes in the region of 0.3–2.8 nm for the different solid catalysts.

Mean particle diameters were calculated from the analysis of different STEM images and mean values of 0.7, 1.0, and 2.8 nm were obtained for Pd/C, Pd/MgO, and Pd/Al_2O_3, respectively. The authors obtained a very narrow particle size distribution, especially for Pd/C. The Pd/C catalyst displayed a higher selectivity toward the formation of ethylene in the catalytic process in comparison with the other catalysts. The authors showed with X-ray absorption near edge structure or XANES experiments that the electronic properties of Pd are modified by the carbon. The interaction of carbon with Pd, together with the low Pd coordination of the small sized particles, greatly enhances the selectivity.

A second example describes of a more complex system as it uses a CNT based catalyst for the selective oxidation of 5-hydroxymethylfurfural (HMF) to 2,5-diformylfuran (DFF) [66]. This reaction is important in the context of the use of renewable resources as alternative feedstock for fuels and chemical reagents. DFF is useful as a monomer and can be applied in the production of pharmaceuticals and fungicides. The starting product HMF is a biomass based compound and selective oxidation to DFF without byproducts is still challenging. These byproducts can be: 5-hydroxymethyl-2-furancarboxylic acid (HMFCA), and 5-formyl-2-furancarboxylic acid (FFCA). The selective oxidation needs an (preferably green) oxygen source (e.g. molecular oxygen) and a catalyst. For the latter, many options have been attempted including salts, organic compounds, and transition metals (V, Ru, Mn). It was an organometallic ruthenium complex that was anchored in this example on a modified CNT.

The CNT was first very easily functionalized with poly(4-vinylpyridine) (PVP) by adding the monomer 4-vinylpyridine to the CNT in the presence of the radical initiator azobisisobuytronitrile (AIBN). PVP is a nitrogen-containing polymer with known affinity for transition metals and provides an excellent support for a $RuCl_3$ catalyst, combined with the strong CNT as mechanical support (Figure 6.24). An amount of 2.2 wt% of ruthenium was attached to the PVP/CNT.

A full HMF conversion toward DFF was observed with a DFF yield of 94% in DMF at 120 °C, after 12 h and with 2.0 MPa of O_2. The idea of developing a catalytic system

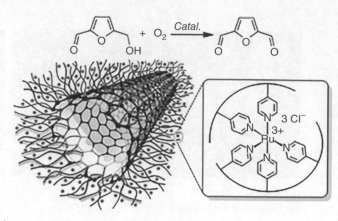

Figure 6.24
The carbon nanotube modified with poly(4-vinylpyridine) is used as support material for the catalyst
$RuCl_3$. The heterogeneous catalyst is used in the conversion of 5-hydroxymethylfurfural (HMF) to
2,5-diformylfuran (DFF) with molecular oxygen as oxidant. Source: Reproduced with permission of the
RSC [66].

with the combination of the affinity of pyridine for Ruthenium (under the form of
PVP) and the structural strength of the CNT can lead to a successful catalytic system.
However, it is always of foremost importance to study the recyclability of the catalytic
system. In this example, reuse of the catalyst showed a decrease of the conversion and
yield in the second run. The authors attributed the decrease to various effects: loss of
catalyst during filtration, slight decrease in Ru content, adsorption of polymerized
compounds on the catalyst. This shows that every aspect must be considered during
the evaluation of a catalytic system, including the leftover of products on the catalytic
system.

The latter is an issue often observed after catalysis. It can lead to deactivation or
poisoning of the catalytic species, and also to the formation of coke [61f]. Any left-
overs in the material can also block micropores and physically limit the reachability
of the catalytic sites.

In previous example, a rather simple Ru complex ($RuCl_3$) was anchored on a rather
complex support (PVP modified CNT). The next case, by Van Der Voort, shows the
anchoring of a bulky manganese complex on a mesoporous polymer (Figure 6.25)
[67].

The envisaged complex in this research was a manganese (III) salen complex, also
known as the homogeneous Jacobsen catalyst, N,N′-bis(3,5-di-tert-butylsalicylidene)
-1,2-cyclohexanediaminomanganese (III) chloride. This commercial catalyst has
a proven track-record for asymmetric alkene epoxidation reactions. Its special
conformation allows stereochemical control during the epoxidation process. As a
homogeneous catalyst, the complex is easily deactivated due to the formation of
dimeric μ-oxo-manganese (IV) species. The heterogenization of this complex would
prevent this.

A mesoporous phenolic polymer was used as a support. It was synthesized via the
soft-templated procedure with resorcinol and formaldehyde in the presence of the
template F127 in acidic conditions. The surface of the OMP was activated with 0.03 M
of sodium hydroxide (step [i] in Figure 6.25) and subsequently exchanged by the manganese center of the Jacobsen catalyst (step [ii]). This resulted in 0.09 mmol g^{-1} of
the Manganese salen complex on the OMP. The resulting material exhibited a BET

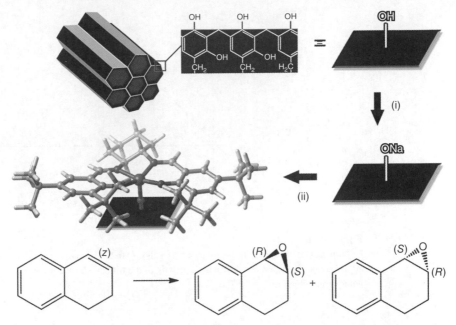

Figure 6.25
Schematic representation of the anchoring of the Jacobsen catalyst on the surface phenol groups of a mesoporous phenolic resin. The anchoring itself is achieved by (i) exchange of hydrogen by sodium; and (ii) exchange of sodium by the manganese center of the Jacobsen catalyst. The catalyst is used in the epoxidation of 1,2-dihydronaphthalene (1,2-dialin) toward 1,2-(R,S)-dihydronaphthalene oxide and 1,2-(S,R)-dihydrophthalene oxide. Source: Reproduced with permission of Elsevier [67].

surface area of $397 \, m^2 \, g^{-1}$, a pore volume of $0.52 \, ml \, g^{-1}$ and a BJH pore diameter of $8.2 \, nm$. The heterogeneous catalyst was tested in the asymmetric epoxidation reaction as depicted in Figure 6.25. Conversions of 62% were detected in the first run with *ee* values of 80% toward 1,2-(S,R)-dihydronaphthalene oxide. The enantioselectivity was maintained during the second and third run and a hot filtration experiment showed that no leaching of the manganese salen complex took place during catalysis (Figure 6.26).

6.5.2.2 *Carbocatalysis: Carbon-Based Materials Acting as Catalysts*

Carbon catalysis or also called carbocatalysis is metal-free catalysis and a valuable and sometimes successful alternative for metal oxide-based catalysts that are used in certain transformations such as dehydrogenation and selective oxidation [61c]. It is the carbon material itself that acts as a catalyst. It was already reported in 1925 by Rideal and Wright [68]. They described the aerobic oxidation reaction of oxalic acid with the aid of the surface of charcoal.

Carbocatalysis in the gas phase is already extensively researched. Oxidative dehydrogenation (ODH) reactions are normally performed with metal oxide containing catalysts. In the conversion of *n*-butane to butadiene, it is especially difficult to avoid further oxidation of the butadiene toward CO_2. Conventional materials, such as activated carbon, often undergo deactivation by coke formation or combustion. It was shown that CNTs with a modified surface are stable enough and can catalyze this process metal free, without the formation of coke [69].

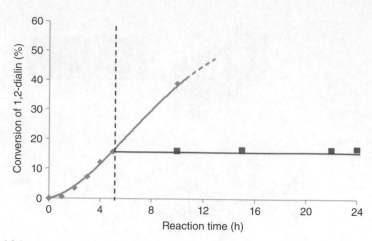

Figure 6.26

Hot filtration experiment. The heterogeneous catalyst (Mn Salen complex anchored on OMP) is removed after 5 hours of reaction and the catalytic activity of the filtrate is further examined. Source: Reproduced with permission of Elsevier [67].

The surface of the CNT is either modified with oxygen rich function groups (e.g. peroxo based groups) by refluxing the CNT in concentrated HNO_3 (oCNT) or it is functionalized with additional phosphorus containing groups (P-oCNT) [69]. The surface of the latter materials also contains peroxo, diol, and dicarbonyl groups. The introduction of PO_x significantly enhanced the selectivity whereas the oCNT material mostly has led to an overoxidation and the formation of carbon monoxide, carbon dioxide, and water. P-oCNT with 0.5 ± 0.1 wt% of P showed an alkene yield of 13.8%. An impressive result for this type of reaction and comparable with results of V/MgO catalysts.

The applicability in liquid-phase reactions of carbon-based materials as a catalyst itself was less elaborately researched in the past than its use as a catalytic support [61c]. This was mainly due to low activity, low stability, and low oxidation resistance issues with the materials. However, with the current development of the nanomaterials, including nanocarbons such as fullerenes and CNT, this has drastically changed and the amount of research performed on this topic has grown during the latest years [61c, g, 70]. Nevertheless, currently, a lot of research still needs to be done to prove its applicability.

The hydroxylation of benzene to phenol is still a challenging process that now proceeds via cumene and cumene hydroperoxide. Avoiding this multistep procedure and performing a direct hydroxylation was attempted with a defective multiwalled CNT (Figure 6.27) [61g, 71]. The observed benzene conversion was still low (~6%), but a remarkably high selectivity toward phenol was noted (~99%). The authors showed that the number of defects in the MWCNTs determines the catalytic performance of the material. A reaction mechanism was proposed and although the authors could not provide any direct evidence, it showed some similarities with a previously reported mechanism. A monoactive oxygen atom is formed by H_2O_2 decomposition on the defect of the MWCNT with the release of water. This active oxygen will subsequently attack benzene and form the end-product phenol.

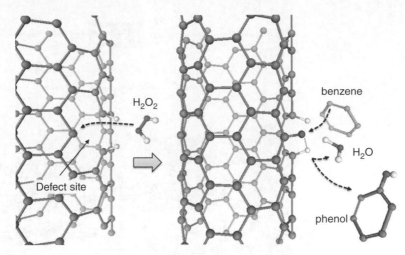

Figure 6.27
The hydroxylation of benzene on a defective site of a carbon nanotube. Source: Reproduced with permission of Elsevier [71].

6.5.3   Electrochemical Applications: Energy Storage

Also in the broad field of electrochemistry, these carbon-based nanomaterials are heavily researched and it includes many specialized topics such as energy conversion (e.g. solar cells, fuel cells), energy storage (e.g. batteries, supercapacitors), electrochromism (a reversible process that occurs when an applied voltage can alter the optical characteristics of a material such as transmittance, absorbance and reflectance), and electrochemical sensing and biosensing [8b, 33a, 72]. The materials of interest for electrochemical applications are currently mainly fullerenes, CNTs, graphene and OMCs. Especially, the latter have attracted a lot of attention lately for energy conversion/storage and electroanalysis due to their interconnected pore system and high specific surface area that is conductive [72a, b]. Here, we will focus on the topic of energy storage and the use of carbon-based materials as electrodes in batteries.

Basically, a rechargeable battery consists of an anode, a cathode, a separator and an electrolyte and a schematic representation of a typical Li-ion battery is shown in Figure 6.28. Electrochemical reactions will take place at the electrodes when the battery is discharging. Consequently, electrons will flow via an external circuit. When the battery is charging, an external electromotive force will be applied across the electrodes. This type of battery is the most widespread rechargeable battery. Their performance is mainly dictated by the used electrode materials. More information can be found in the following references [72b], [74].

Although it is mostly the cathode that currently limits the performance of the battery [75], many attempts are made to change the graphite in the anode by materials with a higher capacity and in the meantime still have a good electronic conductivity [8b].

In adsorption (Section 6.5.1) and catalytic (Section 6.5.2) applications, the exact morphology of the material is less important and many adsorbents or catalysts are used in powder form. In contrast, electrochemical applications mostly require a certain morphology, which adds another special condition to the list of properties that the material must possess. This can also complicate the synthetic process as the

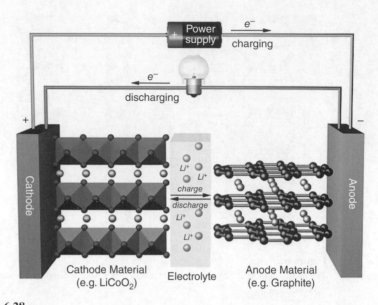

Figure 6.28
Representation of a standard commercial Li-ion battery. The charge-discharge intercalation mechanism is also shown. Source: Reproduced with permission of the RSC [73].

regular synthetic procedure must be altered or an alternative must be developed. Two morphologies are important when considering electrode configurations; that is, films-coated and bulk composite electrodes. Film-coated electrodes can be obtained by either the direct synthesis of a continuous system or the use of a coating of the electrode with the particles by using a powder in suspension. Bulk composite electrodes are easier to prepare. The particles are mixed and a conductive composite matrix is created. This method does not require any special morphology, is easy to perform and all components are in close contact with each other [8b].

In general, the following properties are important for electrochemical applications and, in particular, energy storage:

- High surface areas: this allows to have or to immobilize a high number of active sites on the surface;
- Porosity: an open and interconnected pore system increases mass transport in the pores;
- Good conductivity;
- Intrinsic electrocatalytic properties;
- High mechanical stability;
- Suitable morphology if applicable.

A few examples are given next where carbon-based materials are investigated for their use in batteries.

Lithium-sulfur batteries have a higher theoretical energy density which is 3–5 times higher than the normal Li-ion batteries described previously. However, sulfur possesses a very low electronic conductivity. If the sulfur can be contained in a conductive "host," this could solve the electronic conductivity issue. Porous conductive carbon materials can be useful for this [76]. Researchers prepared a hierarchical OMC mate-

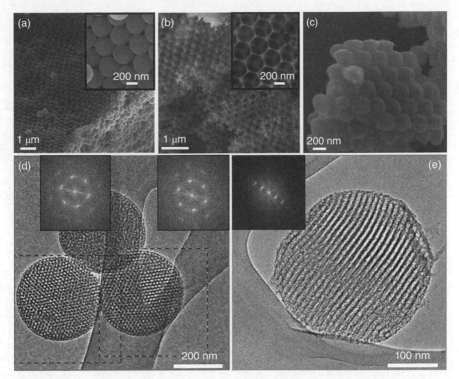

Figure 6.29

(top) SEM images of (a) polymer (PMMA) spheres used as first hard template; (b) Silica inverse opal structure, prepared with the polymer spheres and used as second hard template; (c) ordered mesoporous carbon prepared with the silica inverse opal template. (bottom) TEM images of the OMC nanoparticles: (d) projected along the columns, and (e) tilted out of the columnar projection. The insets in (d) and (e) are Fast Fourier transforms of the squares. Source: Reproduced with permission of John Wiley & Sons, Ltd [76].

rial with spherical morphology (diameter ~300 nm) by using a specially designed synthesis method based on the hard-templated method described in Section 6.2.1 (Figure 6.29) and with a very high mesoporosity. The materials exhibited an impressive specific surface area of 2445 m^2 g^{-1}, a pore volume of 2.32 ml g^{-1}, a bimodal pore size distribution with pore diameters of 6.0 and 3.1 nm and a 2D-hexagonal symmetry. The pores were filled via a melt-diffusion strategy with ~ 50, ~60, and ~70 wt% of S, which was homogeneously distributed in the pores of the material. The resulting C/S composites were investigated as cathode material in Li—S batteries. Figure 6.30 shows that the composites exhibit good initial discharge capacity values (up to 1200 mAh g^{-1}) and good cyclability. The spherical bimodal mesoporous carbons filled with S are denoted as S-BMC/S. The graph also shows a comparison between these spherical nanoparticles and regular bulk material (BMC-1 filled with ~50 and ~ 60 wt% of S). Clearly the use of nanoparticles increased the specific capacity.

Other cases investigate the inclusion of Mn_3O_4 nanoparticles [77], CoO particles [78], CuO [79], and so on in OMCs. More examples can be found in some excellent reviews of A. Walcarius who is an expert in the application of nanomaterials the field of electrochemistry [8b, 72a, b, 80].

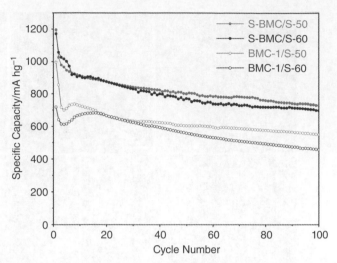

Figure 6.30

Comparison of the electrodes prepared from BMC-1 and S-BMC (Spherical Bimodal Mesoporous Carbon) with two different amounts of S included, that is, ~50 and ~60 wt%. All cells were operated at a current rate of 1 C (1675 mAg^{-1}) at room temperature. Source: Reproduced with permission of John Wiley & Sons, Ltd [76].

Alain Walcarius is born in 1967 in Belgium and studied Chemistry at Namur University (Belgium). After obtaining his Master's degree in 1989, he carried out his Ph.D. study in Analytical Chemistry in the field of zeolite modified electrodes at the same university. Subsequently, he performed several post-doctoral stays at New Mexico State University (USA) and Nancy University (France). In 2000 he obtained his Habilitation in Molecular Chemistry and Physics at Nancy University. Currently Walcarius is research director at CNRS and director of the Laboratory of Physical

Chemistry and Microbiology for the Environment (LCPME) at Lorraine University. He won several awards, including the Fellow of the International Society of Electrochemistry in 2016 and has more than 260 publications and 23 patents. His field of expertise is situated in electroanalytical chemistry and sensors, combined with the development of (porous) nanomaterials.

Photograph: https://www.researchgate.net/profile/Alain_Walcarius

Next to OMCs, nanocarbons such as graphene (sheets) and CNTs have also been investigated for this type of application. One very particular interesting material is the graphitic carbon nanosheet (CNS) [81]. Researchers have found that vertically aligned CNS showed great potential for energy storage applications due to their large electrode – electrolyte interface and strong interaction with the current collector [82]. Doping these vertically aligned CNS with nitrogen and metal carbides ($MoC_{0.654}$, WC, TaC, NbC) even improved their electrochemical properties.

The authors coated a commercially available nickel foam support with a drop of ionic liquid carbon precursors (Figure 6.31). The precursor consisted of 1-ethyl-3-methylimidazolium dicyanamide mixed with a salt mixture ($ZnCl_2$ and KCl). After a thermal treatment under inert gas atmosphere, the vertically aligned nitrogen doped CNS were obtained. They also prepared nitrogen doped CNS modified with metal carbides by adding precursor metal salts. The result is a unique material that exhibits a strong interaction with its support, and has a porous, but robust structure with small particle size. Furthermore, it has a large electrode-electrolyte contact area and short electron-ion diffusion pathways. These characteristics make this material ideal as electrode material and were investigated as anode material for lithium ion batteries. They showed a high specific capacity and

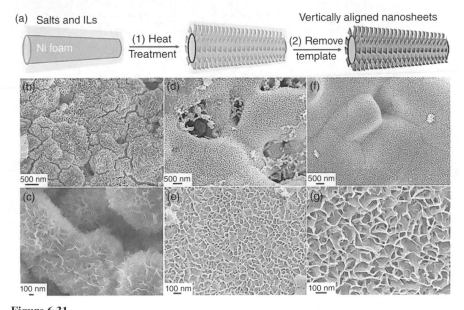

Figure 6.31

(a) Schematic representation of the synthesis process of the vertically aligned 2D nanosheets. Field emission scanning electron microscopy (FESEM) of the nitrogen doped graphitic CNS: (b–c) the CNS; (d–e) $MoC_{0.654}$@CNS1, and (f–g) $MoC_{0.654}$@CNS2. Source: Reproduced with permission of ACS [81].

a long lifespan as anode. A first discharge capacity of 1667 mAh g^{-1} was observed for the MoC$_{0.654}$@CNS2 material.

The introduction of nitrogen atoms in nanocarbons is an interesting pathway to improve their properties. The nitrogen doped variants have a higher charge capacity in comparison with the undoped materials when they are used as electrode in Li-ion batteries [83]. Different nitrogen-containing precursors can be used. This was also briefly discussed in Section 6.2.3.1 where nitrogen was introduced into OMPs via the polymerization reaction of melamine. Figure 6.32 shows how N-doped carbon

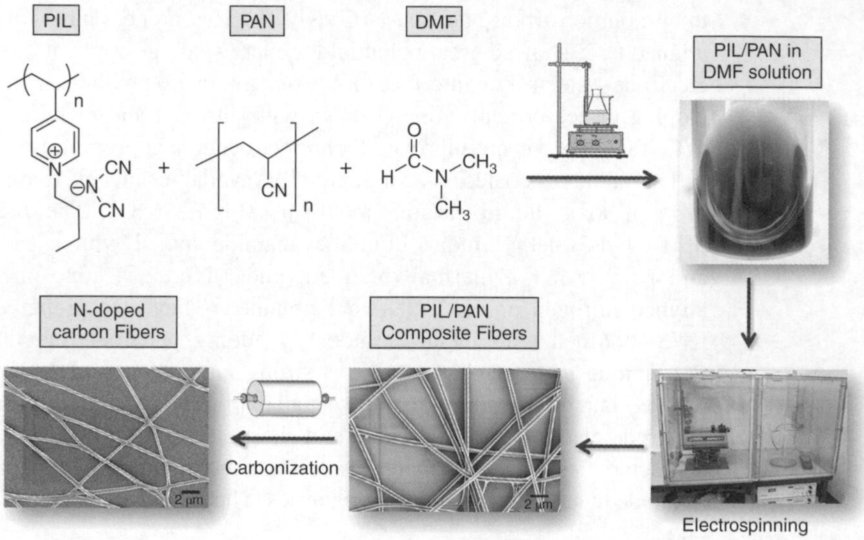

Figure 6.32
Overview of the synthetic pathway to create nitrogen doped carbon fibers using the precursors PIL and PAN dissolved in DMF which are subsequently electrospun to PIL/PAN composite fibers and carbonized into N-doped carbon fibers under inert atmosphere. Source: Reproduced with permission of John Wiley & Sons, Ltd [83].

1. nitro
2. amine
3. pyrolic
4. pyridinic
5. graphitic

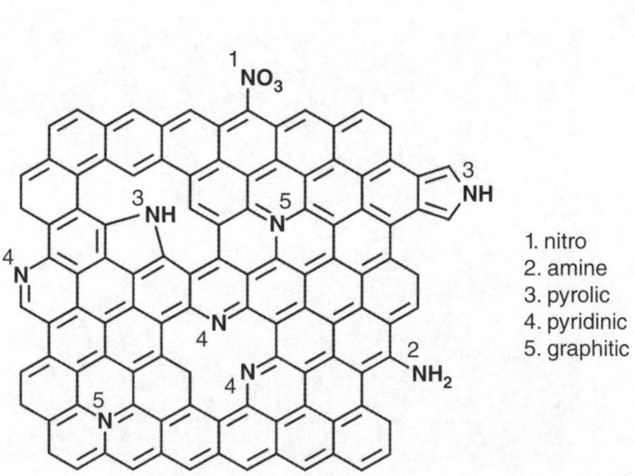

Figure 6.33
The different nitrogen doped sites with a carbon network. Nitrogen can be incorporated in the following forms as a nitro or amine group or can be included as a pyrolic, pyridinic and graphitic functional group. Source: Reproduced with permission of John Wiley & Sons, Ltd [83].

fibers can be synthesized from poly(ionic liquid) (PIL) and poly(acrylonitrile) (PAN) in DMF via an electrospinning process. The nitrogen atoms can be incorporated in the structure in different configurations as shown in Figure 6.33.

Graphitic carbon nitride or g-C_3N_4 is related to this type of materials and shows many similarities with the nanocarbon materials discussed previously. It is the most stable allotrope of the carbon nitride family represented by $(C_3N_3H)_n$. g-C_3N_4 is a defect-rich poly(tri-s-triazine) and can also be considered as a Covalent Organic Framework or COF, which are elaborated on further in Chapter 9. Graphitic carbon nitride is also used in electrochemical applications due to its interesting electronic structure as shown by the research of M. Antonietti [61h, 81, 83, 84] and B. Lotsch [85] who has also performed research in the field of nanocarbons and are very active in energy related applications. Their profiles are discussed in the Chapter 9 on Covalent Organic Frameworks.

Exercises

1 What type of precursors are used to synthesize activated carbon? What are the advantages and disadvantages by using this type of precursors?
2 What type of methods exist to activate activated carbon? What is the biggest difference between both methods? Why is the activation step important?
3 Explain the hard-templated and soft-templated synthesis methods to prepare an ordered mesoporous polymer. What are the disadvantages of each method?
4 Which parameters are important for the mesostructure of Ordered Mesoporous Polymers prepared via the soft-templated method?
5 Write down the reaction mechanism of the prepolymerization of the resol and Novolac prepolymer. Why is the reaction mechanism different?
6 What are the different methods to synthesize OMPs? Explain briefly.
7 How would you introduce nitrogen into an ordered mesoporous polymer?
8 Why have OMPs/OMCs got such a high mechanical and hydrothermal stability? Which other material, which has some similarities with OMPs and OMCs, also exhibits a very high stability?
9 Which synthesis method would you choose to synthesize multiwalled carbon nanotubes?
10 When you must use OMPs or OMCs for the adsorption of bulky organic pollutants. Which important characteristics of the pollutant/adsorbent do you have to consider before you start any experiments?

Answers to the Problems

1 Activated carbon can be synthesized from any agricultural byproduct. It is mainly prepared from waste. Typical precursors are: wood, bamboo, coal, shells from coconuts, palm trees and cacao, coffee beans, seaweed, and so on. The advantage of using these precursors is their low-cost and wide availability as it is a waste product. A huge disadvantage is the variation in precursor. Two batches of coconuts shells will always have slightly different properties.

2 Answer:

1. *Thermal or physical activation (PA) method: this method involves two steps with the first step being a carbonization procedure. Secondly, a gasification is performed with a mild oxidizing agent.*

2. *Chemical activation (CA) method: The precursor is treated with an inorganic base which degrades during carbonization.*

   The CA method normally uses lower temperatures and shorter reaction times in comparison with the PA method. However, the CA method uses acids, bases or salts as inorganic additive and sometimes the salts or metals (e.g. K, Na, Ca, Zn) remain in the activated carbon. This can be a drawback depending on the envisaged application.

3 The *hard-templated method* (also called exocasting or nanocasting method) makes use of a carbon precursors (e.g. sucrose) that are polymerized in a suitable solid porous template. After removal of the template, an ordered mesoporous polymer is obtained. The disadvantages of this method are: it is time-consuming as multiple steps are required; it is sometimes difficult to completely fill the entire template with precursor; a suitable template must be found: it must be stable during the impregnation procedure, but not too stable as it must be removed after polymerization, it must have a continuous pore system otherwise a non-porous system will be formed; unsuitable for up-scaling; hazardous or corrosive reagents are required to remove the template; tuning of the wall thickness and pore diameter is limited to the available templates. The *soft-templated method* makes use of a surfactant (thus "soft") structure-directing agent. The carbon precursors (e.g. phenol and formaldehyde) are polymerized around the surfactant. After formation of the solid, the template is removed to obtain the ordered mesoporous polymer. The disadvantages of this method are: a suitable precursor-template system must be found: suitable interactions between the carbon precursors and template must be present; the carbon precursors might be toxic, for example, formaldehyde; control over the synthesis conditions (pH, temperature, etc.) is crucial as this determines the final properties of the OMP.

4 Important parameters are: the choice of structure-directing agent, precursor to SDA ratio, and synthesis conditions (pH).

5 The prepolymerization mechanism of resol: Scheme 6.1; The prepolymerization mechanism of Novolac: Scheme 6.3; The reaction mechanism is different due to the different conditions during reaction. The prepolymerization of resol occurs under basic conditions. This means that the phenol molecule will be activated as a proton will be subtracted. It is the conjugated phenol that will attack formaldehyde. The prepolymerization reaction of Novolac takes place under acidic conditions. Here formaldehyde is activated via the protonation of the carbonyl group with the formation of a carbocation.

6 Hydrothermal synthesis; Using the EISA procedure; Two-phase system-based method; see Section 6.2.3.4 and Table 6.4.

7 Several methods are applicable. One can use a nitrogen-containing precursor such as melamine, dicyandiamide or urea. This precursor can be combined with phenol or resorcinol and polymerized using the standard reaction conditions. Another method is grafting a nitrogen-containing silane on the surface of an OMP. Here, the grafting procedures, usually applied on silica-based materials, are used that are described in Chapters 4 and 5. Also, a post-treatment with different reagents such as ammonia is possible.

8 OMPs and OMCs have a three-dimensional covalently bonded network, which is entirely made up by carbon atoms. They lack hydrolyzable groups that can occur in silica materials (silane bridges; Si-O-Si bonds) or ether/ester bridges in polymers (R-O-R or R[C=O]-O-R' bonds); Diamond also entirely consists of carbon atoms structured in a 3D network, all covalently connected.

9 The chemical vapor deposition method is an appropriate choice as it uses softer synthetic conditions and can thus be well-controlled. Also, it can produce MWNT on a relatively large-scale, still with high purity.

10 Size of the organic pollutant: you must first verify if the size of the bulky organic pollutant is not too large for the pore system of the OMPs or OMCs. If necessary, you will need to adapt the adsorbent / Charge of the organic pollutant: depending of the pH of your solution, the pollutant can carry a charge or be neutral. This must be compatible with the charge of your adsorbent. If necessary, you will need to adapt the pH of the solution to bring the pollutant in the "right form." / Interactions between the organic pollutant and the adsorbent: are the right interactions present? Hydrophobicity, aromaticity, the affinity for certain heteroatoms can play a vital role. If necessary, you will need to incorporate additional heteroatoms to boost the interactions.

References

1. (a) Gupta, V.K. and Saleh, T.A. (2013). Env. Sci. Pollution Res. 20: 2828–2843. (b) Allen, S.J., Whitten, L., and McKay, G. (1998). Dev. Chem. Eng. Mineral Proc. 6: 231–261; (c) Kwiatkowski, J.F. (2011). *Activated Carbon: Classifications, Properties and Applications*. Nova Science Publishers.

2. Yahya, M.A., Al-Qodah, Z., and Ngah, C.W.Z. (2015). Renewable Sust. Energy Rev. 46: 218–235.

3. ASTM D5158-98(2013) (2013). *Standard Test Method for Determination of Particle Size of Powdered Activated Carbon by Air Jet Sieving*. West Conshohocken, PA: ASTM International www.astm.org.

4. Ma, T.Y., Liu, L., and Yuan, Z.Y. (2013). *Chem. Soc. Rev.* 42: 3977–4003.

5. Ryoo, R., Joo, S.H., and Jun, S. (1999). *J. Phys. Chem. B* 103: 7743–7746.

6. Jun, S., Joo, S.H., Ryoo, R. et al. (2000). *J. Am. Chem. Soc.* 122: 10712–10713.

7. Lu, A.H. and Schuth, F. (2006). *Adv. Mater.* 18: 1793–1805.

8. (a) Van Der Voort, P., Vercaemst, C., Schaubroeck, D., and Verpoort, F. (2008). *Phys. Chem. Chem. Phys.* 10: 347–360. (b) Walcarius, A. (2017). *Sensors* 17: 1863. (c) Liang, C.D., Li, Z.J., and Dai, S. (2008). *Angew. Chem. Int. Ed.* 47: 3696–3717; (d) Sayari, A. and Yang, Y. (2005). *Chem. Mater.* 17: 6108–6113; (e) Wu, Z.X., Meng, Y., and Zhao, D.Y. (2010). Microporous Mesoporous Mat. 128: 165–179.

9. Lu, A.H., Schmidt, W., Taguchi, A. et al. (2002). *Angew. Chem. Int. Ed.* 41: 3489–3492.

10. Roggenbuck, J. and Tiemann, M. (2005). *J. Am. Chem. Soc.* 127: 1096–1097.

11. Kang, M., Kim, D., Yi, S.H. et al. (2004). *Catal. Today* 93, 695–95, 699.

12. (a) Kresge, C.T., Leonowicz, M.E., Roth, W.J. et al. (1992). *Nature* 359: 710–712. (b) Zhao, D.Y., Huo, Q.S., Feng, J.L. et al. (1998). *J. Am. Chem. Soc.* 120: 6024–6036.

13. Meng, Y., Gu, D., Zhang, F.Q. et al. (2005). *Angew. Chem. Int. Ed.* 44: 7053–7059.

14. Meng, Y., Gu, D., Zhang, F.Q. et al. (2006). *Chem. Mater.* 18: 4447–4464.

15. L.H. Baekeland, US Patent, Method of making insoluble products of phenol and formaldehyde, 1909, No. US 942699 A.

16. Muylaert, I., Verberckmoes, A., De Decker, J., and Van der Voort, P. (2012). *Adv. Colloid. Interfac.* 175: 39–51.

17. (a) Libbrecht, W., Verberckmoes, A., Thybaut, J.W. et al. (2017). *Carbon* 116: 528–546. (b) Enterria, M. and Figueiredo, J.L. (2016). Carbon. 108: 79–102.

18. Schlienger, S., Graff, A.L., Celzard, A., and Parmentier, J. (2012). *Green Chem.* 14: 313–316.

19. Nelson, K.M., Mahurin, S.M., Mayes, R.T. et al. (2016). *Microporous Mesoporous Mater.* 222: 94–103.

20. Liu, S.X., Huang, Z.H., and Wang, R. (2013). *Mater. Res. Bull.* 48: 2437–2441.

21. Saha, D., Warren, K.E., and Naskar, A.K. (2014). *Carbon* 71: 47–57.

22. Long, D.H., Qiao, W.M., Zhan, L. et al. (2009). *Microporous Mesoporous Mater.* 121: 58–66.

23. (a) Huang, Y., Yang, J.P., Cai, H.Q. et al. (2009). *J. Mater. Chem.* 19: 6536–6541; (b) Li, J.G., Lin, Y.D., and Kuo, S.W. (2011). *Macromol.* 44: 9295–9309.

24. Li, M. and Xue, J.M. (2014). *J. Phys. Chem. C* 118: 2507–2517.

25. Wei, J., Zhou, D.D., Sun, Z.K. et al. (2013). *Adv. Funct. Mater.* 23: 2322–2328.

26. Chen, A.B., Yu, Y.F., Zhang, Y. et al. (2014). *Carbon* 80: 19–27.

27. Wickramaratne, N.P. and Jaroniec, M. (2013). *Carbon* 51: 45–51.

28. Deng, Y.H., Yu, T., Wan, Y. et al. (2007). *J. Am. Chem. Soc.* 129: 1690–1697.

29. Deng, Y.H., Cai, Y., Sun, Z.K. et al. (2010). *Adv. Funct. Mater.* 20: 3658–3665.

30. Garcia, A., Nieto, A., Vila, M., and Vallet-Regi, M. (2013). *Carbon* 51: 410–418.

31. Wan, Y., Shi, Y.F., and Zhao, D.Y. (2007). *Chem. Commun.* 897–926.

32. (a) Wang, X.Q., Liang, C.D., and Dai, S. (2008). *Langmuir* 24: 7500–7505. (b) Gao, P., Wang, A., Wang, X., and Zhang, T. (2008). *Chem. Mater.* 20: 1881–1888.

33. (a) Xin, W. and Song, Y.H. (2015). *RSC Adv.* 5: 83239–83285. (b) Xu, J.M., Wang, A.Q., and Zhang, T. (2012). *Carbon* 50: 1807–1816; (c) Liu, R.L., Shi, Y.F., Wan, Y. et al. (2006). *J. Am. Chem. Soc.* 128: 11652–11662. (d) Tanaka, S., Nakatani, N., Doi, A., and Miyake, Y. (2011). *Carbon* 49: 3184–3189.

34. Trick, K.A. and Saliba, T.E. (1995). *Carbon* 33: 1509–1515.

35. Li, X.X., Larson, A.B., Jiang, L. et al. (2011). *Microporous Mesoporous Mater.* 138: 86–93.

36. (a) Otero, R., Esquivel, D., Ulibarri, M.A. et al. (2014). *Chem. Eng. J.* 251: 92–101. (b) F, Q., Zhang, Y., Meng, D.G. et al. (2005). *J. Am. Chem. Soc.* 127: 13508–13509; (c) Deng, Y., Liu, C., Gu, D. et al. (2008). *J. Mater. Chem.* 18: 91–97.

37. Lin, J.M. and Ma, C.C.M. (2000). *Polym. Degrad. Stab.* 69: 229–235.

38. Wan, Y., Shi, Y.F., and Zhao, D.Y. (2008). *Chem. Mater.* 20: 932–945.

39. Tian, Y., Wang, X.F., and Pan, Y.F. (2012). *J. Hazard. Mater.* 213: 361–368.

40. (a) Mangun, C.L., Benak, K.R., Economy, J., and Foster, K.L. (2001). *Carbon* 39: 1809–1820. (b) Z, H., Xiang, D.P., Cao, L. et al. (2014). *Adv. Mater.* 26: 3315–3320; (c) Deng, Q.F., Liu, L., Lin, X.Z. et al. (2012). *Chem. Eng. J.* 203: 63–70; (d) Hao, G.P., Li, W.C., Qian, D., and Lu, A.H. (2010). *Adv. Mater.* 22: 853–857.

41. Wu, Z.X., Webley, P.A., and Zhao, D.Y. (2010). *Langmuir* 26: 10277–10286.

42. (a) Georgakilas, V., Perman, J.A., Tucek, J., and Zboril, R. (2015). *Chem. Rev.* 115: 4744–4822. (b) Li, Z., Liu, Z., Sun, H.Y., and Gao, C. (2015). *Chem. Rev.* 115: 7046–7117.

43. Geim, A.K. and Novoselov, K.S. (2007). *Nat. Mater.* 6: 183–191.

44. Kroto, H.W., Heath, J.R., Obrien, S.C. et al. (1985). *Nature* 318: 162–163.

45. Kroto, H.W. (1987). *Nature* 329: 529–531.

46. Wilson, R.J., Meijer, G., Bethune, D.S. et al. (1990). *Nature* 348: 621–622.

47. Kratschmer, W., Lamb, L.D., Fostiropoulos, K., and Huffman, D.R. (1990). *Nature* 347: 354–358.

48. (a) Hirsch, A. (1993). *Angew. Chem. Int. Ed.* 32: 1138–1141; (b) Hirsch, A. (1999). *Fullerenes and Related Structures*, vol. 199, 1–65. Berlin, Heidelberg: Springer.

49. (a) Bhattacharya, S. and Samanta, S.K. (2016). *Chem. Rev.* 116: 11967–12028. (b) Zhang, J., Terrones, M., Park, C.R. et al. (2016). *Carbon* 98: 708–732.

50. (a) Iijima, S. (1991). *Nature* 354: 56–58. (b) Iijima, S. and Ichihashi, T. (1993). *Nature* 363: 603–605.
51. Prasek, J., Drbohlavova, J., Chomoucka, J. et al. (2011). *J. Mater. Chem.* 21: 15872–15884.
52. Ajayan, P.M. (1999). *Chem. Rev.* 99: 1787–1799.
53. Journet, C., Maser, W.K., Bernier, P. et al. (1997). *Nature* 388: 756–758.
54. (a) Mubarak, N.M., Yusof, F., and Alkhatib, M.F. (2011). *Chem. Eng. J.* 168: 461–469. (b) Kruusenberg, I., Alexeyeva, N., Tammeveski, K. et al. (2011). *Carbon* 49: 4031–4039; (c) Hou, P.-X., Liu, C., and Cheng, H.-M. (2008). *Carbon* 46: 2003–2025.
55. Wu, Z.X. and Zhao, D.Y. (2011). *Chem. Commun.* 47: 3332–3338.
56. Wang, S.B. and Zhu, Z.H. (2007). *Dyes Pigm.* 75: 306–314.
57. (a) Libbrecht, W., Verberckmoes, A., Thybaut, J.W. et al. (2017). *Langmuir* 33: 6769–6777. (b) Liu, F.L., Xu, Z.Y., Wan, H.Q. et al. (2011). *Environ. Toxicol. Chem.* 30: 793–800.
58. Ariga, K., Vinu, A., Miyahara, M. et al. (2007). *J. Am. Chem. Soc.* 129: 11022–11023.
59. Vinu, A., Miyahara, M., Sivamurugan, V. et al. (2005). *J. Mater. Chem.* 15: 5122–5127.
60. (a) Tran, H.N., You, S.J., Hosseini-Bandegharaei, A., and Chao, H.P. (2017). *Water Res.* 120: 88–116. (b) Peng, W., Li, H., Liu, Y., and Song, S. (2017). *J. Mol. Liq.* 230: 496–504; (c) Santhosh, C., Velmurugan, V., Jacob, G. et al. (2016). *Chem. Eng. J.* 306: 1116–1137; (d) Ihsanullah, A.A., Al-Amer, A.M., Laoui, T. et al. (2016). *Sep. Purif. Technol.* 157: 141–161.
61. (a) Perez-Mayoral, E., Calvino-Casilda, V., and Soriano, E. (2016). *Catal. Sci. Technol.* 6: 1265–1291. (b) P. Serp, J. L. Figueiredo, Carbon Materials for Catalysis. John Wiley & Sons, 2009; (c) Su, D.S., Wen, G., Wu, S. et al. (2017). *Angew. Chem. Int. Ed.* 56: 936–964;(d) Yan, Y.B., Miao, J.W., Yang, Z.H. et al. (2015). *Chem. Soc. Rev.* 44: 3295–3346.(e) Matos, I., Bernardo, M., and Fonseca, I. (2017). *Catal. Today* 285: 194–203.(f) Rodríguez-Reinoso, F. (1998). *Carbon* 36: 159–175.(g) Su, D.S., Perathoner, S., and Centi, G. (2013). *Chem. Rev.* 113: 5782–5816.(h) Navalon, S., Dhakshinamoorthy, A., Alvaro, M. et al. (2017). *Chem. Soc. Rev.* 46: 4501–4529.
62. (a) Li, M., Xu, F., Li, H., and Wang, Y. (2016). *Catal. Sci. Technol.* 6: 3670–3693; (b) Benavidez, A.D., Burton, P.D., Nogales, J.L. et al. (2014). *Appl. Catal., A* 482: 108–115.
63. Restivo, J., Soares, O.S.G.P., Órfão, J.J.M., and Pereira, M.F.R. (2017). *Chem. Eng. J.* 309: 197–205.
64. Rankel, L.A. (1993). *Energy Fuels* 7: 937–942.
65. Illán-Gómez, M.J., Brandán, S., Linares-Solano, A., and Salinas-Martínez de Lecea, C. (2000). *Appl. Catal., B* 25: 11–18.
66. Chen, J., Zhong, J., Guo, Y., and Chen, L. (2015). *RSC Adv.* 5: 5933–5940.
67. De Decker, J., Bogaerts, T., Muylaert, I. et al. (2013). *Mater. Chem. Phys.* 141: 967–972.
68. Rideal, E.K. and Wright, W.M. (1925). *J. Chem. Soc. Trans.* 127: 1347–1357.
69. Zhang, J., Liu, X., Blume, R. et al. (2008). *Science* 322: 73–77.
70. Figueiredo, J.L. and Pereira, M.F.R. (2010). *Catal. Today* 150: 2–7.
71. Song, S., Yang, H., Rao, R. et al. (2010). *Catal. Commun.* 11: 783–787.
72. (a) Walcarius, A. (2012). *TrAC, Trends Anal. Chem* 38: 79–97; (b) Walcarius, A. (2013). *Chem. Soc. Rev.* 42: 4098–4140; (c) Titirici, M.M., White, R.J., Brun, N. et al. (2015). *Chem. Soc. Rev.* 44: 250–290; (d) El-Kady, M.F., Shao, Y.L., and Kaner, R.B. (2016). *Nat. Rev. Mater.* 1: 1–14; (e) Sun, M.H., Huang, S.Z., Chen, L.H. et al. (2016). *Chem. Soc. Rev.* 45: 3479–3563. (f) Zhang, Q.F., Uchaker, E., Candelaria, S.L., and Cao, G.Z. (2013). *Chem. Soc. Rev.* 42: 3127–3171.
73. Liu, R., Duay, J., and Lee, S.B. (2011). *Chem. Commun.* 47: 1384–1404.
74. (a) Su, D.S. and Schlögl, R. (2010). *Chem. Sus. Chem.* 3: 136–168; (b) Lotsch, B.V. and Maier, J. (2017). *J. Electroceram.* 38: 128–141.
75. Bruce, P.G. (2008). *Solid State Ionics* 179: 752–760.
76. Schuster, J., He, G., Mandlmeier, B. et al. (2012). *Angew. Chem. Int. Edit.* 51: 3591–3595.
77. Li, Z.Q., Liu, N.N., Wang, X.K. et al. (2012). *J. Mater. Chem.* 22: 16640–16648.
78. Zhang, H.J., Tao, H.H., Jiang, Y. et al. (2010). *J. Power Sources* 195: 2950–2955.
79. Yang, M. and Gao, Q.M. (2011). *Microporous Mesoporous Mater.* 143: 230–235.
80. Walcarius, A., Minteer, S.D., Wang, J. et al. (2013). *J. Mater. Chem. B* 1: 4878–4908.
81. Zhu, J.X., Sakaushi, K., Clavel, G. et al. (2015). *J. Am. Chem. Soc.* 137: 5480–5485.
82. (a) Miller, J.R., Outlaw, R.A., and Holloway, B.C. (2010). *Science* 329: 1637–1639; (b) Bo, Z., Zhu, W.G., Ma, W. et al. (2013). *Adv. Mater.* 25: 5799–5806.

83. Einert, M., Wessel, C., Badaczewski, F. et al. (2015). *Macromol. Chem. Phys.* 216: 1930–1944.
84. (a) Liu, J., Wang, H.Q., and Antonietti, M. (2016). *Chem. Soc. Rev.* 45: 2308–2326; (b) Xu, J.S., Antonietti, M., and Shalom, M. (2016). *Chem. Asian J.* 11: 2499–2512.
85. (a) Podjaski, F., Kroger, J., and Lotsch, B.V. (2018). *Adv. Mater.* 30. (b) Lau, V.W.H., Moudrakovski, I., Botari, T. et al. (2016). *Nat. Commun.* 7; (c) Schwinghammer, K., Mesch, M.B., Duppel, V. et al. (2014). *J. Am. Chem. Soc.* 136: 1730–1733.

7 The Era of the Hybrids – Part 1: Periodic Mesoporous Organosilicas or PMOS

7.1 Introduction

The basic silica materials described in Chapter 5, such as MCM-41 (Mobil Composition of Matter-41) and SBA-15, are pure silica materials without any organic functionalization. For many high-end applications, such as catalysis and adsorption, the presence of an appropriate organic functionality is required. In Chapter 4, two different routes are described and discussed to introduce organic functional groups on silica materials; that is, via post-modification of the prepared silicas or via a one-pot-synthesis procedure. However, a third and very appealing route to create ordered materials with organic functionalities exists and the resulting materials are called Periodic Mesoporous Organosilicas or PMOs. They consist of a silica network containing organic functionalities that are homogeneously distributed throughout the entire material, including the inside of the pore walls (Figure 7.1). The organic groups are directly and covalently linked to the silicon atoms of the network, resulting in hybrid materials that combine both organic and inorganic properties in one material.

This new class of nanomaterials was first and almost simultaneously introduced in 1999 by three independent research groups, headed by G.A. Ozin [2], S. Ingaki [3], and A. Stein [4]. A short biography of these three pioneers in the field of PMOs is presented next. Since then, this research field boomed and a lot of research has been performed on the formation and eventual properties of solids, and, additionally, many possible applications have been explored such as heterogeneous catalysis, adsorption of gases, metals, and organics, their use as insulating layers in low-k material, controlled drug release system, stationary chromatographic packing, and so on.

Introduction to Porous Materials, First Edition.
Pascal Van Der Voort, Karen Leus and Els De Canck.
© 2019 John Wiley & Sons Ltd. Published 2019 by John Wiley & Sons Ltd.

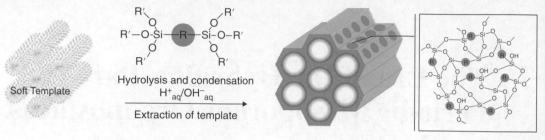

Figure 7.1
Schematic overview of the synthesis of a PMO. Source: Reproduced with permission of John Wiley &
Sons, Ltd [1].

*Shinji Inagaki is senior researcher at Toyota Central R&D labs., Inc. in Japan and
known for his pioneering work in the field of Periodic Mesoporous Organosilicas.
He received his Master's degree in 1984 from the Department of Synthetic Chemistry
at the Faculty of Engineering, Nagoya University in Japan. Afterward, he joined the
Toyota Central R&D Labs., Inc. Inagaki obtained his doctoral degree in Engineering
(Nagoya University) in 1998 entitled "Synthesis and Structure of Mesoporous Crystal, FSM-16." In 2009, he was appointed as Senior Fellow at Toyota Central R&D
Labs., Inc. He is author of many high impact papers, patents, and several books. He
won several awards such as Promotion Award of the Japan Society on Adsorption
(2001), the Chemical Society of Japan Award for Creative Work for 2004, the Min-
ister Award of Education, Culture, Sport, Science and Technology (2005), and the
Japan Society on Adsorption Award (2008). His field of interest lies in the synthesis and application of mesoporous materials for photocatalysis and as optical and
electrical-responsive material.*

Photograph: https://www.tytlabs.com/sflabinagaki/profile.html

Geoffrey A.S. Ozin *received his first-class honors degree in Chemistry from the King's College London in 1965 and his doctoral degree in Inorganic Chemistry at Oriel College, Oxford University in 1967. Afterward, he performed a Postdoctoral Fellowship at Southampton University. Now he is Distinguished University Professor at the University of Toronto (Canada) and Tier 1 Canada Research Chair in Materials Chemistry. Additionally, he is Honorary Professor at The Royal Institution Great Britain and University College London, External Advisor for the London Centre for Nanotechnology, Alexander von Humboldt Senor Scientist at the Max Planck Institute for Surface and Colloid Science Potsdam, and Guest Professor at the Centre for Functional Nanostructures at Karlsruhe Institute of Technology (KIT). He won several awards such as the Meldola Medal and Prize (1972), the Rutherford Memorial Medal (1982), and the Albert Einstein World Award of Science (2011). He has performed pioneering work in the field of Nanochemistry.*

Photograph: https://alchetron.com/Geoffrey-Ozin

Andreas Stein *is currently Professor in the Department of Chemistry at the University of Minnesota. He obtained his bachelor's degree in 1986 and his Master's degree*

in 1988 at the University of Calgary and the University of Toronto, respectively. In 1991, he obtained this doctoral degree with his dissertation entitled "Synthesis, Characterization and Opto-Electronic Properties of Sodalite-Encapsulated Insulators, Semiconductor-Components and Metals" under the guidance of Geoffrey Ozin at the University of Toronto. After several postdoctoral fellowships, he became Assistant Professor in 1994 and Professor in 2003 at the University of Minnesota. He won many awards during his scientific career such as the Distinguished McKnight University Professorship (2008–2013), Merck Professorship in Chemistry (2007–2008), and Highly Cited Materials Scientist in Thomson Scientific's ISIHighlyCited.com *(2007). His interests lie in the field of porous materials and nanocomposites.*

Photograph: http://www.adamamaterials.com/andreas-stein/

Many similarities can be found between the syntheses of PMOs and the ordered silicas (MCM and SBA type). PMOs are also synthesized using a structure-directing agent (SDA) to create porosity and a silane that first hydrolyzes and condenses around the SDA (see Chapter 5). A significant difference between the preparation of both types of material is the silicon source, also called the "PMO precursor", that is added at the beginning of the synthesis (see Figure 7.1). Polysilsesquioxanes that include an organic functional linker or "organic bridge" are employed as precursors to build up the hybrid silica network. This contains both the desired organic functionality and the silicon atoms with hydrolyzable alkoxy groups (e.g. methoxy, ethoxy, etc.) necessary for the hydrolysis and condensation process during synthesis. Most often, a bis-silane is used that is represented by $(R'O)_3Si-R-Si(OR')_3$, with OR' being the hydrolyzable group and R the organic functional linker.

Figure 7.1 shows a brief overview of the most important steps in the synthesis of a PMO material. In this figure, a soft template acts as a SDA that forms micelles (e.g. the triblock copolymer $EO_{20}PO_{70}EO_{20}$; Pluronic P123) and a bis-silane (with R = an organic functionality) that hydrolyzes and subsequently polycondenses with the formation of an organosilica, which are combined in an acidic (H^+) or basic (OH^-) environment.

The organosilica material is formed around the micelles of the SDA and in the case of the template P123 the walls of the micelles are perforated with the hydrophilic ethylene oxide chains of the surfactant. This mechanism is completely identical to the formation of an SBA-15 material (see Chapter 5) [5]. After a certain aging period where the organosilica further condenses, the SDA is removed by an extraction or mild heat treatment under an inert atmosphere, and a highly porous PMO material remains. In general, the synthesis conditions can be hydrothermal, which is most commonly used, via the Evaporation Induced Self-Assembly (EISA) process [6] or with the use of microwaves [7]. These methods are all comparable to the different synthesis methods of regular ordered silica materials.

A variety of SDAs can be used in different conditions (acidic, basic, or neutral) just like MCM and SBA type materials. This creates materials with all kinds of different pore structures and morphologies. Additionally, a plethora of polysilsesquioxanes is either commercially available or can be synthesized. The organic bridges of the precursors can be very simple or very complex structures. They can be aliphatic or aromatic, inert, or very active, with or without heteroatoms, rigid, or very flexible. This results in a vast amount of different PMO materials, each with its specific properties originating from the used organic bridge. This extends the applicability of PMOs to very different fields. The materials with more complex structures such as metal complexes and chiral bridges were developed for high-end applications, such as chromatography or even chiral catalysis [8].

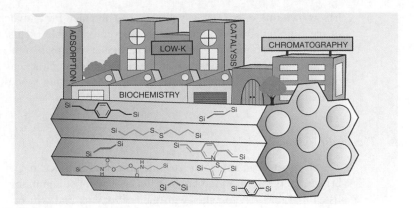

Figure 7.2
Schematic representation of the different application fields of PMOs. Source: Reproduced with permission of the RSC [8].

There are some significant differences between a PMO and a functionalized SBA material, which is, for example, obtained by grafting with a functional silane. The major difference is that the functional groups are not only at the surface, but they are everywhere in the walls. Every silicon atom has one or two bonds to a carbon atom. This means that many of the functional groups are in fact not accessible. These organic groups contribute to the rigidity or flexibility of the walls and to the general structural characteristics of the materials. Moreover, a regular grafting process (see Chapter 5) anchors the organic functionalities via sometimes weak siloxane bridges (Si—O—Si). These bonds are very sensitive to hydrolysis in acidic and alkaline environments, which results in less stable materials.

Since their discovery in 1999, many reviews have appeared on PMOs written by some leading groups in this field of expertise [1, 9]. We also contributed with a comprehensive review of PMOs, discussing in-depth the functionalities, their morphology, and their applications [8]. In this chapter, we try to explain the basics of PMO synthesis, their most important and interesting properties, and introduce the diverse variety of applications of PMOs; that is, catalysis, adsorption, chromatography, and so on [8] (Figure 7.2).

7.2 Synthesis of PMOs

Many parameters influence the synthesis of a PMO and the eventual morphology, structure, and properties of the material. One of the most predominant factors is the synthesis conditions [10] that highly influence the resulting material. Additionally, many other factors influence the PMO, such as the type of polysilsesquioxane, the SDA, additives, pH, temperature, synthesis time, and the solvent, First, some general aspects of PMO synthesis are discussed. Then, PMO materials with different types of (small) organic bridge are presented.

7.2.1 General Aspects of PMO Synthesis

Here, a brief overview of the most important parameters during PMO synthesis is given. It must be noted that many parameters are important and that each reagent has its own influence on the synthesis. The obtained PMO material is thus the result of the simultaneous interplay between several variables. This makes it difficult to determine at certain moments and, in particular, distinguish the exact role of certain reagents, such as the additives or the reaction temperature, as they influence each other greatly. The optimization of a recipe for a certain type of PMO material can therefore be tedious.

7.2.1.1   Type of Precursor

The precursors can be categorized into aliphatic, olefinic, aromatic, and multi-organic bridges, depending on the organic linker. The most commonly used precursors are listed in Table 7.1, together with their acronyms that will be used throughout this chapter.

   Some PMO materials prepared with these bridges will be discussed in more detail in Sections 7.2.2 and 7.2.3.

7.2.1.2   Type of Structure-Directing Agent

The SDAs used in PMO synthesis are completely similar to those used for the synthesis of mesoporous silicas (see Chapter 5, Table 5.3 for a more elaborate discussion).

7.2.1.3   Addition of Inorganic Salts and Organic Molecules During Synthesis

The addition of inorganic salts or organic molecules can greatly influence the synthesis and thus affect the resulting PMO material.

   When using non-ionic SDAs such as triblock copolymers, inorganic salts will exhibit a special effect during the PMO assembly by influencing the interaction between several parts of the polymer [11]. The salt causes a dehydration of the hydrated ethylene oxide units of the polyethylene oxide (PEO) chain, which is located next to the polypropylene oxide (PPO) chain. This results in an increased hydrophobicity of the PPO chain and significantly decreases the hydrophilicity of the PEO. Many salts have shown to improve the hydrothermal stability of during the crystallization process (e.g. NaF, NaCl, KF, KCl, Na_2SO_4, etc.) [12]. It also allows us to prepare highly ordered materials in a wide range of acidic concentrations.

   The addition of KCl significantly increases the interaction between the SDA and the polysilsesquioxane and weakens the disordering of the organic units. It especially improves the interaction between the non-ionic block copolymer and the oligomers of the precursor that are formed during the hydrolysis and condensation.

   Organic molecules can also function as additives during PMO synthesis. They can either act as swelling agents (see Section 7.3.1) or as co-solvents. The latter influences the formation of the PMO material. The addition of butanol, for example, when using the SDA Pluronic P123 in an acidic environment, results into PMOs with a narrow pore size distribution and a higher degree of order [13].

7.2.1.4   Synthesis Conditions: Temperature, pH, and Reaction Time

The reaction temperature can influence the pore size of the PMO and plays a big role during the formation of PMOs. For example, it was found that higher reaction

Table 7.1 List of frequently used precursors in PMO synthesis.

Abbreviation	Full name	Structural formula
BTME	1,2-bis(trimethoxysilyl)ethane	$(MeO)_3Si$ ⌒ $Si(OMe)_3$
BTEE	1,2-bis(triethoxysilyl)ethane	$(EtO)_3Si$ ⌒ $Si(OEt)_3$
BTEENE	1,2-bis(triethoxysilyl)ethylene	$(EtO)_3Si$ ═ $Si(OEt)_3$
BTEM	1,2-bis(triethoxysilyl)methylene	$(EtO)_3Si$ ⌒ $Si(OEt)_3$
BTEB	1,4-bis(triethoxysilyl)benzene	$(EtO)_3Si$ —⬡— $Si(OEt)_3$
BTEEB	1,4-bis(triethoxysilylethyl)benzene	$(EtO)_3Si$ ⌒—⬡—⌒ $Si(OEt)_3$
BTEVB	1,4-bis-((*E*)-2-(triethoxysilyl)vinyl)benzene	$(EtO)_3Si$ ═—⬡—═ $Si(OEt)_3$

(Continued)

Table 7.1 *(Continued)*

Abbreviation	Full name	Structural formula
BTEVS	4,40-bis((*E*)-2-(triethoxysilyl)vinyl)stilbene	
BTET	2,5-bis-(triethoxysilyl)thiophene	
BTEPUE	bis[3-(triethoxysilyl) propyl urethane]ethane	
BTEPDS	bis[3-(triethoxysilyl)propyl]disulfide	
BTEPTS	1,4-bis(triethoxysilyl)propane tetrasulfide	
ICS	tris[3-(trimethoxysilyl)propyl]isocyanurate	

temperatures will lead to larger pore sizes [14]. Also, an increase in crystallization temperature will increase the condensation degree of the pore walls. Furthermore, the pH of the reaction mixture will determine the specific interaction between the SDA and the precursor. An improvement will result in better ordered materials [15]. Additionally, the reaction time and specifically the stirring time will also influence the structural properties of the PMOs [16].

7.2.2 PMOs with Aliphatic Bridges

In this section, focus is placed upon PMO materials with relatively small and simple organic bridges and a brief discussion on the synthesis, properties, or some peculiarities concerning the ethylene-bridged ($-CH_2CH_2-$) and methylene-bridged ($-CH_2-$) is given. A more detailed discussion with an overview of many different synthetic procedures and additional references can be found in Reference [8].

7.2.2.1 *The Ethane PMO*

The ethane PMO (also referred to as an ethylene-bridged PMO) is well-described in literature as it is the most investigated PMO material. The precursor 1,2-bis(trimethoxysilyl)ethane (BTME) or 1,2-bis(triethoxysilyl)ethane (BTEE) is solely employed in synthesis, or it is polycondensed with other organosilanes. Several types of surfactant (cationic and non-ionic) are employed and will influence the ordering of the mesophase, and they can also influence the external morphology of the particles.

The first reports on PMOs in 1999 by Stein et al. and Inagaki et al. described the synthesis of ethane PMOs with the cationic SDAs cetyl trimethyl ammonium bromide (CTAB [7]) and octadecyltrimethylammonium chloride (OTAC [8]), respectively. Different symmetries were obtained depending on the synthetic composition and conditions. By using cationic surfactants, different external morphologies or particle shapes could also be synthesized: rope- and gyroid-shaped particles [3, 16, 17] or even spherical [17], decaoctahedral [18], and dodecahedral [19] shapes. Gemini surfactants are also used [20]. These SDAs contain two charged ammonium heads and, depending on the type of gemini surfactant used, a different phase is obtained such as 2D hexagonal, 3D hexagonal, lamellar, cubic, and so on. Cationic SDA normally leads to materials with rather small pore sizes.

In contrast to cationic surfactants, neutral polymer or non-ionic based templates can significantly increase the pore diameter of the PMO materials. Two types of surfactants are used: block copolymers with Pluronics P123 [21] and F127 [22] being the most famous examples, and the poly(alkylene)oxides such as Triton-X100 [23] and Brij-56. The latter surfactant tends to provide excellent quality PMO materials with a 2D-hexagonal symmetry, not only for the ethane bridges but also for other silanes (e.g. methane, ethene, and benzene) with an extremely robust synthesis method [24].

Different external morphologies can be obtained. In particular, hollow spheres [25] and nanotubes [26] are of great interest in biological/biomedical related applications (Figure 7.3). A more extensive overview of different methods to prepare ethane PMOs, properties, and references can be found in Reference [8].

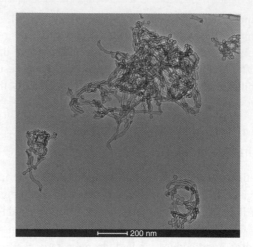

Figure 7.3

TEM image of hollow nanotubes prepared with the ethane PMO precursor. Source: Reproduced with permission of ACS [25d].

7.2.2.2   The Methane PMO

Only a few reports can be found on the methane or methylene-bridged PMO materials. It was Ozin's group that initially reported on them and the materials were prepared with 1,2-bis(triethoxysilyl)methylene (BTEM) and CTAB in a basic environment [27]. However, only materials with small pores were obtained which limits their applicability significantly. PMOs with larger pores were synthesized using P123 in an acidic aqueous medium and with the addition of the salt NaCl [28]. The very small organic bridge and the lack of modification possibilities of the methane group makes the methane PMO less applicable than other PMO materials.

7.2.3   PMOs with Olefinic and Aromatic Bridges

An overview is given next of some interesting features concerning olefinic and aromatic bridged PMO materials. Here the ethene bridged (—CH=CH—) and heavily investigated benzene bridged (—C_6H_4—) PMOs are presented.

7.2.3.1   The Ethene PMO

The ethene PMO or ethenylene-bridged PMO offers more chemical modification possibilities (see Section 7.4) due to the presence of the C=C bond. It was again Ozin's group that published one of the earliest reports in *Nature* on their synthesis [2]. The cationic surfactant CTAB in a basic medium (NH_4OH) was used as SDA and 1,2-bis(triethoxysilyl)ethylene (BTEENE) (see Table 7.1) as a bis-silane. These materials with a hexagonal symmetry showed a specific surface area of 637 m^2 g^{-1} and a pore diameter of almost 4 nm. At the same moment, Stein and coworkers[4] also synthesized ethene PMOs. The same SDA was used, but in an acidic environment (H_2SO_4). The precipitation was subsequently induced by raising the pH with NaOH. It resulted in materials with a high specific surface area, 1210 m^2 g^{-1}, and pores of 2.4 nm were observed. This clearly shows the effect of the different synthesis conditions on the resulting properties of the PMO although the same PMO precursor was used in both procedures.

Mokaya et al. reported on the synthesis of ethene PMO materials with very different morphologies. They prepared PMOs with various alkyltrimethylammonium bromide surfactants (C_nTMABr, n = 14, 16, 18), which resulted in different type of particle [29]. Materials with different morphologies (see Figure 7.4) were obtained, including monodisperse spheres, rod- or cage-like particles, and rope-like particles. The same group also observed molecular-scale periodicity. This phenomenon is even more clearly observed in benzene PMO materials [30]. Next to the expected intense long range (100) reflection and two other well-resolved signals in the range of 3–10°, an additional reflection at $2\theta = 16.5°$ appeared (see Figure 7.5). This indicates molecular-scale periodicity of the PMO material.

Robert Mokaya received his B.Sc. in Chemistry from the University of Nairobi in 1988 after which he spent a year working for Unilever in Kenya. He was awarded his Ph.D. from the University of Cambridge in 1992. In 1992 he was elected to a Research Fellowship at Trinity College, Cambridge and in 1996 was awarded an EPSRC Advanced Fellowship. He was appointed to a lectureship in Materials Chemistry at Nottingham in 2000, was promoted to Reader in Materials Chemistry in 2005, and to Professor of Materials Chemistry in 2008. His research group's interests lie in design, synthesis, and characterization of novel porous inorganic materials and the study of their structure-property relations.

Photograph: http://blogs.nottingham.ac.uk/talkingofteaching/2015/ttp-school-of-chemistry-improving-the-student-experience-of-assessment-and-feedback/

Non-ionic block copolymers [13, 31] and poly(alkylene oxides) [13] were used as templates in an attempt to enlarge the pores. A series of PMOs was synthesized by using the surfactants Brij-56, -58, -76, and -78 under acidic conditions [13]. Specific surface areas of 847–981 m^2 g^{-1} and pore sizes up to 5.1 nm were noted. The PMO materials prepared with Brij-56 and particularly Brij-76 exhibited a well-ordered

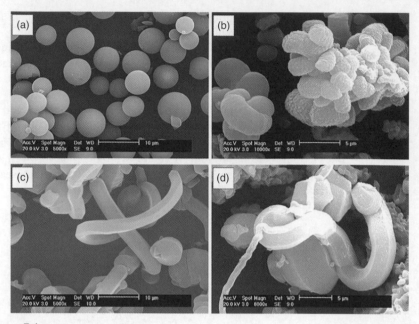

Figure 7.4
SEM images of ethene PMO materials prepared with different surfactants: (a) C_{12}TMABr, (b) C_{14}TMABr, (c) C_{16}TMABr, and (d) C_{18}TMABr. Source: Reproduced with permission of ACS [29].

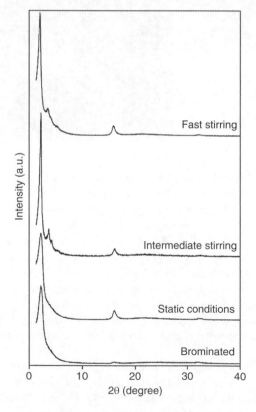

Figure 7.5
X-ray diffraction patterns for ethene PMO materials prepared under different synthesis conditions and after bromination of the double bond. Source: Reproduced with permission of the RSC [30a].

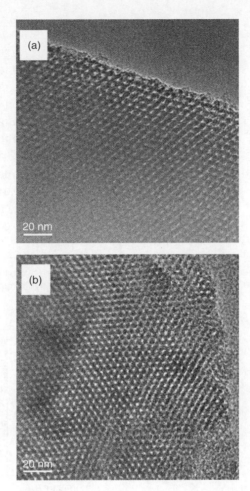

Figure 7.6

TEM images for an ethene PMO material prepared with the surfactants Brij-76 (a) and Brij-56 (b). (Both projected along the [001] zone axes). Source: Reproduced with permission of ACS[13].

structure with a hexagonal symmetry (see Figure 7.6). The use of the Pluronic P123 led to the increase of pore sizes up to 8.6 nm and also a large pore volume of 0.91 ml g^{-1}. The authors also showed that the addition of 1-butanol during the synthesis is very beneficial for pore size distribution. It significantly narrowed the distribution in comparison to the material prepared without additives, thus improving the overall quality of the PMO.

Van Der Voort and coworkers [32] expanded the research on ethene PMO materials further. The authors were the first to report the synthesis of a 100% pure trans ethene PMO material [32a]. They could considerably reduce the synthesis time and still retain excellent structural properties of the PMO material with a specific surface area of >1000 m^2 g^{-1}. Additionally, it was also possible to vary the ratio of *cis* and *trans* configurations [32f]. Distinct differences between the resulting PMOs were detected by examining the porosity; that is, surface area, micropore volume, and pore wall thickness (Figure 7.7) [32d, f].

A few studies have investigated the control of the morphology of the ethene PMO. For example, spherical particles with molecular-scale periodicity were obtained by

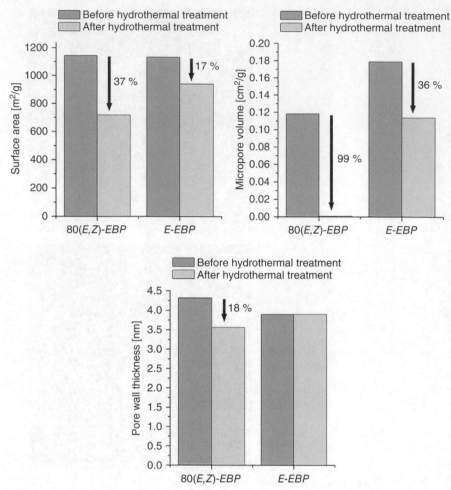

Figure 7.7

Influence of the isomeric configuration on the hydrothermal stability of the ethene PMO: comparison between a PMO prepared with 80% *trans-* and 20% *cis-* (80(*E,Z*)-ethenylene bridged PMO (EBP)) and 100% *trans* (E-EBP) ethene bonds. Source: Reproduced with permission of ACS[32f].

employing C_{12}TMABr in highly basic conditions [33]. Furthermore, it is possible to achieve other morphologies by adjustment of the synthesis conditions and composition (Figure 7.8) [30a].

7.2.3.2   *The Benzene PMO*

Benzene PMOs or phenylene bridged PMO materials have been extensively investigated by various research groups. Investigations are still on-going since the moment that Ozin's group reported this type of PMO material with an aromatic bridge [34]. This PMO is normally very easy to prepare and offers functionalization possibilities (see Section 7.4). Additionally, high specific surface areas can be obtained. For example, Ozin et al. could produce PMO materials with a high specific surface area of 1365 m^2 g^{-1} and pore sizes of 2 nm. They used quite mild acidic conditions and the ionic surfactant CPCl next to the precursor BTEB. Large pore materials were

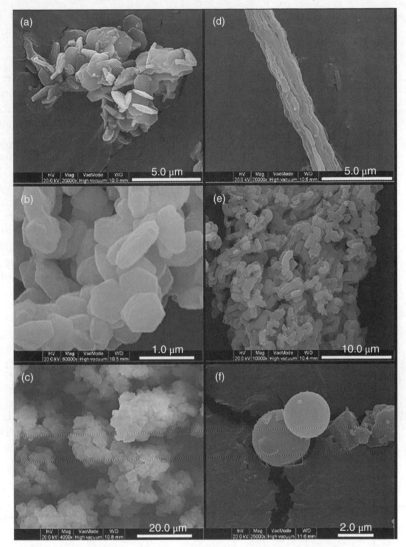

Figure 7.8
SEM images of different ethene PMO particles obtained depending on the synthesis conditions and composition. Source: Reproduced with permission of the RSC [32e].

accomplished using P123 for a 2D-hexagonal and F127 for 3D-cubic systems with pore sizes of 6.0–7.4 nm [35] and 8 nm [36], respectively.

The benzene bridges of this PMO give the material some special properties. For example, one publication briefly mentioned the occurrence of π-π stacking between the benzene moieties and describes some degree of ordering in the pore walls [37]. Up until that moment, it was assumed that PMOs are completely amorphous materials without any crystallinity. One year later, Inagaki et al. reported the first benzene PMO material with crystal-like pore walls with the aid of OTAC, (see Table 5.3) under basic conditions [38].

The material exhibited a S_{BET} of 818 m^2 g^{-1} and pore size of 3.8 nm. Moreover, it displayed a structural periodicity of 7.6 Å along the direct channels caused by the alternation of silica and benzene moieties. This alternation created hydrophobic and hydrophilic layers within the same material. It is said that this molecular-scale periodicity exists due to hydrophobic and hydrophilic or π-π interactions of the BTEB, which helps self-assembly during synthesis [39].

As the example was set up with the benzene PMO, the research was extended by the preparation of PMO materials with different aromatic bis-silanes. Precursors such as 1,4-bis(triallylsilyl)benzene [40], 1,3-bis(triethoxysilyl)benzene [41], 1,4-bis(diallylethoxysilyl)-benzene, and 1,4-bis-(triallylsilyl)benzene [42] gave rise to PMOs, with and without molecular-scale periodicity. The presence of the benzene moiety raises interesting functionalization opportunities (see Section 7.4).

7.2.4 PMOs with Multi-Organic Bridges

A multi-organic bridged PMO material was first reported by Ozin and coworkers [43] and later more studied in detail by our research group [44]. This PMO was based on the precursor 1,1,3,3,5,5-hexaethoxy-1,3,5-trisilacyclohexane (Scheme 7.1; compound **1**), which is commercially available at the moment and can be synthesized according to the reaction presented in Scheme 7.2. The resulting PMO material contains $[Si(CH_2)]_3$ rings [45].

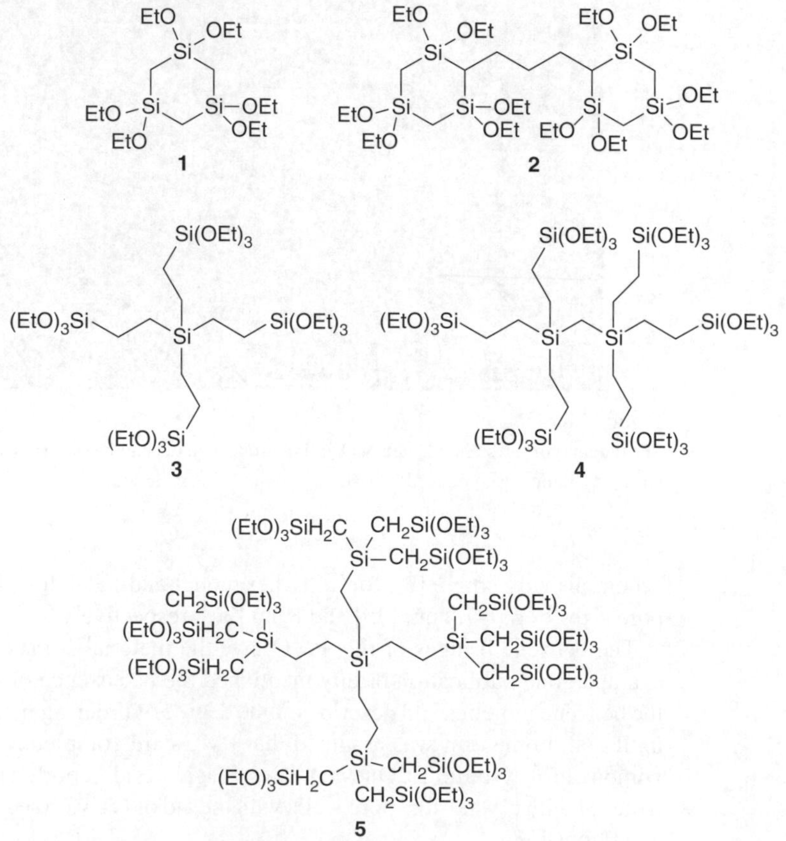

Scheme 7.1
Overview of different PMO precursors with multi-organic bridges.

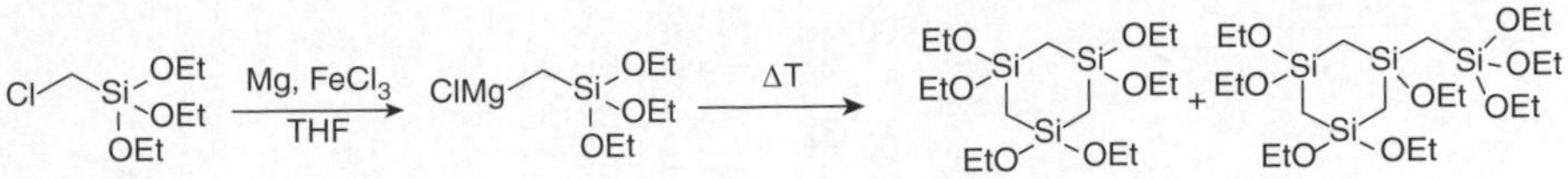

Scheme 7.2
Schematic overview of the synthesis of the precursor 1,1,3,3,5,5-hexaethoxy-1,3,5-trisilacyclohexane.

The PMO synthesis with the resulting silicon oil and the surfactant Brij-76 in acidic conditions produced highly ordered materials with specific surface areas and pore volumes of $1000–1120\,m^2\,g^{-1}$ and $0.83–1.02\,ml\,g^{-1}$, respectively. During the synthesis, KCl was added to improve the quality of the multi-organic bridged PMO material, also denoted as ring PMO. Moreover, a higher hydrothermal, mechanical, and chemical stability was observed in comparison with ordered mesoporous silicas (see Section 7.3.2) [44a].

These PMO materials (compounds **1–5**) have a huge advantage as they contain a high C:Si ratio with organic groups that are directly incorporated in the silica framework [43, 46]. This results in an enhanced stability and hydrophobicity. Organosilane **2** (Scheme 7.1) can be prepared out of organosilane **1** via a lithiation process with *t*BuLi in THF (tetrahydrofuran) at −78 °C [47]. This material showed a high thermal stability up to 400 °C under an inert atmosphere. Compounds **3–5** are the precursors used for the so-called periodic mesoporous dendrisilicas or PMDs. They are dendrimers with trialkoxy silyl groups and result in very inert organic bridged PMOs. These materials are mainly investigated as drug delivery systems [43].

The lithiation procedure to prepare compound **2** has been further exploited and modified to expand the applicability of these multi-organic bridges with exceptional hydrophobic properties. Organosilane **1** (Scheme 7.2) can be modified with a dangling allyl functionality. The resulting PMO material is often referred to as allyl ring PMO material [45a]. The functionalization is performed as generally described in Scheme 7.3. The resulting organosilane precursor can be used to prepare the very stable allyl ring PMO. The dangling allyl function is very useful as it can be functionalized with a plethora of organic reactions (see Section 7.4.2).

Scheme 7.3
Synthesis of the allyl ring precursor.

7.3 General Properties of PMOs

7.3.1 Pore Size Engineering

An important aspect of tailoring a PMO material is the ability to enlarge the pore size. This is also called *pore size engineering*. Particularly for applications such as the immobilization or adsorption of proteins, enzymes, or drugs, this is of major importance. Large molecules must be able to enter the pores smoothly without any hindrance. The use of block copolymers and poly(alkylene oxides) has already

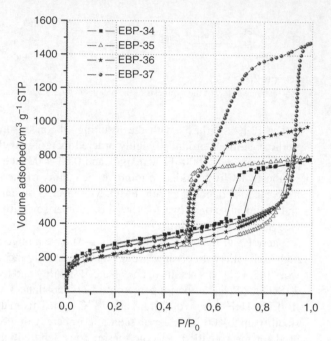

Figure 7.9

Nitrogen adsorption desorption isotherms of EBP-34, EBP-35, EBP-36, and EBP-37 prepared with the addition of 0, 1.73, 2.59, and 7.76 mmol TMB, respectively. Source: Courtesy of Carl Vercaemst.

resulted in larger pores (see Section 7.2). However, values above 10 nm cannot be achieved by purely changing the SDA.

The addition of swelling agents to the reaction mixture can provide an answer for this issue. These additives are typically hydrophobic compounds and will interact with the surfactant by settling in the hydrophobic part of the polymer. An expansion of the core of the surfactant will occur and this results in larger pores. Typical swelling agents are 1,3,5-trimethylbenzene (TMB), 1,3,5-triisopropylbenzene (TPB), and cyclohexane, but xylene, toluene, and benzene are also reported.

This method is applied for ethene PMOs with TMB as swelling agent [32c]. Several PMOs were prepared by the addition of TMB just after completely dissolving the SDA P123. The amount of swelling agent was varied and a pore expansion up to 28.3 nm was accomplished. The effect of TMB can be clearly noticed in the nitrogen sorption isotherms (Figure 7.9) as the shape and the size of the hysteresis changes significantly. Not only did an enlargement of the pores occur, but also the pore volume reached an extremely high value, namely 2.25 ml g^{-1}. Van Der Voort also succeeded in developing PMO materials with a bimodal pore size distribution (Figure 7.10). The resulting PMO exhibited pore sizes of 8.1 and 21.3 nm.

Kruk and coworkers utilized cyclohexane in combination with P123 to expand the pores. They noticed that a decrease of the initial temperature during the synthesis expanded the pores significantly. Therefore, with initial temperatures of 10–15 °C, a pore expansion of 11.9 up to 22.0 nm was achieved [48]. Low temperatures will benefit the penetration of the swelling agent in the micelle cores. Additionally, it also increases the hydrophilicity of the PEO and PPO units of the SDA such as P123 and F127 [49]. He observed that the type of bridge drastically affected the formation of the framework and the unit-cell size.

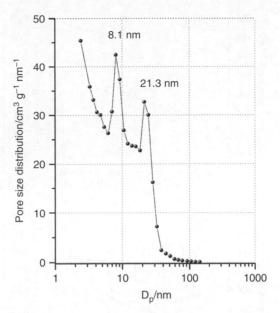

Figure 7.10

Pore size distribution of a bimodal PMO material (calculated from the adsorption branch of the isotherm with the Barrett, Joyner, Halenda [BJH] method). Source: Courtesy of Carl Vercaemst.

Van Der Voort, following the methods to create the PHTS (Plugged Hexagonal Templated Silica) for pure silica (see Chapter 5) was able to repeat this in the case of the ethene PMO. Materials going from fully open materials to fully closed variants could be prepared [32d] (Figure 7.11).

7.3.2 (Hydro)thermal and Chemical Stability

The type of PMO bridge also determines the (hydro)thermal stability of the entire material. The overall stability of PMO is a very important feature that mainly determines the applicability of the material in applications such as catalysis, adsorption, chromatography, and so on. For example, a PMO material that decomposes after one catalytic run is useless.

7.3.2.1 Thermal Stability

The organic bridges can decompose when a heat treatment is applied. The stability highly depends on the type of organic bridge and the atmosphere used during calcination (air or inert). It is also the reason why the removal of the SDA is preferably performed via an extraction or with a very mild heat treatment under inert atmosphere. Generally, benzene PMO materials are the most thermally stable and they can withstand temperatures up to 500–570 °C in air. The thermal stability of ethene and ethane PMO materials are comparable and are only limited to 200–300 °C in air. An overview of the upper temperature limits reported for PMOs with simple organic bridges is presented in Table 7.2. Additionally, it seems that the symmetry of the PMO does not influence the thermal stability [51].

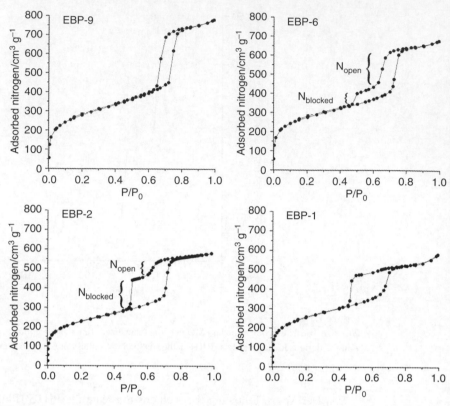

Figure 7.11

N$_2$-isotherms of ethene PMOs with different fractions of blocked mesopores. EBP-9 consists exclusively of open mesopores, EBP-6 has a relatively small fraction of blocked mesopores, EBP-2 has a relatively large fraction of blocked mesopores and EBP-1 consists mainly of blocked mesopores. Source: Reproduced with permission of ACS [32d].

Table 7.2 Overview of the thermal stability of PMOs with simple organic bridges in air.

	T$_{air}$ (°C)	References
Ethane	200–300	[18b]
Ethene	200–300	[30b]
Methane	400	[27]
Benzene	500–570	[50]

7.3.2.2   *Hydrothermal Stability*

Next to the thermal stability, also the hydrothermal stability is important. This is the stability of the PMO material in aqueous environment under elevated temperature. In many cases, the stability is related to the hydrophobicity of a material as hydrophobic materials repel water and this avoids hydrolysis of Si—O—Si bonds on the surface. Thicker pore walls are also beneficial for the hydrothermal stability of a PMO material [3, 4, 38].

Markowitz and coworkers determined the hydrophilicity of several PMOs with simple organic bridges and compared them with silica. They found the following sequence with increasing hydrophilicity: ethane > benzene > methane > silica [52]. The PMO materials underwent a hydrothermal treatment of 18 hours in boiling water and showed no structural degradation. The hydrophobic character of the material improves the mechanical stability as well. The same was observed by other research groups when comparing PMOs and their silica counterparts [50]. The study was extended for ethane PMOs that were treated for 60 days. Here, also, no degradation was observed [53].

The hydrothermal and mechanical stability of ethene and ethane PMO materials was compared with a steam treatment of three days at 105 °C [44a]. The ethane PMO turned out to possess a higher stability as the ethene bridged PMO showed a significant decrease in specific surface area, pore volume and micropore volume. The same trend was observed for the mechanical stability. Hydrophobization with 1,1,1,3,3,3-hexamethyldisilazane (HMDS) greatly improved the stability of the ethene PMOs and no further differences were observed between the two materials. This clearly proved that the hydrophobicity is of paramount importance.

7.3.2.3 *Chemical Stability*

The multi-organic bridged $[Si(CH_2)]_3$ ring PMO is also an example of a very stable material [44a]. The high number of —CH_2— bridges leads to an increased hydrophobicity of the material. It exhibits a pH stability up to 14. Additionally, the allyl ring PMO material introduced in Section 7.2.4, and based on the ring PMO, shows a remarkable stability in a high pH range between 1.75 and 12, and a temperature resistance up to 150 °C in 60 vol% of water [45a] (Figure 7.12).

7.3.3 Metamorphosis in PMOs

The concept of "metamorphosis" in this context comprises of a transformation of the bridging groups of PMOs under the influence of an external factor and it is also referred to as a framework rearrangement. The bridges are converted into terminal organic functionalities. The most important factor is probably temperature.

This phenomenon was first described by Ozin [27]. Methane PMOs were treated at 400 °C in air or inert atmosphere. The Si—C bonds of the bridges were cleaved and end-standing —CH_3 functionalities were formed. It was assumed that a proton transfer from silanols or residual water to the —CH_2— moiety took place. Similar transformations were observed for other PMO materials. The bridges of benzene PMOs were converted into pendant phenyl and oxidized phenyl moieties when treated at 400 °C in humid conditions or an oxidizing atmosphere [54]. The treatment of ethene PMOs resulted in pendant vinyl groups using different thermal conditions [32b, 55]. The structural properties of the material were still good, although the thermal treatment partially destroyed the PMO. A general decrease was observed of the S_{BET}, V_p, and d_p from 1112 to 795 $m^2\,g^{-1}$, 1.17 to 0.85 $ml\,g^{-1}$, and 8.1 to 7.0 nm, respectively. The X-Ray Diffraction (XRD) patterns before and after treatment also showed that the material was hexagonally ordered. Additionally, metamorphosis was also observed for the ring PMO material. Here, dangling methyl functional groups were formed with a temperature treatment of 450 °C under nitrogen atmosphere. This

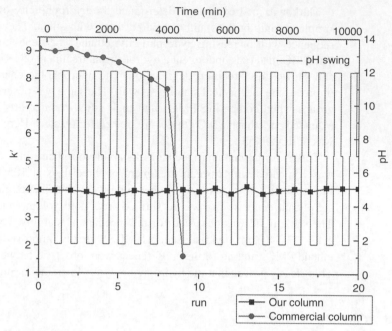

Figure 7.12

Evaluation of the pH stability of the allyl ring PMO material in a HPLC set-up. A pH swing experiment was performed with high pH (pH = 12, 20 mM of triethylamine) and low pH (pH = 1.75, 1 vol% of formic acid). Comparison was performed with a commercial column (ethylene bridged hybrid 2.5 mm X-Bridge C18 column). Source: Reproduced with permission of the RSC [45a].

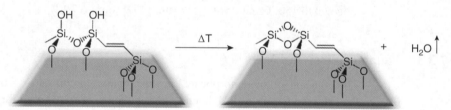

Scheme 7.4

Metamorphosis of an ethene PMO.

procedure is also called auto-hydrophobization or self-hydrophobization due to the addition of hydrophobic methyls to the surface of the material (Scheme 7.4).

These examples, describing the (hydro)thermal and chemical stability, clearly show that the hydrophobicity of a PMO material is of major importance. This relationship originates from the structural degradation by hydrolysis of siloxanes under the influence of water. A more hydrophobic material prevents this and thus exhibits a higher mechanical stability. Furthermore, the hydrophobicity combined with thicker pore walls further improves the stability in comparison with silica materials.

7.4 Post-Modification of PMOs

The PMO bridges described previously contain rather small and inert organic moieties. These materials are very interesting for applications where only their inherent properties (e.g. hydrophobicity, stability) are needed. Chromatography and low-*k* insulators are two examples of this kind of applications. However, for more advanced applications such as catalysis and adsorption, more advanced functionalities with heteroatoms are needed. They can be introduced via the post-modification of a simple organic bridge (see Section 7.4.1) or they can already be integrated during the synthesis. The direct inclusion of heteroatoms or (complex) functionalities in PMOs can be very straightforward when using the appropriate precursor. Many precursors are already commercially available at different chemical suppliers or they can be synthesized via procedures published in literature. Several concrete examples of the latter will be given in Section 7.5. A more extensive overview can be found in Reference [8].

Sometimes, it is more useful to post-functionalize a PMO material. Many precursors that include heteroatoms are too large and flexible to assemble a rigid PMO framework and are thus difficult to incorporate. One way to circumvent this is the addition of a small bis-silane (e.g. BTEE), tetraethoxysilane (TEOS), or tetramethoxysilane (TMOS) to increase the rigidity of the walls. Alternatively, the post-functionalization of the organic bridge is at certain moments a more useful method that allows maintaining the high ordering of the materials while introducing the appropriate function. Different post-functionalization methods are described in the following sections.

7.4.1 Post-Functionalization of the Unsaturated Bridges

The most common unsaturated carbon bridge of PMO materials is the ethene bridge. Less common is the butane bridge [56]. A whole array of organic reactions can be performed to convert the double bond on the surface to a more appropriate functionality.

7.4.1.1 *Bromination of the Double Bond – Nucleophilic Substitution*

One of the most familiar reactions is the bromination of the C=C (Scheme 7.5). The addition of bromine is already elaborately described by different research groups such as Stein and coworkers [4], Ozin and coworkers [2], Sayari and coworkers [13], Mokaya and coworkers [30a], Yoshitake and coworkers [57], and Van Der Voort and coworkers [32f]. The reaction is very straightforward and performed with gaseous bromine or bromine dissolved in a chlorinated solvent (e.g. dichloromethane). Although the procedure is easy, it must be executed with extreme care due to the irritating bromine and the volatile solvents. In general, bromination with gaseous bromine instead of dissolved bromine in dichloromethane results in a higher yield. Bromination yields between 10 and 50% were reported depending on the bromination method, but also on the preparation of the PMO material itself. This procedure

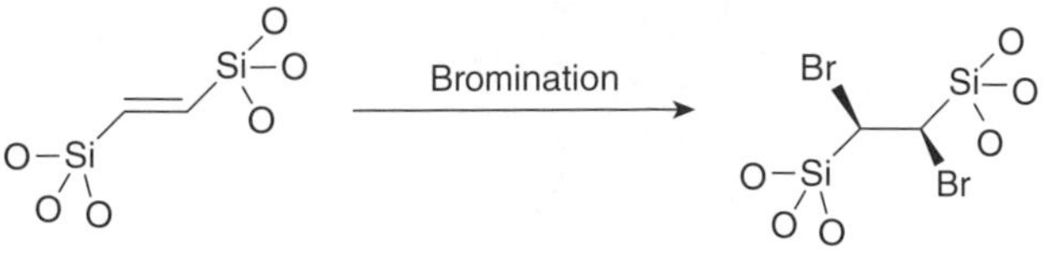

Scheme 7.5
General representation of the bromination of the ethene PMO material. Different synthesis conditions can be used, for example, gaseous bromine or Br_2 dissolved in dichloromethane.

did not have a large effect on the mesostructure of the material. The specific surface area, pore volume, and pore diameter decreased only because of the modification. The general ordering and mesopore structure are maintained in all cases.

Abdelhamid Sayari obtained his Ph.D. in the field of heterogeneous catalysis at the University of Tunis and the Université Claude Bernard in Lyon (France). After this, he performed a postdoctoral fellowship at the University of Pittsburgh (USA) and joined the National Research Council of Canada. Sayari became Professor of Chemical Engineering at Laval University (Quebec City) in 1990. In 2001, he obtained a mandate as Canada Research Chair to set up the Centre for Catalysis Research and Innovation (CCRI) at the University of Ottawa (Canada). He is now Full Professor of Chemistry and Chemical Engineering at the same university. He specialized in inorganic materials such as zeolites and PMOs in the fields of catalysis and adsorption. He has several patents in the field of CO_2 capture and more than 250 research publications and book chapters.

Photograph: https://science.uottawa.ca/chemistry/people/sayari-abdelhamid

   The bromine atom is an excellent leaving group and can be substituted by more useful functionalities via a nucleophilic substitution reaction. One method proposed in literature is the substitution of bromine with a diamine. Yoshitake and coworkers used ethylenediamine as nucleophile to obtain a nitrogen-containing material (Scheme 7.6)

Scheme 7.6

An example of the replacement of the Bromine atom via a nucleophilic substitution reaction with a diamine. EDA = ethylene diamine. Reaction conditions: toluene at 60 °C for 12 hours.

[57]. A total amount of 0.87 mmol of amine groups per gram was achieved and was used as arsenate adsorbent.

The bromination of the ethene bond of a PMO material can also have a different purpose. It can probe the number and chemical accessibility of the ethene bonds of the material.

7.4.1.2 Epoxidation of the Double Bond

Inagaki and coworkers [58] optimized the epoxidation reaction to functionalize the C=C bond of a PMO material (Scheme 7.7). They investigated several oxidation procedures such as temperature, pH, and different oxidants. The authors concluded that the epoxidation is best performed under mild basic conditions with anhydrous tert-butylhydroxyperoxide (THBP) as oxidant.

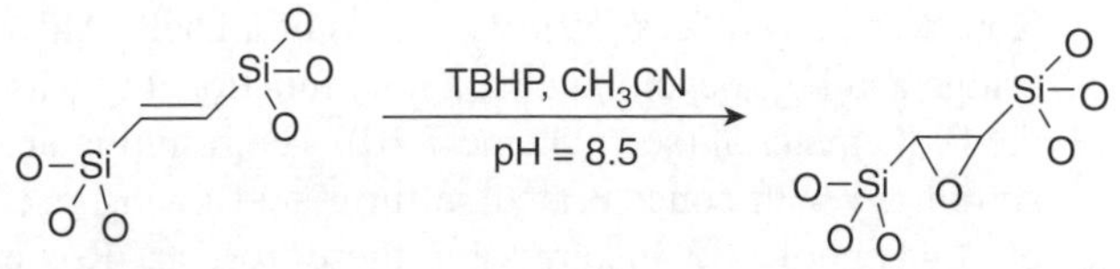

Scheme 7.7
Epoxidation reaction of the ethene PMO material with *tert*-butylhydroperoxide as oxidant.

The resulting oxirane can be further modified via a standard ring-opening reaction. In the research of Inagaki mentioned previously, the PMO was further modified with an aqueous ammonia solution resulting in the introduction of an amine and alcohol function (Scheme 7.8).

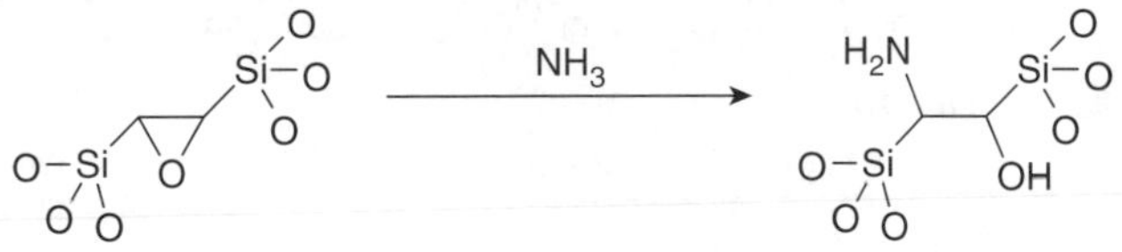

Scheme 7.8
Ring-opening reaction of the oxirane with an aqueous ammonia solution.

7.4.1.3 Ozonolysis of the Double Bond

Ozonolysis can be performed on double bonds to perform a cleavage of the bridge and obtain carboxylic acid functionalities (Scheme 7.9). Unfortunately, this method is rather destructive and a partial breakdown of the organic moieties on the surface occurs. This implies the creation of small indentations inside the pore walls where the

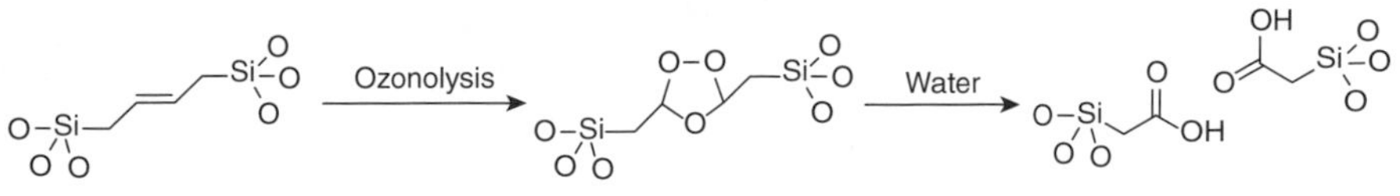

Scheme 7.9
Ozonolysis of the double bond of the butene PMO.

bonds are cleaved. Consequently, the functionalities are embedded in the indentation of the pore wall instead of being added on top of the surface. This leaves the pores open and unhindered. This procedure was reported by the group of Polarz on a butene PMO material with ozone [56].

Unfortunately, the described procedure cannot be applied on the ethene PMO material. It was reported by Polarz et al. that attempts to ozonolyze the C=C resulted in the complete removal of the organic bridge. The PMO was converted into a material that retains its hexagonal ordering, but a general shrinkage of the material was observed. It was concluded that the ozone treatment was impossible due to the α-position of the siloxane moieties in position to the double bond.

7.4.1.4   Diels–Alder Reaction with the Double Bond

An interesting and versatile modification of ethane PMOs was first reported by Kondo and coworkers [59]. By means of a Diels–Alder reaction between the ethene bridge and benzocyclobutene, a new functionality was successfully incorporated in the PMO material (see Scheme 7.10). The resulting aromatic ring was subsequently sulfonated with concentrated sulfuric acid giving rise to materials with an acidity of $1.44\,\text{mmol}\,\text{g}^{-1}$. Inspired by these results, Romero-Salguero and coworkers significantly extended this method with other dienes; for example, antracene and cyclopentadiene [60].

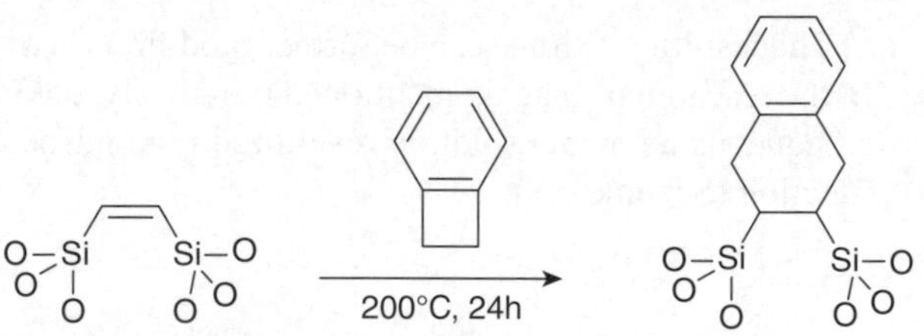

Scheme 7.10
Diels–Alder reaction with benzocyclobutene and the ethene PMO material.

7.4.1.5   Friedel–Crafts Reaction

Another post-modification method performed on the ethane bridge was developed by Dubé et al. and involved a Friedel–Crafts reaction [61]. The ethene bond was arylated with benzene and $AlCl_3$ as a catalyst (see Scheme 7.11). This material was subsequently sulfonated to produce an acid catalyst.

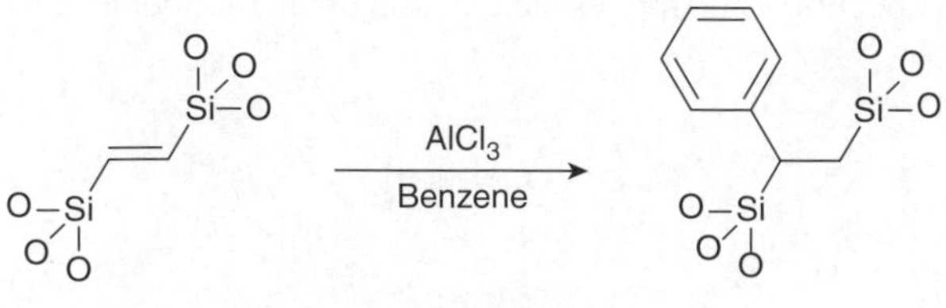

Scheme 7.11
Friedel–Crafts reaction with benzene and the ethene PMO material.

7.4.1.6 *Thiol-Ene Click Reaction*

A very interesting post-modification is the reaction between thiols and alkenes or alkynes. This reaction is also called a thiol-ene or thiol-yne click reaction, respectively. Click reactions are typically very efficient reactions, with high yields, mild reaction conditions, and produce no or a limited amount of waste [62]. They fit perfectly in the well-known concept of Green Chemistry [63]. Van Der Voort described the photo-initiated reaction between an ethene PMO and a thiol group recently. Cysteamine and the amino acid cysteine are anchored via the C=C of the ethene PMO resulting in a thioether linkage (Scheme 7.12) [64]. These materials were applied in catalysis for the aldol condensation of 4-nitrobenzaldehyde and acetone.

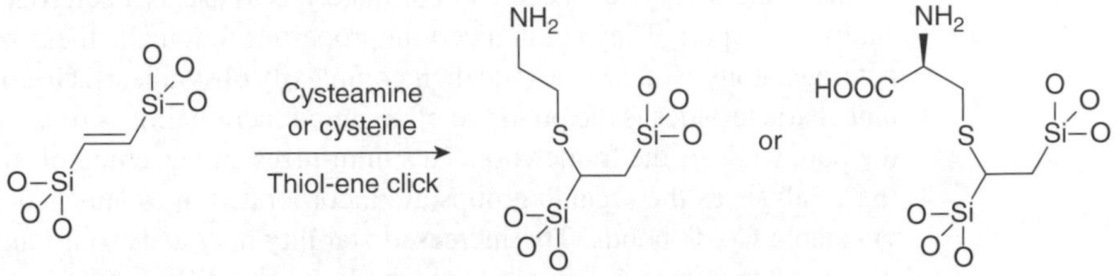

Scheme 7.12
Thiol-ene click reaction on an ethene PMO material.

A similar reaction is also performed on the allyl ring PMO with 2-mercaptoethanol and 1,2-ethanedithiol [45b] for the green oxidation of cyclohexanol and (±)-menthol.

7.4.2 Post-Modification of the Aromatic Ring

The post-functionalization methods, described previously, all concern the modification of olefinic bonds. However, aromatic bridges have also been functionalized. For example (Scheme 7.13), benzene PMO material was sulfonated to obtain sulfonic acid groups [38, 54]. Moreover, amines were introduced by two subsequent functionalizations; that is, a nitration of the benzene moiety followed by a reduction toward an amine [65]. Bromination of the benzene ring has also been attempted [66]. The authors noticed unwanted side reactions as breakage of the Si—C bonds occurred when Br_2 and $NaBrO_3$ was used on the PMO material.

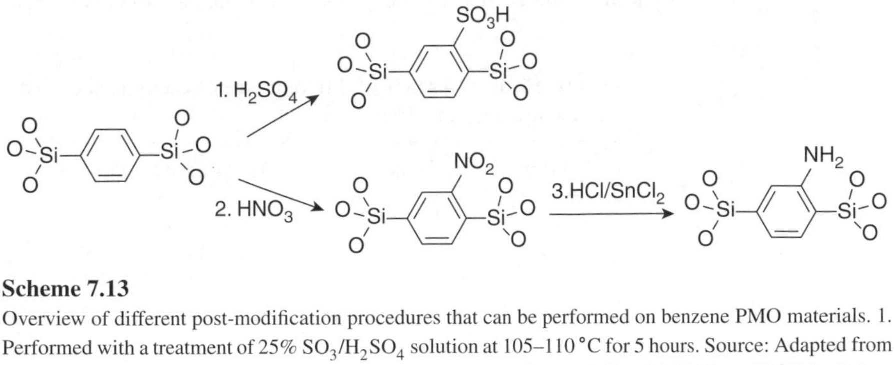

Scheme 7.13
Overview of different post-modification procedures that can be performed on benzene PMO materials. 1. Performed with a treatment of 25% SO_3/H_2SO_4 solution at 105–110 °C for 5 hours. Source: Adapted from Reference [38]. 2. Performed with a treatment of 96 vol% H_2SO_4 and 69 vol% HNO_3 at 27 °C for 3 days [65]. 3. Performed with 37 vol% HCl and $SnCl_2$ at 27 °C for 3 days.

7.5 Applications of PMOs

7.5.1 As Heterogeneous Catalysts

7.5.1.1 Synthesis Methods

One of the most researched applications for PMOs is catalysis. The number of literature reports discussing the development and application of PMOs in this specific research area has increased enormously in previous years. The need and interest for heterogeneous catalysts with high activity and selectivity is growing tremendously for economic and ecological reasons. Especially now, when *green and sustainable chemistry* is becoming very important.

PMOs are a very advanced class of materials to use as a heterogeneous catalyst or catalytic support. They contain certain properties that make them very interesting as heterogeneous catalysts, despite their complexity of synthesis. One of the most important characteristics is the presence of organic functionalities completely embedded in the pore walls of the framework. This minimizes any leaching of functionalities during catalysis as the organic groups are incorporated in or attached to the framework via stable C—C bonds. The increased stability also adds to a longer lifetime. Furthermore, they are mesoporous, which allows the diffusion of larger substrates and products in and out of the pores. On the downside, the catalytic activity of heterogeneous catalysts is lower than their homogeneous counterpart most of the time, as reactions take place at the liquid–solid or gas–solid interface. This can be countered by a large surface area. The latter in combination with a high amount of available bridges ensures that many functionalities are accessible. Additionally, PMOs also exhibit a certain hydrophobicity due to organic bridges.

Catalysis can only be performed when a catalytic site or function is present. One can distinguish several types of catalytic sites and some are more complex or advanced than others (see Table 7.3). In particular, the (chiral) organometallic complexes and (chiral) organocatalysts require more attention due to their complexity and special conformation.

Different methods can be used to develop a PMO-based heterogeneous catalyst. They are all based on the well-known principles to introduce organic functionalities in a silica material: co-condensation and grafting. An overview is presented in Figure 7.13 and each type is briefly discussed next.

Method 1: The synthesis of PMOs using polysilsesquioxanes with bulky and flexible organic bridges is often an issue. It is difficult to obtain a well-ordered and highly porous material as the polysilsesquioxane cannot easily organize itself around

Table 7.3 Overview of different type of catalytic sites that can be introduced in a PMO material.

Non-metal-based catalyst	Metal-based catalyst
Acidic	Metal sites
Basic	Metal nanoparticles
Organocatalysts	Organometallic complexes
Chiral organocatalysts	Chiral organometallic complexes

A difference between metal containing and non-metal containing catalysts is made. A more detailed discussion per type of catalytic site can be found later in this chapter.

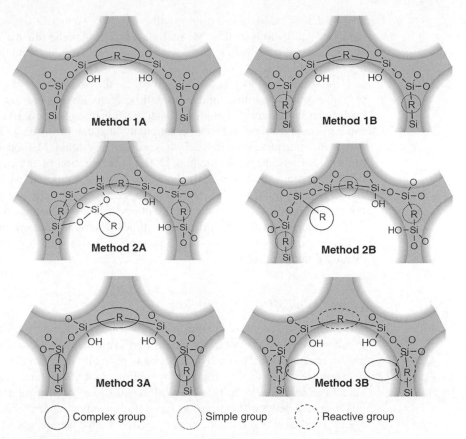

Figure 7.13

Overview of different methods to introduce (catalytic) functionalities in a PMO material. Method 1A:
complex bis-silane co-condensed with TEOS/TMOS; Method 1B: complex bis-silane co-condensed with
a simple bis-silane; Method 2A: grafting of dangling organosilane on a PMO with simple bridges; Method
2B: co-condensation of dangling organosilane with a simple bis-silane; Method 3A: PMO prepared with
complex bis-silane; and Method 3B: PMO prepared with a simple bis-silane that can be further adapted
into the catalytic function.

the SDA. Therefore, the co-condensation with a simple precursor can be used. This
can be the pure silica source TEOS or TMOS (Method 1A), but also the precursors
BTME, BTEE, or BTEB (Method 1B). Strictly speaking, materials prepared via the
co-condensation of a bis-silane with TEOS or TMOS are not PMOs. They possess
the typical silica structure with some organic functionalities distributed throughout
the material. They mainly lack the increased stability and hydrophobicity of true
PMO materials. Additionally, the catalytic function is incorporated inside the pore
walls, which can hinder the conformation of the catalyst. As a result, the catalytic
activity and/or selectivity can be decreased or even lost.

Method 2: The catalytic functionality can also be anchored via the grafting
of an organosilane on a PMO prepared with simple bridges (e.g. ethane, ethene,
benzene) (Method 2A). Or the same dangling catalytic moiety can be obtained via
the co-condensation of the simple bis-silane with the organosilane (Method 2B). The
resulting materials can be called PMOs and the catalytic function dangles inside
the pores. This is probably the biggest advantage of using this preparation method.
The catalysts are unhindered by the surface or have rotational freedom. There is also

a huge drawback when using this synthesis route. Method 2A materials are more vulnerable to hydrolysis than Method 2B materials, as the organic functionality is attached via siloxane bonds. However, in quite mild reaction conditions this should not be a problem.

Method 3: The last method is probably the most interesting one. Here a polysilsesquioxane with a rigid and small bridge is used as the sole precursor. The bridge can immediately contain the catalytic function (Method 3A). Or, the bridge can also contain a rigid and small functionality that can be further modified (Method 3B). Both materials can be called PMOs without any doubt. Method 3A suffers from the same disadvantages as Method 1. The incorporation of the catalytic function inside the pore walls can force the functionality in to a different conformation. This can negatively influence the catalytic activity and selectivity. Method 3B, on the other hand, is probably the most important one and has gained much interest lately.

It must be noted that the different methods described here are not only limited to the preparation of catalysts. These methods are also valid for other applications discussed in this chapter.

7.5.1.2 *Acid Catalysis with PMOs*

Mostly Brønsted acid sites are introduced for this type of catalysis. The sulfonic acid group or -SO_3H is the most investigated one and different routes can be used to obtain a solid acid catalyst. The catalytic activity is generally influenced by an entire range of parameters, such as hydrophobicity/hydrophilicity, structure, acid strength, the amount of acid site, stability of the PMO, and so on. A clear relationship between all these parameters and the activity has unfortunately not been established yet [67]. So, trial and error is still a widespread practice when developing acid PMO catalysts for a certain catalytic reaction.

Figure 7.14 gives a non-exhaustive overview of sulfonic acid containing PMOs. Normally, Method 2 (A and B) and Method 3B are commonly used. The preparation of the materials a and b (Method 2B) is probably the easiest as it only requires the co-condensation of the right precursors. In some cases, the oxidation of a thiol function is needed that originates from the organosilane 3-mercaptopropyltrimethoxysilane or 3-mercaptopropyltriethoxysilane. Materials c–e are prepared via Method 2A and again sometimes require an additional oxidation step. The materials prepared via Method 2B are generally more stable than the materials prepared via Method 2A [68]. The synthesis and post-functionalization of the materials in f–m (Method 3) is more complicated as, first, a well-structured PMO with the right properties must be synthesized. Afterward, the functionalization with the sulfonic acid function must be achieved with retainment of the structure of the PMO. Proper and in-depth characterization of the material must be performed to verify this. However, once the PMO is prepared, a stable acid catalyst is usually obtained.

The synthesis and subsequent post-modification can be very tedious, especially when a high number of reaction step is required. A successful attempt is reported to limit this. A PMO was synthesized with the home-made 1-thiol-1,2-bis(triethoxysilyl)ethane where the thiol group was oxidized in situ [69]. This was performed by merely adding H_2O_2 during the synthesis. Another advantage of this procedure is the tunability of the functional loading. An adjustment of the thiol to ethane ratio simply changes the loading. For example, an acidity of 0.85 mmol H^+ g^{-1} was obtained by mixing 50–50% thiol and ethane precursor. The material was applied in the esterification of acetic acid and benzyl alcohol and compared to the catalytic performance of the benchmark material Amberlyst 15

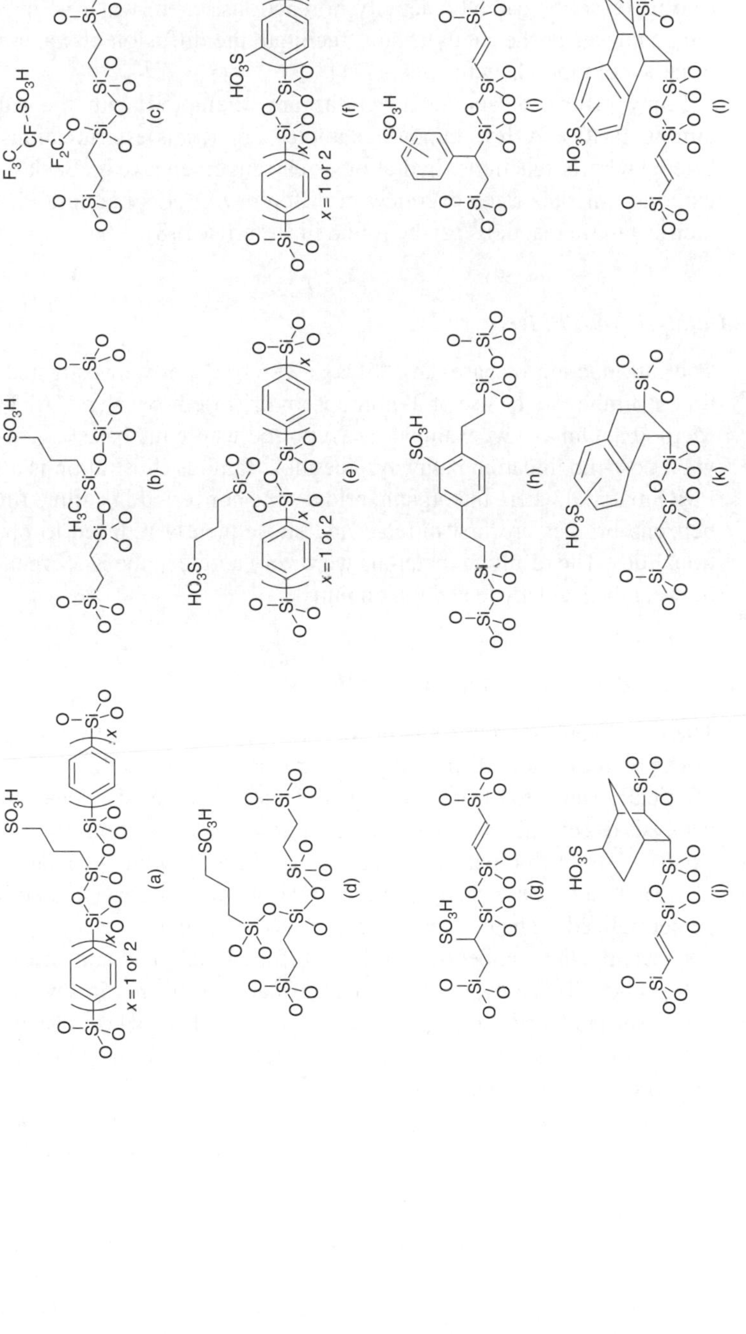

Figure 7.14

Overview of different sulfonic acid group containing PMO materials. The materials are prepared either via Method 2 (A and B) or Method 3B.

(acidity = 4.8 mmol H$^+$ g^{-1}). The PMO showed an impressive turn over frequency (TOF) value of 334 h^{-1} while the benchmark only reached 72 h^{-1}. The authors attributed this high activity to the fact that PMOs do not present any swelling problems in contrast to polymers like Amberlyst 15. Moreover, the SO$_3$H PMO prepared in situ was very stable during catalysis and could be reused for five runs.

Next to their non-swelling behavior, their hydrophobicity also plays a key role. First, this hydrophobic property leads to an enhanced stability in comparison with their silica counterparts. Many examples can be found in literature where the PMO is structurally stable for several catalytic runs [59a, 67a, 70]. The hydrophobicity can also increase the catalytic activity in reactions where water is involved. Hydrophobic patches near the catalytic sites facilitate the diffusion of water out the pores and organic compounds in the pores [71].

Many different catalytic reactions are attempted with the sulfonic acid containing PMOs. A few famous examples are (trans)esterifications, etherifications, Friedel–Crafts reactions, hydrations, rearrangements (e.g. Beckmann), and dimerizations. An elaborated overview of different SO$_3$H containing PMO structures and their catalytic reactions can be found in Reference [8].

7.5.1.3 Base Catalysis with PMOs

The introduction of bases in PMOs is less frequently investigated and most of the time is limited to the use of 3-aminopropyltri(m)ethoxysilane (APTMS, APTES) via Method 2. Only a few examples are reported where the base (e.g. —NH$_2$) is inserted via a post-modification pathway. The most famous illustration is a benzene-bridged PMO material where the organic bridges are converted in aniline functions [65]. The benzene bridges are first nitrated and subsequently reduced to obtain amine functionalities. The resulting materials were very active in the Knoevenagel condensation between benzaldehyde and malononitrile.

7.5.1.4 Combined Acid and Base Catalysis with PMOs

The heterogenization of catalytic sites results in a material where the catalytic functions are completely dispersed over the surface of the solid. The functions are thus physically separated from each other. This is a huge advantage over homogeneous catalysis as several types of catalytic sites can be present on one material that cannot exist next to each other in homogeneous phase. In homogeneous catalysis, an acid (e.g. HCl) and a base catalyst (e.g. NaOH) would just react and both catalysts would be neutralized. In heterogeneous catalysis, this is not the case.

Currently, the number of reports on bifunctional (acid and base) PMO catalysts is still limited. Typically, the functionalities are introduced in several steps during the synthesis and mostly some protection steps and close pH control are involved. These precautions are necessary to avoid any unwanted neutralization reactions of the acids or based during the synthesis.

This is illustrated in Figure 7.15 where an ethene PMO is modified via two different routes [58a]. The eventual materials contain SO$_3$H and NH$_2$ functions, 1.1 and 1.3 mmol g^{-1}, respectively, and are obtained by the clever combination of co-condensation and grafting. The catalyst is employed in the one-pot deacetalization/nitroaldol condensation [72].

Another example (Figure 7.16) shows the necessity of using protection groups [73]. The sulfonation of the aromatic moiety on a PMO material can only be achieved

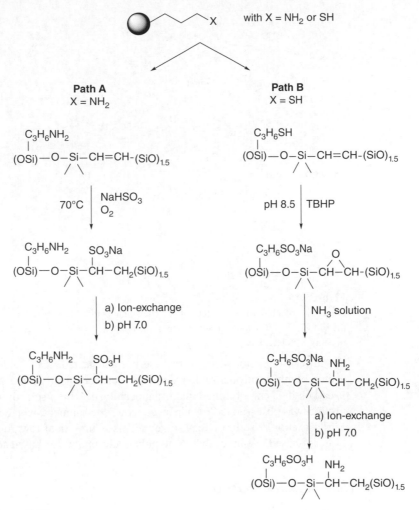

Figure 7.15

Preparation of a bifunctional PMO material containing both acid and base functions. Two different routes can be followed starting from the ethene PMO.

if the dangling amine group is protected by, for example, a *tert*-butoxycarbonyl group. Without protection, the amine group will be protonated and the sulfonation will not succeed. However, deprotection is sometimes difficult and harsh conditions are needed as can be seen in Figure 7.16.

It is clear from these examples that much attention should be given to the design of a bifunctional catalyst and especially to the sequence of introducing the different functionalities during the synthesis route.

7.5.1.5 *Organocatalysts Supported on PMOs*

The amount of research performed on the development of organocatalysts supported on PMOs is only scarce. Some amino acids (e.g. proline, serine) are widely known as organocatalysts. The anchoring of L-serine for example is performed on a benzene bridged PMO materials via a post-modification route [74] (see Figure 7.17). First, the benzene bridge is transformed in an aniline moiety (see Section 7.4.2). Afterward,

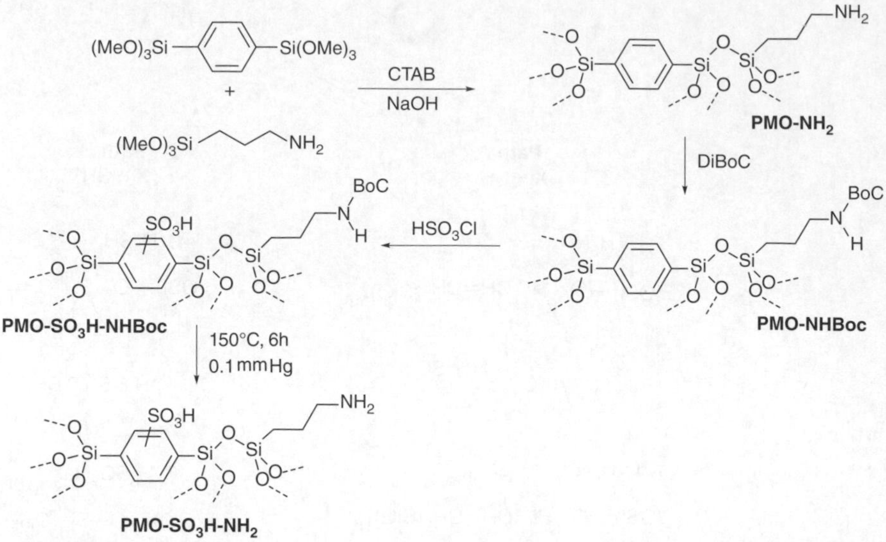

Figure 7.16

Preparation of a bifunctional PMO material containing both acid and base functions. The introduction of the sulfonic acid function, via sulfonation of the aromatic ring, can only succeed when the amine functionality is protected.

the PMO is functionalized with L-serine and tested in the aldol condensation of acetone with 4-nitrobenzaldehyde. The presence of an amine and alcohol on the β-carbon delivers a bifunctional organocatalyst. A high selectivity toward the aldol product was observed (>96%) in an aqueous environment, which is quite remarkable as most reports use hexane as solvent. This is important toward the valorization of biomass-based resources that preferably occur in a water-based solvent.

7.5.1.6   Metal Sites Supported on PMOs

Most metal sites are incorporated in PMO materials to perform redox processes. Isomorphic substitution is a very straightforward and easy method. It involves the use of hydrolyzable metal precursors that are added during the synthesis and results in the incorporation of isolated tetrahedral metal sites in the PMO. Many metals (Ti [75], Sn [76], V [77], Nb [78], Cr [79], Al [80]) are already included in PMOs, particularly ethane PMOs. The catalytic activity per metal center is generally higher due to the hydrophobic microenvironment created by the organic bridges. This positive effect is also seen on the selectivity. Most of these PMO materials were also more stable. Unfortunately, in certain cases, metal leaching of, for example, V and Cr was observed. Figure 7.18 shows a graph with the catalytic results of an ethane PMO (S6) and HMDS hydrophobized ethane PMO (S5), both with titanium metal sites. The catalytic activity in the epoxidation of 1-octene benefited greatly from the hydrophobic environment [81].

7.5.1.7   Metal Complexes Supported on PMOs

The anchoring or inclusion of a homogeneous metal complex on a solid support is probably one of the most investigated topics regarding PMOs and catalysis. Currently, many homogeneous catalysts exist for the most important reactions in fine

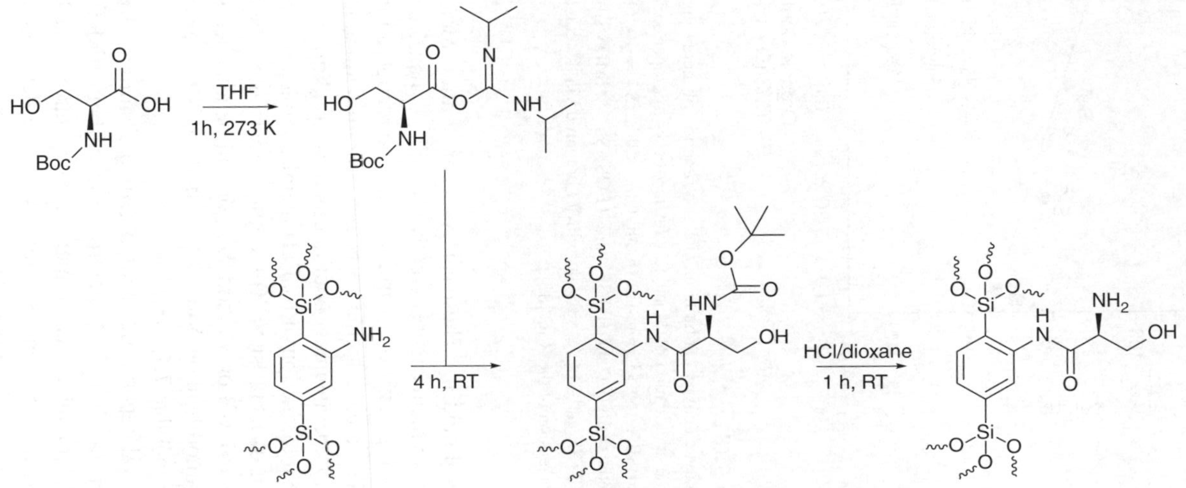

Figure 7.17

Schematic representation of the anchoring of L-serine on a benzene bridged PMO material. The material is afterwards used as a catalyst in the aldol condensation.

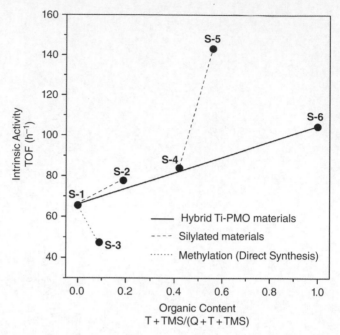

Figure 7.18

The catalytic activity per titanium site as a function of increasing organic content for the different materials in the epoxidation of 1-octene with *tert*-butyl hydroperoxide (TBHP). S1 = SBA-15; S2 = HMDS silylated SBA-15; S3 = in-situ methylated silica; S4 = PMO prepared via the condensation of 1,2-bis-(triethoxysilyl)ethane (BTSE) and TEOS; S5 = HMDS silylated PMO prepared via the condensation of BTSE and TEOS; S6 = pure PMO prepared via the condensation of BTSE. Source: Reproduced with permission of the RSC [81].

chemical industry. These catalysts can contain very selective and even chiral ligands that make them enantioselective but also expensive. The most important motives for the heterogenization, therefore, are: the recuperation of the expensive or toxic metal complex, thus reducing costs and contamination of the end-products; and the possibility to operate under continuous flow conditions.

The introduction of (chiral) metal complexes is more complicated than the introduction of metal sites. The complex can either be part of the organic bridge itself (Method 1A-B or Method 2A), or can be attached to an existing organic bridge via a post-modification process (Method 3). A few examples of both strategies are presented in Table 7.4.

Several reports are available in literature where the metal complex is part of the organic bridge. N-heterocyclic carbene (NHC) moieties are very interesting ligands for all kinds of metals. They have already proven in the past that very successful homogenous catalysts can be developed with NHCs. Now attempts are being made to heterogenize these types of ligands/catalysts. For example, palladium NHC complexes are promising catalysts for Suzuki–Miyaura coupling reactions [82]. A NHC containing bis-silane was first synthesized via several steps with IMes (N,N'-bis(2,6-dimethylphenyl)imidazol-2-ylidene) (Figure 7.19) as the organic bridge. As this bridge is relatively large, the authors chose to co-condense the precursor with BTEE, P123 and KCl in an acidic environment. Up to 20 mol% of the IMES-precursor was added to the reaction mixture and this led to PMO materials

Table 7.4 Non-exhaustive overview of (chiral) metal complexes incorporated or attached to PMO materials as heterogeneous catalysts.

Functionality	Catalytic Reaction	References
	Suzuki–Miyaura couplings of various aryl halides and phenylboronic acid	[82]
	Hydration of diphenylacetylene	[83]

(Continued)

Table 7.4 (*Continued*)

Functionality	Catalytic Reaction	References
	Cyanosilylation of aldehydes	[84]
	Asymmetric addition of diethylzinc to aldehydes	[85]

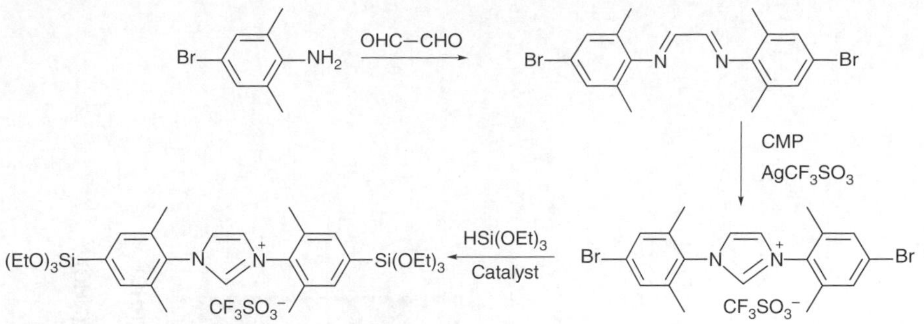

Figure 7.19

Preparation of a bis-silane with the NHC ligand IMES as organic bridge. CMP = Chloromethyl pivalate; Catalyst = Rh(cod)(CH$_3$CN)$_2$BF$_4$. Source: Reproduced with permission of ACS [82].

with a loading of maximum 0.7 mmol g^{-1} of IMES ligand. A part of the IMES was destroyed during the formation of the PMO due to the acidic environment. Afterward, Pd(OAc)$_2$ was coordinated to the material. The PMO catalyst was tested for the Suzuki–Miyaura coupling of some challenging aryl and benzyl halides (e.g. chlorobenzene) and phenylboronic acids.

The incorporation of this catalyst in a PMO not only resulted in an active catalyst (see Table 7.4), but it was also recyclable (up to eight cycles), as it avoids the formation and growth of Pd(0) nanoparticles. This normally leads to a decrease in catalytic activity. The stable PMO framework combined with the stable and bulky IMES ligand ensures that the Pd does not cluster. The latter is another important advantage of the use of PMO materials as catalytic support.

A different strategy for the heterogenization of NHC ligands was followed by Van Der Voort and De Canck. They focused on a gold(I) NHC complex for hydration reactions of alkynes [83]. Here, the NHC complex is anchored via a post-modification process after synthesizing the PMO support first, which is an ethane bridged PMO bearing sulfonic acid functions [69]. The SO$_3$H functions are well-known to have a high affinity for Au and react with the OH function of the [Au(OH)(NHC)] complex via an acid–base reaction. The resulting PMO material is shown in Figure 7.20 and is a perfect example of Method 3 functionalization. This approach has several advantages over the method used to produce the Pd-NHC-PMO. First of all, a well-ordered material with high functional loading (SO$_3$H) can be obtained due to the small and rigid organic bridge. By using a post-modification, the loading of the NHC complex can be perfectly chosen without wasting any expensive NHC complex during synthesis. Additionally, the resulting catalyst is a boomerang-type catalyst; that is, it allows the active species to detach from the PMO surface to perform catalysis and recombines with the solid after all the starting material is consumed. This exceptional design thus combines the high activity and selectivity of homogeneous catalysts with the recyclability and ease of separation of heterogeneous catalysts. The IPr*-based Au catalyst showed high activity for the hydration of diphenylacetylene and could be easily recovered and reused in several catalytic runs. No decomposition or degradation of the complex or PMO support was observed.

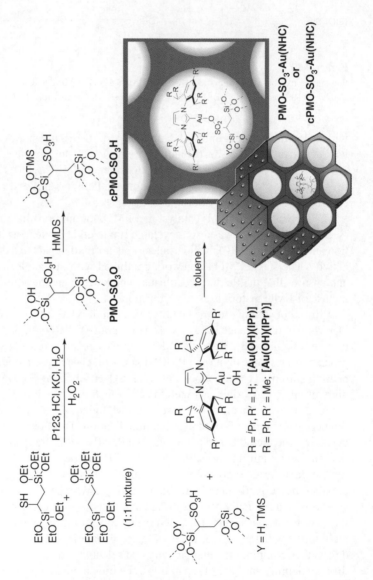

Figure 7.20

Preparation of the ethane PMO bearing sulfonic acid functions and subsequent attachment of the [Au(OH)(IPr)] and [Au(OH)(IPr*)] complexes. Source: Reproduced with permission of John Wiley & Sons, Ltd [83].

7.5.1.8 Chiral Complexes on a PMO

The examples described before did not possess any chiral complexes. However, the introduction of chiral metal complexes and chirality is another interesting route [84, 86]. Chirality is also of major importance for the pharmaceutical industry in particular. The most challenging is the retention of the enantioselectivity of the homogeneous variant after heterogenization. A small distortion of the complex conformation or any steric hindrance can already negatively influence the enantioselectivity.

A very famous example by Avelino Corma of a chiral metal catalyst is the salen-type catalyst, which is based on an N,N'-bis(salicylidene)ethylenediamine moiety. First, a chiral bis-silane is synthesized that already coordinates the vanadium metal. Subsequently, a PMO material is prepared via co-condensation with TEOS in the presence of CTAB. Only relatively low loadings could be achieved ($0.014 \, \mathrm{mmol \, V \, g^{-1}}$) as much vanadium was lost during the SDA removal. The catalyst was used in the cyanosilylation reaction of several aldehydes with trimethyl silyl cyanide (TMSCN). A conversion of ~82% and a selectivity of >98% was observed. Moreover, the chiral catalyst was recyclable for four times and no leaching was observed. The enantioselectivity was, however, much lower than expected; that is, an enantiomeric excess of ~30% was obtained, most likely due to an effect of the PMO walls [86, 87]. Only slightly better enantioselectivity was observed for R-(+)-BINOL based PMO catalysts, obtained via the co-condensation with BTME, for the asymmetric addition of diethylzinc to aldehydes with Ti as metal center [85]. They showed a high catalytic activity (99%) with only moderate enantioselectivity (~40%), probably due to conformation differences in comparison with the homogeneous counterpart.

These examples clearly show that, although some homogeneous chiral metal complexes show high selectivity and enantioselectivity, caution is needed when developing their heterogeneous variants. The conformation of the chiral center and the accessibility must be carefully monitored ensuring that the heterogenization does not influence these two parameters. The use of the chiral complex as an organic bridge adds to the risk of losing enantioselectivity. A post-modification route is more advisable.

7.5.1.9 Bifunctional Metal Modification of a PMO

A final interesting concept in the heterogenization of metal complexes is the development of bifunctional catalysts similar to the already discussed acid/basic catalysts. These materials can perform two consecutive reactions in a one-pot synthesis without any work-up steps between the two catalytic reactions. The intended catalytic reactions are also called cascade, tandem, or relay reactions. Only a few examples are available.

Two site-isolated metal catalysts were combined on a single PMO material to perform Pd-catalyzed Suzuki cross-coupling and Ru-catalyzed asymmetric transfer hydrogenation in two consecutive steps [88]. The authors prepared a PMO material by co-condensing three silanes (Figure 7.21), that is, (S,S)-4-((trimethoxysilyl)ethyl)phenylsulfonyl-1,2-diphenylethylene-diamine

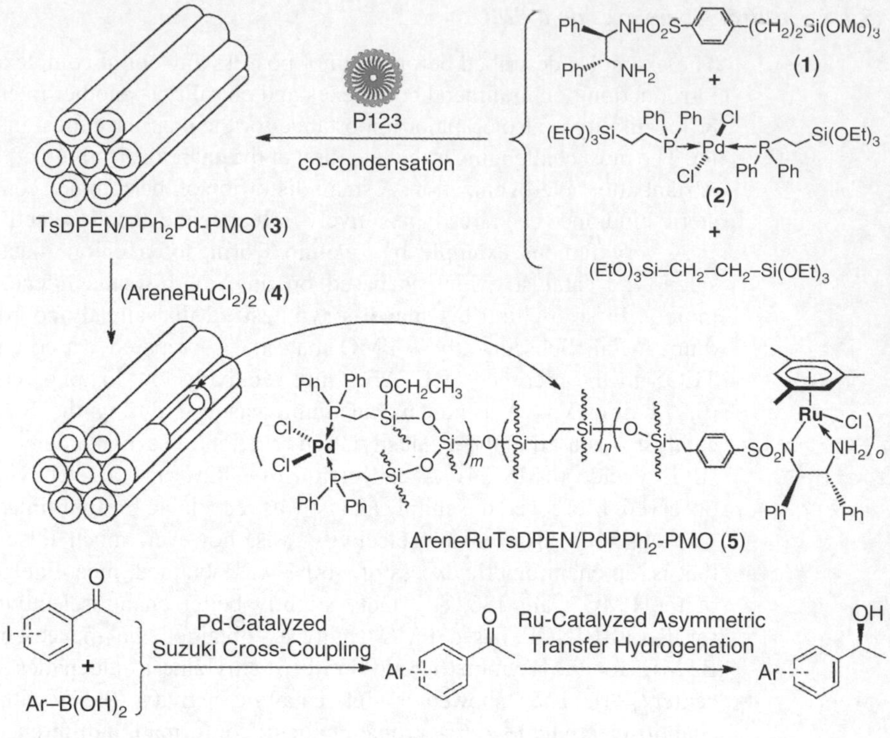

Figure 7.21
Synthesis of a bifunctional PMO material, containing a ruthenium and palladium catalyst. Source: Reproduced with permission of John Wiley & Sons, Ltd [88].

(TsDPEN; **1**), bis[(diphenylphosphino) ethyltriethoxy-silane)] palladium dichloride (**2**) and 1,2-bis(triethoxysilyl)ethane for rigidity. Once the PMO (**3**) is obtained, a ruthenium complex – (areneRuCl$_2$)$_2$ (**4**) – is attached to the solid. This results in the bifunctional catalyst (**5**), which is evaluated for the reaction of haloacetophenones and arylboronic acid. The resulting product is a chiral biaryl alcohol (Figure 7.21). The material could be reused for eight times without any loss of catalytic activity. Similar research was performed for the asymmetric hydrogenation – ATH-Sonogashira coupling one-pot tandem reaction of 4-iodoacetophenone and ethynyl-benzene. The catalyst could be recycled up to seven times [89].

7.5.2   As Adsorbents of Metals, Organic Compounds, and Gases

7.5.2.1   *Adsorption of Toxic Metals*

A large amount of work has been done on the development of adsorbents for environmental applications and also for the recovery of precious compounds. Many diverse types of pollutants exist that must be removed from, for example, waste waters. This includes toxic metals (cations, oxyanions, and radionuclides.), gases (e.g. CO$_2$), volatile organic compounds (VOCs) (e.g. aldehydes), and toxic organic species. Also, expensive metals such as Au, Pt, and rare earth elements (La, Eu, etc.) are interesting to recuperate from an economical point of view. It is very challenging to outperform the typical cheap and commercial adsorbents such as silica and activated carbon. PMOs are more expensive and therefore only target very specialized high-end applications where the intrinsic PMO properties are an added-value (hydrophobicity, incorporation of very special and expensive functional groups, etc.). Parameters such as adsorption capacity and selectivity are important to assess the performance of an adsorbent. The first parameter really deals with the amount of adsorbate that can be adsorbed on the material. The selectivity gives an idea on the affinity for certain pollutants or precious compound and the ease of separation of the target element/molecule versus others that are present. Additionally, the reuse of adsorbent and recovery of adsorbate are also key parameters that should not be neglected but are often not discussed in detail. When designing an adsorbent, one should in the first place consider the target compound. The introduction of functionalities (e.g. heteroatoms, aromaticity) with a known affinity for the target will significantly improve the adsorption capacity and selectivity.

Jaroniec's group reported the development of PMOs for the adsorption of the toxic metal mercury(II) [90]. It is well-known that the soft and large Hg atoms have a high affinity for nitrogen and particularly sulfur containing functionalities [91]. In their research, they combined three different silanes with a high content of N and S (Figure 7.22). By changing the loading of the organic functionality, they could fine-tune the adsorption capacity.

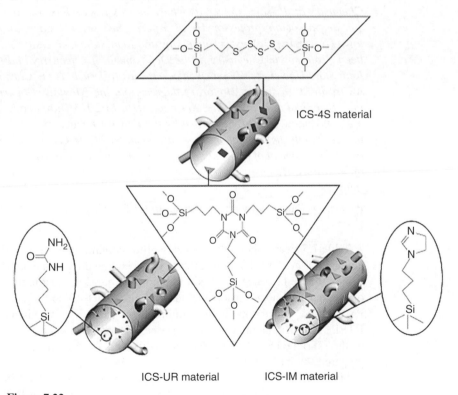

Figure 7.22

A representation of three materials developed as adsorbent for mercury(II). The materials are prepared with TEOS, Tris[3-(trimethoxysilyl)-propyl]isocyanurate (ICS), Bis[3-(triethoxysilyl)propyl]tetrasulfide (4S), N-(3-Triethoxysilylpropyl)4,5-dihydroimidazole (IM), and Ureidopropyltrimethoxysilane (UR). Source: Reproduced with permission of ACS [90].

Mietek Jaroniec obtained his masters at the M. Curie-Sklodowska University (Lublin) in Poland in 1972. In 1976, he obtained his Ph.D. degree and in 1979 a Doctor of Science degree. Afterward, he started working at the Department of Theoretical Chemistry at the same university where he was research assistant, assistant professor, and associate professor. He also performed several stays abroad, at Georgetown University, McMaster University, and Kent State University. Since 1991, he has been Professor in Chemistry at the Department of Chemistry and Biochemistry at Kent State University. He received several awards such as the Carbon Hall of Fame Award in 2001 and an Honorary degree from the Nicolaus Copernicus and Military Technical University in 2009 and 2010. In 2016, Jaroniec won the Medal of Marie Sklodowska-Curie awarded by the Polish Chemical Society for his outstanding work in the field of nanoporous carbon materials. In general, his field of expertise is the development of nanomaterials for environmental, catalytic, and energy storage applications. He has written more than 1000 scientific publications in international journals.

Photograph: https://www.kent.edu/chemistry/profile/mietek-jaroniec

Table 7.5 clearly shows that the tetrasulfide containing materials adsorb the highest amounts of mercury(II) as the affinity between S and Hg is the highest.

Next to the presence of certain functional groups, the loading of adsorption sites on the material also impacts the adsorption. Van Der Voort and Esquivel prepared PMO-based adsorbents with 1-thiol-1,2-bis(triethoxysilyl)ethane (TBTEE) and BTEE [92]. The amount of sulfur in the material was controlled via the adaption of the ratio TBTEE:BTEE. Materials with 100, 75, 50, and 25% of TBTEE were prepared (molar %) (Table 7.6).

Materials with a higher amount of sulfur present will adsorb more Hg(II). An impressive 1183 mg per gram of PMO material was observed for 100SH-PMO. Additionally, the thiol PMO materials were very selective for Hg in presence of Zn, Pb, Cu, and Cd.

The concept of *molecular imprinting* is also very interesting when designing adsorbents. This idea was reported by Burleigh et al. and is best explained

Table 7.5 Overview of the number of organic functionalities in the PMO materials and the mercury adsorption capacity [90].

Sample	$C_{\text{N-lig}}$ (mmol g^{-1})^a	$C_{\text{S-lig}}$ (mmol g^{-1})^b	Hg^{2+} (mmol g^{-1})
ICS-4S-1	0.68	0.13	0.91
ICS-4S-2	0.73	0.33	1.37
ICS-IM-1	0.39	—	0.40
ICS-IM-2	0.58	—	0.51
ICS-UR-1	0.27	—	0.76
ICS-UR-2	0.78	—	0.60

^{a}Concentration of surface ligands containing nitrogen atoms; calculated from elemental analysis data.
^{b}Concentration of surface ligands containing sulfur atoms; calculated from elemental analysis data.
Source: Reproduced with permission of ACS.

Table 7.6 Maximum mercury adsorption capacities for the thiol PMO materials, prepared with different amounts of TBTEE [92].

Material	S_{total} (mmol g^{-1})	Hg adsorption capacity (mg g^{-1})	Hg/S_{total} ratio
100SH-PMO	4.28	1183 ± 2	1.38
75SH-PMO	3.03	833 ± 2	1.37
50SH-PMO	1.93	410 ± 8	1.06
25SH-PMO	0.96	297 ± 1	1.54

Source: Reproduced with permission of the RSC.

by means of an example [93]. Adsorbents for the removal of nickel(II), copper(II), and zinc(II) were developed using the precursors BTEE and N-[3-(trimethoxysilyl)propyl]ethylenediamine ((TMS)en) and cetyltrimethylammonium chloride (CTAC) as SDA. During synthesis Ni(II), Cu(II), and Zn(II) salts are added and will form complexes with the amine containing precursor. After formation of the PMO materials, the metals are removed with nitric acid. This results in a material that possesses cavities shaped like the metal with an identical chemical environment. This significantly increases the affinity of a certain adsorbent for the metal ion. When using the material during adsorption, it will bind selectively to that one metal ion.

Many reports describe the adsorption of metal cations such as Hg, Cd, Cu, Cr, and Cu. However, only a limited amount of research [8] has been done on anionic metal species such as rhenates and arsenates and many possibilities can still be researched.

7.5.2.2 Adsorption of VOCs

A few studies have been done on the use of PMOs in VOC capture, a domain that is hugely dominated by the active carbons. The presence of aromaticity also greatly influences the adsorption process. A study compared the adsorption of three phenolic compounds on an arylene base and ethane PMO [94]. The PMO materials were prepared with the precursors 1,4-bis(trimethoxysilylethyl)benzene

Table 7.7 Uptake of 4-nitrophenol on the PMO adsorbents, using a batch test set-up [94].

Initial concentration (mol l^{-1})	Arylene PMO (%)	Ethane PMO (%)
1.0×10^{-4}	99.4	16.5
2.5×10^{-4}	99.1	16.3
5.0×10^{-4}	98.7	15.8
7.5×10^{-4}	98.2	15.3
1.0×10^{-3}	96.9	15.2

Source: Reproduced with permission of ACS.

or 1,2-bis(triethoxysilyl)ethane and CTAC as SDA in basic environment. The adsorption of 4-nitrophenol (4-NP), 4-chlorophenol (4-CP) and 4-methylphenol (4-MP) were investigated, both in batch and in a column set-up. The results for 4-NP are presented in Table 7.7. The arylene PMO adsorbs ~97–99% depending on the initial concentration of the pollutant. On the other hand, the ethane PMO only adsorbs ~15–16%, even though the surface area of this PMO material is twice as larger as the S_{BET} of the arylene PMO (1300 vs. 550 m^2 g^{-1}, respectively).

The interactions between the aromatic systems of the pollutant and the adsorbent play a significant role. It is believed that π-π interactions are the reason that the arylene bridged PMO adsorbs 14 times more 4-NP than the ethane PMO (based on S_{BET}).

The PMO structure (e.g. 3D cubic versus 2D hexagonal) is also important for adsorbent design and an example is given for the removal of the VOC acetaldehyde. Sulfonated benzene bridged PMOs were prepared with a three-dimensional cubic (*Pm3n*) and a two-dimensional hexagonal (*P6mm*) mesophase [40]. The sulfonation process was performed via an analogous procedure as described in Section 7.4.2. The results of the elimination of gaseous acetaldehyde are presented in Table 7.8 and the sulfonated materials with a 3D structure clearly removes more acetaldehyde in comparison with the other materials.

7.5.3 As Solid Chromatographic Packing Materials

PMO materials with simple organic bridges such as methane, ethane, and benzene clearly possess an intrinsic quality that is of great importance in some applications; that is, the hydrophobicity induced by the organic moieties and the resulting increased

Table 7.8 The removal of acetaldehyde using sulfonated benzene PMOs with different mesophases [40].

	Acetaldehyde adsorbed			
	25 °C; 30 min		60 °C; 90 min	
	mg g^{-1}	%	mg g^{-1}	%
3D cubic	0.314	14.89	0.558	26.47
2D hexagonal	0.045	2.27	0.124	6.27
3D cubic – SO$_3$H	0.763	38.6	1.369	69.22
2D hexagonal – SO$_3$H	0.314	15.91	0.472	23.93

Source: Reproduced with permission of the RSC.

stability. This property is especially useful in chromatography when the PMO is applied as normal or reversed phase.

Currently, silica is mostly used due to the possibility to fine-tune its morphology, porosity, and surface, and its mechanical stability (see Chapter 4). Spherical particles are needed and this is obtained by using the well-known Stöber method [95], which is further adapted to obtain porous spherical particles [96]. A large amount of literature is available on this type of synthesis and applicability of highly porous silica spheres in chromatography. Additionally, for reversed phase high-performance liquid chromatography (HPLC) applications, the silica surface must be hydrophobized, typically with a grafting procedure of C8 or C18 chains to obtain retention, but also to remove any silanols, as these have a negative effect on the peak shape.

Silica as a chromatographic packing material has one major drawback; it possesses a poor hydrolytic stability [97]. Even when the surface is hydrophobized, the working range in reversed phase HPLC applications is limited to pH = 11 at the maximum [98]. At higher pH values, the structure will degrade. In the lower pH range (<2), the grafted C8 or C18 chains and any end capping agents will be removed from the surface due to hydrolysis of the siloxanes. Also, the use of elevated temperatures is problematic with silica materials [98b]. PMO can be a suitable reversed-phase HPLC packing material with a broader pH and temperature working range due to its increased stability and hydrophobicity.

The amount of work performed now on introducing PMOs as chromatographic phase is rather limited. Inagaki and coworkers [99] and Fröba and coworkers [100] reported in the same year the first attempts with a benzene PMO with spherical morphology. Based on the Stöber method, they prepared uniform spheres under basic conditions with cationic surfactants and an alcohol as co-solvent. Particles with an average size between 0.4 and 0.5 μm were obtained [100]. Larger benzene PMO spheres were acquired with particle diameters between 3 and 15 μm by using a combination of two surfactants, CTAB and P123, in acidic environment (Figure 7.23) [101].

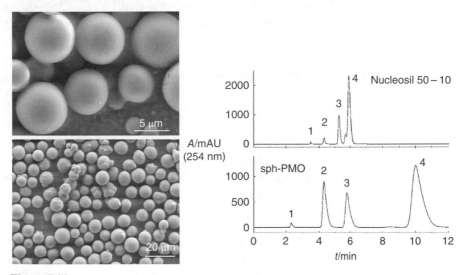

Figure 7.23

SEM images of a spherical benzene PMO (sph-PMO). Chromatograms of mixture 1 (benzene (1), naphthalene (2), biphenyl (3), and phenanthrene (4)) separated with *n*-hexane at a flow rate of 1 ml min^{-1} on the Nucleosil 50-10 column or a flow rate of 2 ml min^{-1} on the sph-PMO column. Source: Reproduced with permission of John Wiley & Sons, Ltd [101].

Michael Fröba (Froeba) was born in 1962 in Lübeck (Germany). He finished his Chemistry studies in 1989 at the Universities of Würzburg and Hamburg in Germany. Fröba subsequently obtained his Ph.D. degree at the University of Hamburg in 1993 in the field of graphite intercalation, after which he performed a postdoctoral study at the Lawrence Livermore National Laboratory (LLNL, USA). In 1996, he started his Habilitation for Inorganic Chemistry at the University of Hamburg (Institute of Inorganic and Applied Chemistry). After this, in 2000, Fröba became Associate Professor of Inorganic Chemistry at the University of Erlangen-Nuremberg and in 2001 Full Professor of Inorganic Chemistry at Justus-Liebig-University Gießen. Since 2007 Fröba is Full Professor of Inorganic Chemistry at the University of Hamburg. He is specialized in the development of nanoporous materials for energy storage applications and nanoporous hybrid materials applied in the fields of sorption, separation science, and catalysis.

Photograph: https://www.chemie.uni-hamburg.de/ac/froeba/mitglieder/Froeba.html

Chromatographic tests were performed to evaluate the spherical benzene PMO and comparison was made with the commercial Nucleosil 50-10. Different test mixtures were selected and the PMO could separate the mixtures in contrast to the Nucleosil material. Although these first results were promising, the particle size was still too broad.

Another approach uses the technique spray-drying to obtain spherical particles. An attempt was made by Van Der Voort. Here the allyl ring PMO is spray-dried and post-modified via thiol-ene click chemistry with C3 and C18 chains to obtain a reversed-phase packing material [45a]. Particles with an average particle size of 2 μm were obtained, which is still too broad for commercial use. However, their chromatographic performance could be tested and already promising retention and selectivity could be observed. On top of that, the material showed unprecedented stability in a high pH and temperature range (Figure 7.24).

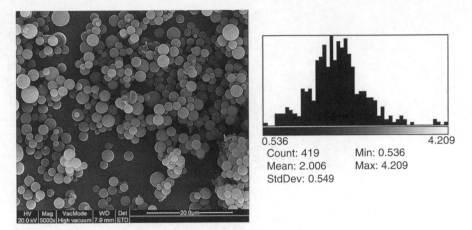

Figure 7.24
(left) SEM picture of spray-dried allyl ring PMO and (right) particle size distribution expressed in micrometer. Source: Reproduced with permission of the RSC [45a].

Overall, it is especially difficult to obtain particles that meet all the conditions. They must be spherical, monodisperse ($D_{90/10} < 1.6$, where $D_{90/10}$ statistically describes the dispersion of the particle size), and possess a large surface area and pore size. They also must be prepared without any metal ions and contain no or a limited number of micropores. Using the described Stöber and spray-drying methods will probably not allow us to meet all these conditions at the same time. Different methods must be explored, such as the core-shell and emulsification procedures.

7.5.4   As Low-*k* Films

PMO materials are not only potential chromatographic packing materials, they can also be applied as insulating interlayer or low-*k* material. A low-*k* material possesses a low dielectric constant (k) and reduces the RC-delay that is induced between wires that are near each other. The *k*-value needs to preferably be below 2 to act as an insulating interlayer. The intrinsic properties of PMOs (e.g. hydrophobicity) make them very promising in micro-electronic devices. Their high porosity or plainly stated cavities filled with air can reduce the *k*-value as significantly as $k_{air} \approx 1$. Furthermore, the organic bridges (methane and ethane) are less polarizable than SiO_2 ($k = 3.9$), which diminishes the *k*-value even more [1].

To function as an interlayer, PMO thin films must be prepared. This can be achieved by two methods: (i) the deposition on a substrate using the well-known EISA method [102] in combination with spin- or dip-coating or (ii) the preparation by hydrothermal synthesis [103]. The latter procedure is not frequently used, but can result in the growth of a free standing PMO film at the liquid-reaction polyethylene bottle interface.

Lu et al. reported the first PMO thin films for this application [104], although the majority of the work was performed by the research groups of Ozin and Van Der Voort. Ozin and coworkers elaborately described the preparation of thin films consisting of methane, ethane, ethene, and ring PMOs. They thoroughly investigated their dielectric, mechanical, and hydrophobic properties [43, 105]. In particular, the hydrophobicity of the ring PMO is a key factor as fewer silanol groups are present and

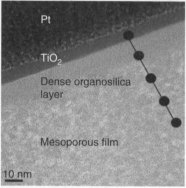

Figure 7.25

(top) SEM and TEM image of a ring PMO thin film, prepared via spin-coating with Brij-76 and ethanol in acidic medium. The length of the scale bar on the TEM image represents 20 nm. (bottom) High Resolution Transmission Electron Microscopy image of the ethane PMO sealed with a dense ring PMO layer. Source: Reproduced with permission of the RSC [44b].

the adsorption of water is reduced. The presence of water can significantly increase the dielectric constant as $k_{water} \approx 80$.

The group of Van Der Voort extended the applicability of the ring PMO as low-k film (Figure 7.25) [44b]. By employing Brij-76 instead of the surfactant CTAB, they improved the chemical stability of the thin films. This was explained by the presence of thicker pore walls in combination with the —CH_2— moieties. The ring PMO thin films also survived a treatment with 1 M NaOH. Moreover, they developed a method to modify the top surface of an ethane based low-k film to seal the pores [106]. By applying this procedure, the k-value was lowered and the mechanical and chemical stability was improved [106b].

Another type of interesting PMO material for low-k applications was presented by Ozin and coworkers [107]. A thin film was spin-coated by using a polyhedral oligomeric silsesquioxane or POSS precursor. The resulting PMO possesses air pockets in the pore walls, which is beneficial for the dielectric constant. A k-value of 1.7 was achieved (Figure 7.26).

7.5.5   As Biomedical Supports

The adsorption or immobilization of proteins, enzymes, amino acids, and drugs is of great interest in the field of biomedical applications. The adsorption process itself

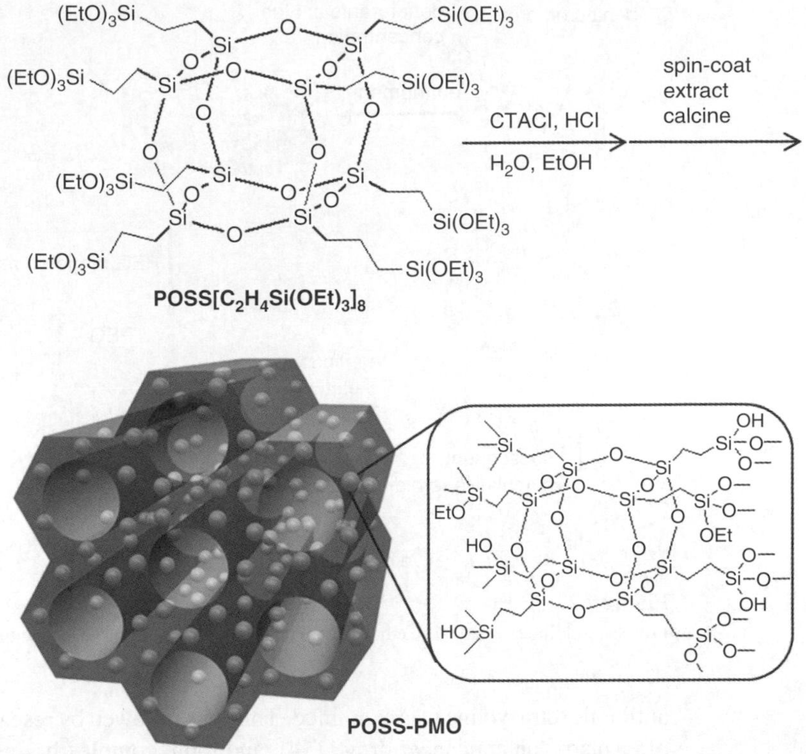

Figure 7.26
A POSS PMO thin film is prepared with a EISA spin-coating procedure. Source: Reproduced with permission of ACS [107].

can be based on not only electrostatic but also π-π interactions, Van der Waals forces, and hydrophobicity. The properties of PMO materials can be fine-tuned so that the material can interact with the adsorbate according to one or more of the mentioned interactions. Additionally, the large mesopores also allow the diffusion of bigger (bio)molecules in the pore network.

One of the first reports presented the adsorption of a protein via electrostatic interactions [108]. Cytochrome c (bovine heart) was adsorbed on large pore ethane PMO materials. The adsorption capacity reached a maximum at the isoelectric point of the protein, unfortunately it did not exceed the values obtained for SBA-15 materials. This research on the adsorption of proteins was expanded by using benzene and biphenyl PMOs for cytochrome c (horse heart) [109], and ethane PMO materials for the adsorption of serum albumin [25a, b] and hemoglobin [110]. The ethane PMO does not only act as a support, it can also assist in protein refolding (Figure 7.27) [111]. Hen egg white lysozyme was entrapped in the pores of the PMO via hydrophilic and hydrophobic interactions and aggregation of the biomolecules was prevented. Continuous elution of the lysozyme triggered by poly(ethyleneglycol) (PEG) resulted in refolding of the protein.

Besides enzymes, also amino acids [112] and drugs were immobilized. The latter is important in controlled drug delivery systems. The use of PMO for this purpose was reported for the first time in 2009 [113]. Ethane PMO materials were prepared as hollow spheres and solid spheres, and their use as drug carriers for the hydrophobic

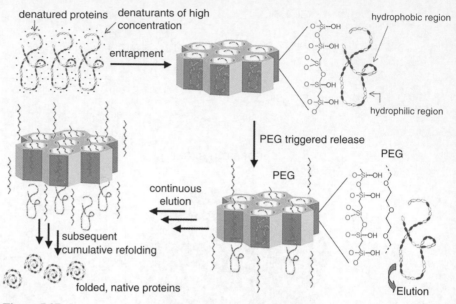

Figure 7.27
Protein refolding assisted by an ethane PMO material. Source: Reproduced with permission of ACS [111].

antibiotic tetracycline was examined. This was followed by research on the adsorption of cisplatin, an anticancer drug [114], and more complex bridges were incorporated to adsorb drugs such as captopril and 5-fluorouracil [8]. It is important to stimulate the interaction between the PMO and the target molecule. The introduction of heteroatoms, in analogy with the development of adsorbents, can increase the potential for a controlled drug release system. Ibuprofen, for example, was adsorbed on PMO materials containing amidoxime [115] and malonamide-based moieties in a report by Van Der Voort [116]. In the latter research, the malonamide bridge (MA) was compared to the N-methyl malonamide bridge (mMA) (Figure 7.28). It was clear that the presence of the hydrogen atoms on the secondary amines plays an important role in the ibuprofen release profile due to the possibility of hydrogen bonding with the drug molecule. Additionally, changing the functional loading affects the release of ibuprofen as well making it possible to optimize the dosage.

Although some initial research is reported with some promising results, the use of PMO materials as a controlled drug delivery system still needs some more investigation and many challenges remain. It is especially important that the rate and the period of delivery are controlled during the treatment. A burst release must be avoided to ensure a long-term effect of the treatment.

Exercises

1 What are the advantages of PMOs in comparison with silicas?
2 Which techniques would you use to evaluate the porosity of a PMO material?
3 Which techniques would you use to examine the external morphology of a PMO material?
4 Why is the amount of research performed on methane PMOs so limited?

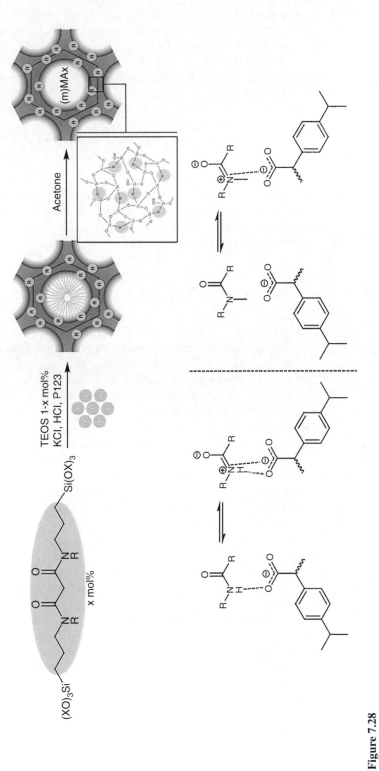

Figure 7.28

The synthesis of the malonamide (R = H; MA) and N-methyl malonamide (R = CH$_3$; mMA) PMO material and the different interactions between ibuprofen and MA (left) or mMA (right).

Source: Courtesy of Sander Clerick.

5 Why is KCl often added during the synthesis of a PMO?

6 How can you tune the pore size of a PMO material?

7 As an assignment, you need to prepare a PMO material in the laboratory using the precursor below (bis[3-(triethoxysilyl) propyl urethane]ethane; BTEPUE). Your tutor gave you a robust recipe that uses a chemically similar, but smaller precursor molecule. Unfortunately, after several attempts you did not manage to prepare a PMO material with a high surface area and mesopores. Can you explain why and how will you try to solve this issue?

8 You need to design an adsorbent for the toxic metal ion cadmium(II). Which properties does your adsorbent need to possess to have a high adsorption capacity? Which PMO materials would you propose?

9 Are PMOs crystalline materials?

Answers to the Problems

1 PMOs can immediately contain the desired organic functionality. The introduction of an organic function on a silica material is normally performed by a grafting or co-condensation procedure. The organic bridges in the PMO material are homogenously distributed in the network.

The organic bridges make the PMO material intrinsic hydrophobic. Particularly for chromatographic applications and the insulation of electric devices (low-*k*), this is an interesting property.

The hydrothermal stability of PMOs is higher than the hydrothermal stability of silicas.

The pore diameter of PMOs is larger, allowing better entrance and diffusion of organic molecules. Additionally, the pores are ordered when comparing with a normal commercial silica, which is also beneficial for diffusion inside the pores.

2 Nitrogen sorption: this technique allows you to determine the specific surface area ($m^2\,g^{-1}$), the pore volume ($ml\,g^{-1}$) and the pore diameter (nm) of the material. Moreover, it gives you an idea of the pore size distribution (narrow, broad, bimodal) and the shape of the pores (cylindrical, slit, inkbottle).

X-ray diffraction: this technique allows you to determine the ordering of the pores and the symmetry. Furthermore, via the position of the reflections, you can calculate the pore wall thickness of the material.

Transmission Electron Microscopy (TEM) allows you to determine the symmetry of the material.

3 Scanning Electron Microscopy (SEM), and for hollow structures TEM can also be interesting.

4 The methane bridge does not contain any interesting functionality. Therefore, the applicability of this PMO material is low and research performed on this type of PMO is limited.

5 The salt KCl can improve the interactions between the precursor molecules and the SDA. This often results in materials with better ordering and structure.

6 One can choose the surfactant carefully. SDAs such as the cationic CTAB will result in relatively small pores (2–3 nm). Selecting a non-ionic and larger surfactant, for example, the SDA P123, will result in large pores (5–8 nm). The addition of additives such as a poreswweller, for example, trichlorobenzene will result in larger pores when using the same surfactant. One can use a post-synthetic expansion procedure with ammonia to etch out the pores and make them bigger.

7 The precursor BTEPUE is a rather large molecule with a high degree of rotational freedom, which means that it is very flexible. It is very difficult to prepare materials with flexible precursors that still exhibit a high porosity (S_{BET}, V_p, d_p) and good ordering. If the recipe already worked with a similar precursor, which is much smaller, then probably this flexibility of BTEPUE is the root cause. To investigate this and to solve the issue, one can attempt to make a PMO material with a mixture of BTEPUE and a small rigid precursor such as BTEE or BTME. The latter precursors will help the formation of the solid during the co-condensation process.

8 Cadmium is a soft metal and has a large affinity for sulfur containing functionalities. The PMO material must therefore possess a high amount of reachable S in the form of thiols (-SH), thioethers (-R-S-R-). The following precursors can be used to prepare PMOs that meet this requirement: 1-thiol-1,2-bis(triethoxysilyl)ethane (TBTEE), bis[3-(triethoxysilyl)propyl]disulfide (BTEPDS) and 1,4-bis(triethoxysilyl) propane tetrasulfide (BTEPTS). Additionally, the molecular imprinting technique can also be used here to enhance the affinity of the PMO material for cadmium(II).

9 PMOs are not crystalline materials as they exhibit no ordering on an atomic scale. For comparison, the lattice of NaCl does exhibit periodicity and thus NaCl is crystalline. However, some molecular-scale periodicity has been observed for ethene and benzene PMO materials. These materials showed a higher degree of ordering that was observed in XRD measurements.

References

1. Hoffmann, F., Cornelius, M., Morell, J., and Fröba, M. (2006). *Angew. Chem. Int. Ed.* 45: 3216–3251.
2. Asefa, T., MacLachlan, M.J., Coombs, N., and Ozin, G.A. (1999). *Nature* 402: 867–871.
3. Inagaki, S., Guan, S., Fukushima, Y. et al. (1999). *J. Am. Chem. Soc.* 121: 9611–9614.
4. Melde, B.J., Holland, B.T., Blanford, C.F., and Stein, A. (1999). *Chem. Mater.* 11: 3302–3308.
5. (a) Zhao, D.Y., Feng, J.L., Huo, Q.S. et al. (1998). *Science* 279: 548–552. (b) Zhao, D.Y., Huo, Q.S., Feng, J.L. et al. (1998). *J. Am. Chem. Soc.* 120: 6024–6036.
6. Brinker, C.J. (2004). *MRS Bull.* 29: 631–640.

7. Smeulders, G., Meynen, V., Van Baelen, G. et al. (2009). *J. Mater. Chem.* 19: 3042–3048.
8. Van Der Voort, P., Esquivel, D., De Canck, E. et al. (2013). *Chem. Soc. Rev.* 42: 3913–3955.
9. (a) Ford, D.M., Simanek, E.E., and Shantz, D.F. (2005). *Nanotechnology* 16: S458–S475; (b) Fujita, S. and Inagaki, S. (2008). *Chem. Mater.* 20: 891–908; (c) Hatton, B., Landskron, K., Whitnall, W. et al. (2005). *Acc. Chem. Res.* 38: 305–312; (d) Hoffmann, F., Cornelius, M., Morell, J., and Fröba, M. (2006). *J. Nanosci. Nanotechnol.* 6: 265–288; (e) Hoffmann, F. and Fröba, M. (2011). *Chem. Soc. Rev.* 40: 608–620; (f) Hunks, W.J. and Ozin, G.A. (2005). *J. Mater. Chem.* 15: 3716–3724; (g) Jaroniec, M. (2006). *Nature* 442: 638–640; (h) Kapoor, M.P. and Inagaki, S. (2006). *Bull. Chem. Soc. Jpn.* 79: 1463–1475; (i) Mizoshita, N., Tani, T., and Inagaki, S. (2011). *Chem. Soc. Rev.* 40: 789–800; (j) Xia, H.S., Zhou, C.H., Tong, D.S., and Lin, C.X. (2010). *J. Porous Mater.* 17: 225–252; (k) F. Hoffmann, M. Fröba, Silica-based mesoporous organic-inorganic hybrid materials, in The Supramolecular Chemistry of Organic-Inorganic Hybrid Materials (Eds K. Rurack, R. Martínez-Máñez), 2010, John Wiley & Sons, Inc., Hoboken, NJ, 39–102.
10. Palmqvist, A.E.C. (2003). *Curr. Opin. Colloid Interface Sci.* 8: 145–155.
11. (a) Zhai, S.-R., Park, S.S., Park, M. et al. (2008). *Microporous Mesoporous Mater.* 113: 47–55; (b) Zhai, S.R., Kim, I., and Ha, C.S. (2008). *J. Solid State Chem.* 181: 67–74.
12. (a) Zhao, D.Y., Sun, J.Y., Li, Q.Z., and Stucky, G.D. (2000). *Chem. Mater.* 12: 275–280; (b) Guo, W.P., Park, J.Y., Oh, M.O. et al. (2003). *Chem. Mater.* 15: 2295–2298.
13. Wang, W.H., Xie, S.H., Zhou, W.Z., and Sayari, A. (2004). *Chem. Mater.* 16: 1756–1762.
14. Xiao, F.S. (2005). *Curr. Opin. Colloid Interface Sci.* 10: 94–101.
15. Bao, X.Y., Zhao, X.S., Li, X. et al. (2004). *J. Phys. Chem. B* 108: 4684–4689.
16. Lee, C.H., Park, S.S., Choe, S.J., and Park, D.H. (2001). *Microporous Mesoporous Mater.* 46: 257–264.
17. Sayari, A., Hamoudi, S., Yang, Y. et al. (2000). *Chem. Mater.* 12: 3857–3863.
18. (a) Guan, S., Inagaki, S., Ohsuna, T., and Terasaki, O. (2000). *J. Am. Chem. Soc.* 122: 5660–5661; (b) Guan, S.Y., Inagaki, S., Ohsuna, T., and Terasaki, O. (2001). *Microporous Mesoporous Mater.* 44: 165–172.
19. Kapoor, M.P. and Inagaki, S. (2002). *Chem. Mater.* 14: 3509–3514.
20. Lee, H.I., Pak, C., Yi, S.H. et al. (2005). *J. Mater. Chem.* 15: 4711–4717.
21. (a) Qiao, S.Z., Yu, C.Z., Hu, Q.H. et al. (2006). *Microporous Mesoporous Mater.* 91: 59–69; (b) Zhu, H.G., Jones, D.J., Zajac, J. et al. (2001). *Chem. Commun.* 2568–2569; (c) Muth, O., Schellbach, C., and Fröba, M. (2001). *Chem. Commun.* 2032–2033; (d) Bao, X.Y., Zhao, X.S., Li, X., and Li, J. (2004). *Appl. Surf. Sci.* 237: 380–386.
22. (a) Zhao, L., Zhu, G.S., Zhang, D.L. et al. (2005). *J. Phys. Chem. B* 109: 764–768; (b) Guo, W.P., Kim, I., and Ha, C.S. (2003). *Chem. Commun.* 2692–2693.
23. McInall, M.D., Scott, J., Mercier, L., and Kooyman, P.J. (2001). *Chem. Commun.* 2282–2283.
24. (a) Sayari, A. and Yang, Y. (2002). *Chem. Commun.* 2582–2583; (b) Burleigh, M.C., Jayasundera, S., Thomas, C.W. et al. (2004). *Colloid Polym. Sci.* 282: 728–733.
25. (a) Li, N., Wang, J.G., Zhou, H.J. et al. (2011). *Chem. Mater.* 23: 4241–4249; (b) Hao, N., Wang, H.T., Webley, P.A., and Zhao, D.Y. (2010). *Microporous Mesoporous Mater.* 132: 543–551; (c) Liu, J., Bai, S.Y., Zhong, H. et al. (2010). *J. Phys. Chem. C.* 114: 953–961; (d) Mandal, M. and Kruk, M. (2012). *Chem. Mater.* 24: 123–132.
26. (a) Mandal, M. and Kruk, M. (2012). *Chem. Mater.* 24: 149–154; (b) Liu, X., Li, X.B., Guan, Z.H. et al. (2011). *Chem. Commun.* 47: 8073–8075.
27. Asefa, T., MacLachlan, M.J., Grondey, H. et al. (2000). *Angew. Chem. Int. Ed.* 39: 1808–1811.
28. (a) Zhang, W.H., Daly, B., O'Callaghan, J. et al. (2005). *Chem. Mater.* 17: 6407–6415; (b) Bao, X.Y., Li, X., and Zhao, X.S. (2006). *J. Phys. Chem. B* 110: 2656–2661.
29. Xia, Y.D. and Mokaya, R. (2006). *J. Phys. Chem. B* 110: 3889–3894.
30. (a) Xia, Y.D. and Mokaya, R. (2006). *J. Mater. Chem.* 16: 395–400; (b) Xia, Y.D., Wang, W.X., and Mokaya, R. (2005). *J. Am. Chem. Soc.* 127: 790–798.
31. Nakajima, K., Tomita, I., Hara, M. et al. (2005). *J. Mater. Chem.* 15: 2362–2368.
32. (a) Vercaemst, C., Ide, M., Allaert, B. et al. (2007). *Chem. Commun.* 2261–2263; (b) Vercaemst, C., Jones, J.T.A., Khimyak, Y.Z. et al. (2008). *Phys. Chem. Chem. Phys.* 10: 5349–5352; (c) Vercaemst, C., de Jongh, P.E., Meeldijk, J.D. et al. (2009).

Chem. Commun. 4052–4054; (d) Vercaemst, C., Friedrich, H., de Jongh, P.E. et al. (2009). *J. Phys. Chem. C* 113: 5556–5562; (e) Vercaemst, C., Ide, M., Friedrich, H. et al. (2009). *J. Mater. Chem.* 19: 8839–8845; (f) Vercaemst, C., Ide, M., Wiper, P.V. et al. (2009). *Chem. Mater.* 21: 5792–5800; (g) Goethals, F., Vercaemst, C., Cloet, V. et al. (2010). *Microporous Mesoporous Mater.* 131: 68–74.

33. Xia, Y.D., Yang, Z.X., and Mokaya, R. (2006). *Chem. Mater.* 18: 1141–1148.
34. Yoshina-Ishii, C., Asefa, T., Coombs, N. et al. (1999). *Chem. Commun.* 2539–2540.
35. Cho, E.B. and Kim, D. (2008). *Microporous Mesoporous Mater.* 113: 530–537.
36. Cho, E.B., Kim, D., Gorka, J., and Jaroniec, M. (2009). *J. Mater. Chem.* 19: 2076–2081.
37. Temtsin, G., Asefa, T., Bittner, S., and Ozin, G.A. (2001). *J. Mater. Chem.* 11: 3202–3206.
38. Inagaki, S., Guan, S., Ohsuna, T., and Terasaki, O. (2002). *Nature* 416: 304–307.
39. Okamoto, K., Goto, Y., and Inagaki, S. (2005). *J. Mater. Chem.* 15: 4136–4140.
40. Kapoor, M.P., Yanagi, M., Kasama, Y. et al. (2006). *J. Mater. Chem.* 16: 3305–3311.
41. Kapoor, M.P., Yang, Q.H., and Inagaki, S. (2004). *Chem. Mater.* 16: 1209–1213.
42. Kapoor, M.P., Inagaki, S., Ikeda, S. et al. (2005). *J. Am. Chem. Soc.* 127: 8174–8178.
43. Landskron, K., Hatton, B.D., Perovic, D.D., and Ozin, G.A. (2003). *Science* 302: 266–269.
44. (a) Goethals, F., Meeus, B., Verberckmoes, A. et al. (2010). *J. Mater. Chem.* 20: 1709–1716; (b) Goethals, F., Ciofi, I., Madia, O. et al. (2012). *J. Mater. Chem.* 22: 8281–8286.
45. (a) Ide, M., De Canck, E., Van Driessche, I. et al. (2015). *RSC Adv.* 5: 5546–5552; (b) Clerick, S., De Canck, E., Hendrickx, K. et al. (2016). *Green Chem.* 18: 6035–6045.
46. Landskron, K. and Ozin, G.A. (2004). *Science* 306: 1529–1532.
47. Landskron, K. and Ozin, G.A. (2005). *Angew. Chem. Int. Ed.* 44: 2107–2109.
48. (a) Mandal, M. and Kruk, M. (2010). *J. Mater. Chem.* 20: 7506–7516; (b) Mandal, M. and Kruk, M. (2010). *J. Phys. Chem. C* 114: 20091–20099.
49. Zhou, X.F., Qiao, S.Z., Hao, N. et al. (2007). *Chem. Mater.* 19: 1870–1876.
50. Cho, E.B. and Char, K. (2004). *Chem. Mater.* 16: 270–275.
51. Kruk, M., Jaroniec, M., Guan, S.Y., and Inagaki, S. (2001). *J. Phys. Chem. B* 105: 681–689.
52. Burleigh, M.C., Markowitz, M.A., Jayasundera, S. et al. (2003). *J. Phys. Chem. B* 107: 12628–12634.
53. Guo, W.P., Li, X., and Zhao, X.S. (2006). *Microporous Mesoporous Mater.* 93: 285–293.
54. Esquivel, D., Jimenez-Sanchidrian, C., and Romero-Salguero, F.J. (2011). *J. Mater. Chem.* 21: 724–733.
55. Asefa, T., Kruk, M., MacLachlan, M.J. et al. (2001). *J. Am. Chem. Soc.* 123: 8520–8530.
56. Polarz, S., Jeremias, F., and Haunz, U. (2011). *Adv. Funct. Mater.* 21: 2953–2959.
57. Nakai, K., Oumi, Y., Horie, H. et al. (2007). *Microporous Mesoporous Mater.* 100: 328–339.
58. (a) Sasidharan, M., Fujita, S., Ohashi, M. et al. (2011). *Chem. Commun.* 47: 10422–10424; (b) Sasidharan, M. and Bhaumik, A. (2013). *ACS Appl. Mater. Interfaces* 5: 2618–2625.
59. (a) Nakajima, K., Tomita, I., Hara, M. et al. (2005). *Adv. Mater.* 17: 1839–1842; (b) Nakajima, K., Tomita, I., Hara, M. et al. (2006). *Catal. Today* 116: 151–156.
60. Esquivel, D., De Canck, E., Jimenez-Sanchidrian, C. et al. (2011). *J. Mater. Chem.* 21: 10990–10998.
61. D. Dubé, M. Rat, F. Beland, S. Kaliaguine, Microporous Mesoporous Mater. 2008, 111, 596–603.
62. H. C. Kolb, M. G. Finn, K. B. Sharpless, Angew. Chem. Int. Ed. 2001, 40, 2004−2021.
63. Anastas, P.T. and Warner, J.C. (1998). *Green Chemistry: Theory and Practice*. Oxford University Press.
64. Ouwehand, J., Lauwaert, J., Esquivel, D. et al. (2016). *Eur. J. Inorg. Chem.* 2144–2151.
65. Ohashi, M., Kapoor, M.P., and Inagaki, S. (2008). *Chem. Commun.* 841–843.
66. Smeulders, G., Meynen, V., Houthoofd, K. et al. (2012). *Microporous Mesoporous Mater.* 164: 49–55.
67. (a) Dhepe, P.L., Ohashi, M., Inagaki, S. et al. (2005). *Catal. Lett.* 102: 163–169; (b) Rac, B., Hegyes, P., Forgo, P., and Molnar, A. (2006). *Appl. Catal. A-Gen.* 299: 193–201; (c) Yang, J., Yang, Q.H., Wang, G. et al. (2006). *J. Mol. Catal. A-Chem.* 256: 122–129.
68. Yang, Q.H., Kapoor, M.P., Inagaki, S. et al. (2005). *J. Mol. Catal. A-Chem.* 230: 85–89.

69. Lopez, M.I., Esquivel, D., Jimenez-Sanchidrian, C. et al. (2015). *J. Catal.* 326: 139–148.
70. Liu, J., Yang, J., Li, C.M., and Yang, Q.H. (2009). *J. Porous Mater.* 16: 273–281.
71. (a) Li, C.M., Yang, H., Shi, X. et al. (2007). *Microporous Mesoporous Mater.* 98: 220–226; (b) Rat, M., Zahedi-Niaki, M.H., Kaliaguine, S., and Do, T.O. (2008). *Microporous Mesoporous Mater.* 112: 26–31.
72. Sayari, A. (2000). *Angew. Chem. Int. Ed.* 39: 2920–2922.
73. Shylesh, S., Wagener, A., Seifert, A. et al. (2010). *Angew. Chem. Int. Ed.* 49: 184–187.
74. Huybrechts, W., Lauwaert, J., De Vylder, A. et al. (2017). *Microporous Mesoporous Mater.* 251: 1–8.
75. Kapoor, M.P., Bhaumik, A., Inagaki, S. et al. (2002). *J. Mater. Chem.* 12: 3078–3083.
76. Sisodiya, S., Shylesh, S., and Singh, A.P. (2011). *Catal. Commun.* 12: 629–633.
77. Shylesh, S. and Singh, A.P. (2006). *Microporous Mesoporous Mater.* 94: 127–138.
78. Feliczak, A., Walczak, K., Wawrzynczak, A., and Nowak, I. (2009). *Catal. Today* 140: 23–29.
79. Shylesh, S., Srilakshmi, C., Singh, A.P., and Anderson, B.G. (2007). *Microporous Mesoporous Mater.* 99: 334–344.
80. Yang, Q.H., Li, Y., Zhang, L. et al. (2004). *J. Phys. Chem. B* 108: 7934–7937.
81. Melero, J.A., Iglesias, J., Arsuaga, J.M. et al. (2007). *J. Mater. Chem.* 17: 377–385.
82. Yang, H.Q., Han, X.J., Li, G.A. et al. (2010). *J. Phys. Chem. C* 114: 22221–22229.
83. De Canck, E., Nahra, F., Bevernaege, K. et al. (2018). *Chem. Phys. Chem.* 19: 430–436.
84. Baleizao, C., Gigante, B., Das, D. et al. (2003). *Chem. Commun.* 1860–1861.
85. (a) Liu, X., Wang, P.Y., Yang, Y. et al. (2010). *Chem. Asian J.* 5: 1232–1239; (b) Wang, P.Y., Yang, J., Liu, J. et al. (2009). *Microporous Mesoporous Mater.* 117: 91–97.
86. Baleizao, C., Gigante, B., Das, D. et al. (2004). *J. Catal.* 223: 106–113.
87. Baleizao, C., Gigante, B., Garcia, H., and Corma, A. (2003). *J. Catal.* 215: 199–207.
88. Zhang, D.C., Xu, J.Y., Zhao, Q.K. et al. (2014). *Chem. Cat. Chem.* 6: 2998–3003.
89. Zhao, Y.X., Jin, R.H., Chou, Y.J. et al. (2017). *RSC Adv.* 7: 22592–22598.
90. Olkhovyk, O. and Jaroniec, M. (2007). *Ind. Eng. Chem. Res.* 46: 1745–1751.
91. Pearson, R.G. (1963). *J. Am. Chem. Soc.* 85: 3533–3539.
92. Esquivel, D., van den Berg, O., Romero-Salguero, F.J. et al. (2013). *Chem. Commun.* 49: 2344–2346.
93. Burleigh, M.C., Dai, S., Hagaman, E.W., and Lin, J.S. (2001). *Chem. Mater.* 13: 2537–2546.
94. Burleigh, M.C., Markowitz, M.A., Spector, M.S., and Gaber, B.P. (2002). *Environ. Sci. Technol.* 36: 2515–2518.
95. Stöber, W., Fink, A., and Bohn, E. (1968). *J. Colloid Interface Sci.* 26: 62–69.
96. Grun, M., Lauer, I., and Unger, K.K. (1997). *Adv. Mater.* 9: 254–257.
97. Vansant, E.F., Van Der Voort, P., and Vrancken, K.C. (1995). *Characterization and Chemical Modification of the Silica Surface.* Elsevier Science.
98. (a) Pettersson, S.W., Collet, E., and Andersson, U. (2007). *J. Chromatogr. A* 1142: 93–97; (b) Claessens, H.A. and van Straten, M.A. (2004). *J. Chromatogr. A* 1060: 23–41.
99. Kapoor, M.P. and Inagaki, S. (2004). *Chem. Lett.* 33: 88–89.
100. Rebbin, V., Jakubowski, M., Potz, S., and Fröba, M. (2004). *Microporous Mesoporous Mater.* 72: 99–104.
101. Rebbin, V., Schmidt, R., and Fröba, M. (2006). *Angew. Chem. Int. Ed.* 45: 5210–5214.
102. Wahab, M.A. and He, C.B. (2009). *Langmuir* 25: 832–838.
103. Park, S.S., Park, D.H., and Ha, C.S. (2007). *Chem. Mater.* 19: 2709–2711.
104. Lu, Y.F., Fan, H.Y., Doke, N. et al. (2000). *J. Am. Chem. Soc.* 122: 5258–5261.
105. (a) Hatton, B.D., Landskron, K., Whitnall, W. et al. (2005). *Adv. Funct. Mater.* 15: 823–829; (b) Wang, W.D., Grozea, D., Kohli, S. et al. (2011). *ACS Nano* 5: 1267–1275; (c) Wang, W.D., Grozea, D., Kim, A. et al. (2010). *Adv. Mater.* 22: 99–102.
106. (a) Goethals, F., Baklanov, M.R., Ciofi, I. et al. (2012). *Chem. Commun.* 48: 2797–2799; (b) Goethals, F., Levrau, E., Pollefeyt, G. et al. (2013). *J. Mater. Chem. C* 1: 3961–3966.
107. Seino, M., Wang, W.D., Lofgreen, J.E. et al. (2011). *J. Am. Chem. Soc.* 133: 18082–18085.
108. Qiao, S.Z., Yu, C.Z., Xing, W. et al. (2005). *Chem. Mater.* 17: 6172–6176.
109. Park, M., Park, S.S., Selvaraj, M. et al. (2011). *J. Porous Mater.* 18: 217–223.
110. Zhu, L., Liu, X.Y., Chen, T. et al. (2012). *Appl. Surf. Sci.* 258: 7126–7134.
111. Wang, X.Q., Lu, D.N., Austin, R. et al. (2007). *Langmuir* 23: 5735–5739.

112. Shin, J.H., Park, S.S., Selvaraj, M., and Ha, C.S. (2012). *J. Porous Mater.* 19: 29–35.
113. Lin, C.X., Qiao, S.Z., Yu, C.Z. et al. (2009). *Microporous Mesoporous Mater.* 117: 213–219.
114. Vathyam, R., Wondimu, E., Das, S. et al. (2011). *J. Phys. Chem. C* 115: 13135–13150.
115. Moorthy, M.S., Park, S.S., Fuping, D. et al. (2012). *J. Mater. Chem.* 22: 9100–9108.
116. Clerick, S., Libbrecht, W., van den Bergh, O. et al. (2015). *Adv. Porous Mater.* 2: 157–164.

8 Era of the Hybrids – Part 2: Metal–Organic Frameworks

8.1 Introduction

Metal–organic frameworks (MOFs) can be considered as an exciting new development in the field of ordered (crystalline) porous materials. The year 1999 was a good one for porous materials; just as the PMOs (Periodic Mesoporous Organosilicas) were invented by three independent research groups (previous chapter), the same holds for the MOFs. They were also reported in 1999 by three independent research groups. Omar Yaghi, Susumi Kitagawa, and Gerard Férey all published their work on MOFs (only Yaghi used the name MOF) in that year.

Omar M. Yaghi is one of the pioneers in the field of MOFs. He was born in Amman, Jordan in 1965 and received his Ph.D. from the University of Illinois-Urbana in 1990. Afterward, he worked at Harvard University, the Arizona State University (1992–1998), the University of Michigan (1999–2006), and at the University of California, Los Angeles (2007–2011). Currently, he is the James and Neeltje Tretter Professor of Chemistry at UC Berkeley and is co-director of the Kavli Energy

Introduction to Porous Materials, First Edition.
Pascal Van Der Voort, Karen Leus and Els De Canck.
© 2019 John Wiley & Sons Ltd. Published 2019 by John Wiley & Sons Ltd.

NanoSciences Institute of the University of California, Berkeley and the Lawrence Berkeley National Laboratory, as well as the California Research Alliance by BASF. He is widely known for inventing several classes of new materials such as MOFs (the Yaghi MOFs have the acronym MOF), ZIFs, and covalent organic frameworks (see Chapter 9). He holds several awards including the U.S. Department of Energy Hydrogen Program Award for outstanding contributions to hydrogen storage (2007), the American Chemical Society Chemistry of Materials Award (2009), the Royal Society of Chemistry Centenary Prize (2010), and the China Nano Award (2013). In 2018 he won the prestigious Wolf Prize in Chemistry. He has published over 200 articles, with an average of 300 citations per paper making him one of the top five most highly cited chemists worldwide.

Photograph: yaghi.berkeley.edu/

Gérard Férey (14 July 1941–19 August 2017) was a French chemist who specialized in the physical chemistry of solids. He founded the Chemistry Department of the Institut Universitaire de Technologies in Le Mans, in 1968, and was a professor at the University of Le Mans during the periods 1967–1988 and 1992–1995. From 1988 until 1992, he was the director of the Chemistry Department of CNRS before he founded the Institut Lavoisier at the new University of Versailles, at the request of CNRS and the Ministry of Education. Although in the beginning, Férey's research interest mainly focused on the chemistry of inorganic fluorides, later he developed many new porous inorganic hybrid materials for several applications. The Férey MOFs have the acronym MIL, derived from his institute. Férey received many international awards, including the CNRS Gold medal in 2010, which is the highest French scientific distinction. He was a member of the French Academy of Sciences, and was the Vice-President of the Société Chimique de France (SCF) in 2007. He is co-author of more than 600 publications.

Photograph: https://lejournal.cnrs.fr/articles/gerard-ferey-architecte-de-la-matiere

Susumu Kitagawa is a Japanese scientist born in 1951. He was Assistant Professor and Associate Professor at the Kindai University (Osaka). From 2007 to now, he is the director of the Institute for Integrated Cell–Material Sciences at Kyoto University. His research interest is focused on coordination chemistry with a special focus on the chemistry of organic–inorganic hybrid compounds, particularly the chemical and physical properties of porous coordination polymers and metal–organic frameworks. He holds several awards including the Humboldt Research Award (2008, Germany), the Kyoto University Shi-Shi Award (2013), Fred Basolo Medal for Outstanding Research in Inorganic Chemistry (2016, Northwestern University), and in 2017 he obtained the Chemistry for the future Solvay Prize. Additionally, he was appointed in 2016 as one of the most highly cited researchers according to the Web of Science having more than 500 publications and over more than 43 000 citations.
Photograph: https://www.solvay.com/en/media/press_releases/20170928-Solvay-prize-professor-Kitagawa.html

MOFs consist of inorganic units forming a network of repetitive building blocks in one or more dimensions. MOFs contain metal ions or metal clusters that are connected with each other by rigid organic linkers to create rigid frameworks (often with permanent porosity) as seen in Figure 8.1.

MOF-5 was one of the first MOFs that was synthesized by Omar Yaghi's group in 1999 [1]. MOF-5 is constructed of Zn_4O tetrahedral clusters that are connected by 1,4-benzenedicarboxylate organic linkers (Figure 8.2).

In recent decades, there has been an exponential growth in the field of MOFs due to the almost unlimited choice of building bricks that can be used during synthesis. For this reason, many acronyms have been introduced in the literature to refer to the synthesized MOF. Sometimes, the acronym is related to the laboratory that first synthesized the new MOF structure followed by a number. Some well-known examples are: MIL (Matériaux de l'Institute Lavoisier), UiO (University of Oslo), HKUST (Hong Kong University of Science and Technology), NU (Northwestern University), and DUT (Dresden University of Technology). In other cases, the acronym refers to the composition or structure of the material such as ZIF (zeolitic imidazole framework), PCP (porous coordination polymers), PMOF (porous metal–organic framework), and MMOF (Microporous Metal–Organic Framework).

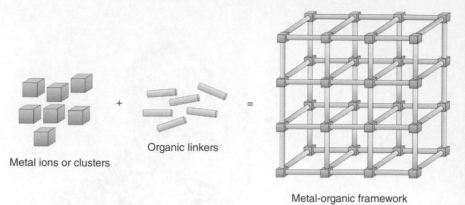

Figure 8.1
Schematic build-up of a metal–organic framework.

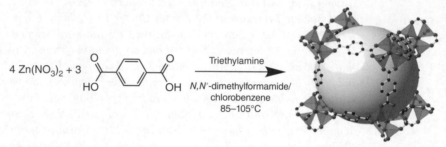

Figure 8.2
Schematic representation of the synthesis of MOF-5.

8.2 Isoreticular Synthesis

MOFs consist of two main components: the organic linkers and the metal ions. The linkers act as "struts" that connect the metal ions, which on their part act as "joints" in the resulting MOF architecture. For this reason, synthesis of MOFs is often based on trial and error techniques. However, the need for "designable" MOFs is high. Within this context, in 2002 O'Keeffe and Yaghi's group introduced the concept of *isoreticular synthesis* (*iso*: the same, *reticular*: forming a net) based on the association of designed rigid secondary building units (SBUs) into predetermined ordered structures (networks) that are held together by strong bonds [2].

Yaghi illustrated this in 2002 by reproducing the octahedral inorganic SBU of MOF-5 *by using similar but other* organic linkers [2]. Several functionalities could be easily built into the framework as —Br, —NH_2, —OC_3H_7, and —OC_2H_{11}. Also, the pore size could be readily expanded by using elongated organic linkers such as biphenyl, terphenyl, and pyrene (Figure 8.3). This allowed the synthesis of large series of isoreticular metal–organic frameworks (IRMOFs) in which the functionality could be tuned, and the pore size could be varied from the microporous until the mesoporous range. They were able to increase the pore size from 0.38 to 2.88 nm without changing the original topology.

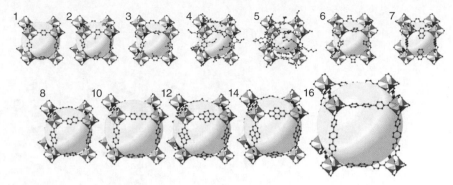

Figure 8.3

Series of isoreticular metal–organic frameworks based on MOF-5. In IRMOF-1 until IRMOF-7 the organic linker had a different functionality whereas in IRMOF-8 to IRMOF-16, the length of organic linker was changed. Source: Reproduced with permission of AAA [2].

Mohamed Eddaoudi was the first author on Yaghi's first papers on the MOF-5 and the isoreticular synthesis. Currently, he is a distinguished professor at KAUST (King Abdullah University of Science and Technology, Saudi Arabia) and the Director of the Advanced Membranes and Porous Materials Research Center. He obtained his Master's and Ph.D. at the Denis Diderot University in Paris. He is highly active in the field of MOFs for hydrogen, CO_2, and water sequestration.

Photograph: http://www.gpcaresearch.com/speaker/prof-mohamed-eddaoudi

8.3 Well-Known MOFs

There are hundreds of MOFs. A few of them have become famous in the field, because of the ease of synthesis, or commercial availability, stability, or applicability. We present some of the best known MOFs here.

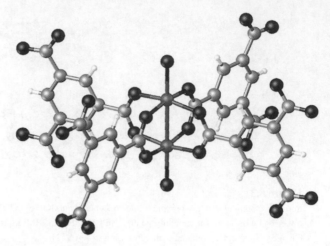

Figure 8.4
Paddle wheel structure adapted by Cu^{2+} cations and BTC anions within the HKUST-1 framework. Source:
Reproduced with permission of the RSC [4].

8.3.1   Cu-BTC

Cu-BTC, also known as HKUST-1, named after the Hong Kong University of Science
and Technology is a 3D channel-based MOF. It was first reported by Williams et al.
[3] in 1999 and consists of Cu-based paddle wheeled SBUs that are linked together
by benzene-1,3,5-tricarboxylic acids (BTCs) linkers (Figure 8.4). The framework
of HKUST-1 consists of Cu dimers in which each copper atom is coordinated by
four oxygen atoms from the BTC linkers and to two water molecules. These can be
removed by gentle heating under low-pressure, creating coordinatively unsaturated
(CUS) metal sites. During this process, the oxidation state of Cu is maintained and
the crystalline structure is preserved making it a versatile material toward its use in
catalysis. Cu-BTC has been examined as a catalyst in, for example, cyanosilylation
of aldehydes and ketones, hydrosilylation of ketones, and in oxidation reactions. It is
commercially available on lab scales and it relatively easily synthesized.

8.3.2   MIL-53

MIL-53, one of numerous MIL members synthesized by Férey's group, is an intrigu-
ing MOF due to its flexible behavior (see Section 8.7.1.4.1) [5]. MIL-53 is built up
by corner sharing $MO_4(OH)_2$ (with M = Al, Fe, V, Cr, Ga, Sc, or In) octahedra that
are connected to each other by 1,4-benzenedicarboxylate organic ligands to form a
1D channel-based MOF (see Figure 8.5). On removal of the solvent molecules, a

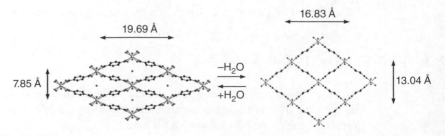

Figure 8.5
Hydration and dehydration process occurring in MIL-53. Source: Reproduced with permission of ACS
[6].

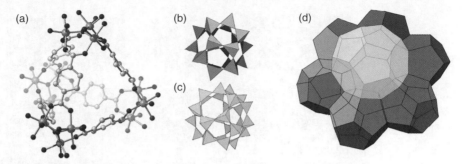

Figure 8.6

(a) MIL-101 super tetrahedron; (b) small and (c) large super cage. (d) MIL-101 structure (MTN zeotype) (Ph.D. thesis, Functionalized Metal–Organic Frameworks as Selective Metal Adsorbents, promoter Pascal Van Der Voort, UGent, 2017). Source: Courtesy of Jeroen De Decker.

reversible structural change occurs. The structure changes from a closed pore toward an open pore form. This change in the pore dimensions is accompanied by a change in the unit cell volume up to 40%. Many factors influence the framework flexibility, such as the employed metal ion, the temperature, the nature of the guest molecules, and the size and functionality of the linkers.

8.3.3 MIL-101

MIL-101(Cr) is another structure, originating from the Férey group [7]. MIL-101 is a mesoporous cage-type MOF (Figure 8.6). Its three-dimensional framework consists of inorganic chromium-oxide clusters, interconnected by terephthalate linkers, which leads to the formation of highly ordered cages of super tetrahedra. MIL-101 possesses two types of mesopores cages with free diameters of 2.9 and 3.4 nm that are accessible through two micropores windows of 1.2 and 1.6 nm. Due to the presence of these mesoporous cages and because of the fact that MIL-101 is a highly porous MOF having a Langmuir surface area of approximately 5900 m^2 g^{-1}, it is a highly attractive structure to incorporate large molecules. This MOF shows excellent resistance to acidic and aqueous media and belongs to the most stable mesoporous MOFs to date.

The MIL-101 topology also exists for other trivalent metals (Fe, Al, Sc, etc.), but the Cr-variant is the most stable of all.

8.3.4 UiO-66

UiO-66, a Zr-based MOF developed by Lillerud et al. [8] at the University of Oslo in 2008, consists of a highly stable inorganic brick of inner $Zr_6O_4(OH)_4$ cores that are connected to 12 carboxylate groups originating from the 1,4-benzenedicarboxylic acid struts to form $Zr_6O_4(OH)_4(CO_2)_{12}$ clusters. The exceptional high thermal stability (up to 540 °C) is due to the stable Zr—O bonds that are formed between the inorganic brick and the carboxylate groups. Its cubic 3D structure consists of an octahedral cage connected to eight corner tetrahedral cages (Figure 8.7). These cages can entrap guest species, but they are usually occupied by solvent molecules that can be removed on heating under vacuum. Because of its high thermal and chemical stability as well as its promising properties for CO_2/CH_4 gas separation, this MOF is one of the most studied in literature. It even shows (some) photocatalytic activity.

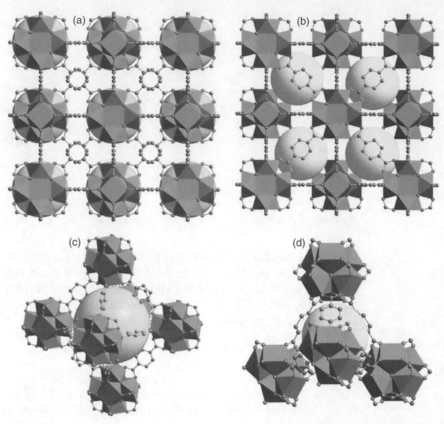

Figure 8.7
Ball-and-stick representations of the 3D cubic framework structure of UiO-66. (a and b) Parts of the framework showing spatial arrangements of the octahedral and the tetrahedral cages. (c and d) Magnified views
of the octahedral and the tetrahedral cages. Source: Reproduced with permission of the RSC [9].

Karl Petter Lillerud (left) and Unni Olsbye from the University of Oslo are two of the
inventors of the UiO-66. Unni Olsbye is holding a molecular model of the UiO-66 in
her hand.
Photograph   Lillerud: https://www.mn.uio.no/kjemi/english/people/aca/kplPhotograph
Olsbye: https://www.ntva.no/unni-olsbye-fikk-arets-innovsasjonspris-ved-uio

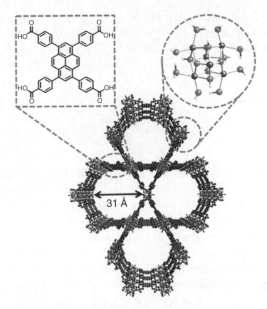

Figure 8.8
Representative structure of the mesoporous Zr-based MOF, NU-1000. Source: Reproduced with permission of Springer Nature [11].

8.3.5 NU-1000

NU-1000 is another Zr-based MOF, synthesized by Hupp and coworkers in 2013 at the Northwestern University [10]. This MOF, obtained via a straightforward solvothermal synthesis method, consists of octahedral Zr_6 nodes and tetratopic pyrene-containing linkers (Figure 8.8) giving rise to exceptionally large mesoporous channels of 3.1 nm. Eight of the 12 octahedral Zr edges are connected to the organic units, while the remaining Zr coordination sites are occupied by terminal −OH ligands. This MOF exhibits a high thermal stability and possesses a good chemical stability in the pH range 1–11 due to the strong bonding between the zirconium nodes and the carboxylate groups of the organic linkers. Because of its high stability and presence of isolated Lewis acid sites on the Zr_6 nodes, this material is interesting for use in catalysis. NU-1000 has also been examined as an adsorbent. For example, it has been observed that the presence of the hydroxyl ligands on the Zr_6 nodes of NU-1000 are important for the adsorptive removal of selenate and selenite from water [12].

Omar Farha (left) and Joseph Hupp (right) awarding an innovation grant to one of their students at Northwestern University (USA). Both professors are highly active in the field of MOFs with a huge focus on the industrial applications of these materials. Fahra produces custom-made MOFs for several applications (see "industrial applications")

Photograph: http://chemgroups.northwestern.edu/hupp/innuvative.html

8.3.6   ZIF-8

ZIF materials represent a subclass of MOFs, and are very similar to the conventional aluminosilicate zeolites in which the Zn^{2+} is the metal node and the imidazolate ions play the role of the oxygen in the zeolite. The angles of these ZIF materials are very close to 109°, creating tetrahedral building blocks very similar to zeolites. Because of this, ZIFs exhibit zeolite-like topologies. ZIF-8, one of the most studied microporous ZIF materials, was reported for the first time by Chen et al. [13] using the solvothermal synthesis method. In ZIF-8, each Zn^{2+} ion is coordinated to four nitrogen atoms from the methylimidazolates moieties (Figure 8.9). ZIF-8 has a sodalite (SOD) zeolite-type structure containing cages with a diameter of 1.16 nm that are accessible through small apertures with diameters of 0.34 nm. As the kinetic diameter of CO_2 is 0.33 nm, CO_2 molecules show a good interaction with the ZIF-8 framework resulting in a high CO_2 adsorption capacity. In addition, the basic sites associated with the imidazole groups in the ZIF-8 structure tend to attract more CO_2 that results not only in a high adsorption capacity, but also with a good selectivity toward CO_2 in the presence of methane, nitrogen, or oxygen.

8.4   Stability of MOFs

Unfortunately, only a few MOFs have a good chemical stability. To demonstrate this, Van Der Voort and Leus evaluated the chemical and hydrothermal stability of several well-known and "believed to be stable" MOFs [15]. The group investigated the hydrothermal and chemical stability toward acids, bases, air, water, and peroxides. Crystallinity and porosity of the fresh materials and the materials after treatment for three days and two months was checked. Most MOF materials showed a good hydrothermal stability due to the high bond strength between the metal oxide and the bridging linker (Table 8.1). The chemical stability toward acids and bases was

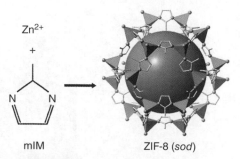

Figure 8.9
Crystal structure of ZIF-8. Source: Reproduced with permission of Elsevier [14].

Table 8.1 Overview of the stability (N_2 and XRD) of the examined MOFs in the various media.

NP, narrow pore.

░ the XRPD pattern and surface area is largely preserved,

▒ the XRPD pattern and surface area is completely destroyed,

▌ the XRPD pattern and surface area is partially decreased/degraded or transformed.

Source: Reproduced with permission of Elsevier [15].

rather disappointing, especially after two months. In hydrogen peroxide (H_2O_2), only the Zr-based MOFs, UiO-66, and NH_2-UiO-66 showed no loss in crystallinity after two months. Many materials that appeared to be stable based on XRPD analysis exhibited a significant decrease in their surface area.

8.5   Preparation of MOFs

Nowadays, there are many methods to synthesize MOFs. In the following, the most commonly employed synthetic approaches will be described briefly: hydro- and solvothermal synthesis, microwave assisted synthesis, electrochemical synthesis, and the high-throughput synthesis method. Additionally, we present some of the approaches used in the preparation of multifunctional frameworks.

8.5.1   Hydro- and Solvothermal Synthesis

In general, MOFs are synthesized using soluble salts as the metal source (e.g. metal sulfates, nitrates, or acetates) with the organic linker in a solvent or mixture of solvents [16]. Hydrothermal synthesis refers to the use of water as solvent whereas solvothermal synthesis involves the use of organic solvents such as alcohols, pyridine, and dialkylformamides. The choice of the solvent depends on its ability to dissolve the organic linker. In both cases, the reaction takes place in a Teflon-lined stainless autoclave in which an autogeneous pressure is built up. The most crucial parameters during the self-assembly are temperature, concentration of metal-salt, the solubility of the reactants in the solvent, and the pH of the solution. The typical temperature range is between 80 and 260 °C for several hours up to several days. Although this procedure can yield good crystals, the long reaction times and the difficult scalability are the mean disadvantages of this method.

8.5.2   Microwave-Assisted Synthesis

Microwave-assisted synthesis of MOFs has attracted growing attention due to rapid synthesis of MOFs and nanoporous materials. The main advantage of microwave heating is its energy efficiency as the energy is generated directly throughout the bulk of the material instead of by conduction from the external surface [17].

As a consequence, microwave heating allows shorter crystallization times, narrow particle size distribution, high yields and selectivities, and a facile morphology control. The very first MOF that was synthesized using a microwave is a chromium-based MOF, MIL-101. The framework was synthesized in 4 hours at 220 °C with a yield comparable to the conventional synthesis method that needed a reaction time of 4 days. Since this first report, many other MOFs have been prepared using microwave heating. Although the MOFs can be obtained in shorter reaction times, in most reports, microcrystalline powders are obtained due to the fast kinetics of crystal nucleation and growth or in other words, the obtained crystals are significantly smaller than the ones obtained by solvothermal heating. For this reason, this synthesis method is sometimes considered unsuitable for the investigation of new phases.

8.5.3 Electrochemical Synthesis Route

The earliest report on the electrochemical synthesis of MOFs was carried out by researchers from BASF in 2006 for the synthesis of HKUST-1 [18]. In this study, bulk copper plates were arranged as the anodes in an electrochemical cell with the carboxylate linker dissolved in methanol and a copper cathode. After a reaction time of 150 minutes, a greenish blue precipitate was formed. Since the pioneering work of BASF, the electrochemical synthesis route has been widely applied in MOF chemistry, including the synthesis of Zn-based MOFs, Cu-based MOFs, and Al-based MOFs.

Interestingly, this synthesis route allows for the preparation of MOF thin films that can be useful in sensing and electrochemical devices. Ameloot et al. [19] modified the electrochemically produced HKUST-1 of BASF to allow the preparation of thin films. By applying an anodic voltage to the Cu-electrode, Cu^{II} ions were introduced into the synthesis solution containing the carboxylate organic linker and methyltributyl-ammonium methyl sulfate as a conduction salt. The crystal size could be tuned by controlling the voltage. Besides offering the advantage of self-completion, the electrochemical synthesis route is also remarkably fast, and the coatings can be formed continuously, which makes them highly applicable in industry.

8.5.4 High-Throughput Analysis

Small changes in reaction parameters can have a significant impact on the resulting framework. For a certain combination of metal-salt and organic linker various crystalline phases can be obtained besides the formation of undesired side products. Unfortunately, each increase in the number of studied parameters raises the number of necessary experiments exponentially. In such cases, high-throughput analysis can offer a solution. Small amounts of reagents and a complete set of parallel reactors can be set to work under identical conditions allowing the efficient discovery of new components, fast optimization of the synthesis conditions and facilitate the observation of certain reaction trends. Most high-throughput studies on MOFs have a goal to elucidate the role of certain parameters such as the influence of the temperature, pH, concentration, and reaction time to allow the discovery of new MOFs and to optimize the synthesis conditions.

8.6 Functionalities in MOFs

MOFs are highly tunable allowing the introduction of several functionalities, and, just as in the case of the PMOs, multiple functionalities can be loaded on one single MOF. Furthermore, their specific hydrophilic/hydrophobic properties can influence the adsorption of specific reagents, products, and intermediates and drive the selectivity of a catalytic process. The highly controlled 3D geometric organization can create a synergetic effect while the nanosized porosity permits shape selectivity.

8.6.1 Active Sites in MOFs

There are several ways to generate catalytically active sites in MOFs. First, the metal or metal cluster connecting points can be used as active functions.

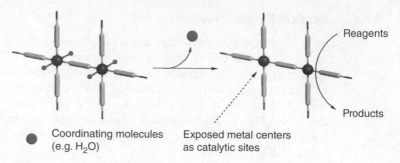

Figure 8.10
Schematic representation illustrating the generation of unsaturated metal connecting points as active sites.
Source: Reproduced with permission of Springer Nature [20].

A metal connecting point with a free coordinating site can be employed as a Lewis acid on removal of the coordinating solvent molecules from the axial positions of the metal center (see Figure 8.10). Nevertheless, in many MOFs the metal centers are completely saturated. Most MOFs do not contain free coordination sites that can be used in catalysis.

A second way to generate catalytic sites in MOFs is by using the active sites generated from the functional groups in a MOF scaffold (see Figure 8.11). In contrast to the traditionally immobilized catalysts, the active sites generated in this fashion are arranged in a predictable and tunable manner due to the periodically ordered nature of MOFs.

Third, the catalytic activity of MOFs can result from encapsulating active species, such as nanoparticles or other metal complexes (Figure 8.12) into the pores/cages of MOFs.

8.6.2 Multifunctional MOFs

Lately, researchers have been particularly interested in the creation of multifunctional MOFs for advanced functionality and applications. Two obvious ways to create multifunctionality are either by using mixed ligands or by using mixed metals as nodes. In both cases, one single structure is created with alternating ligands or metal nodes.

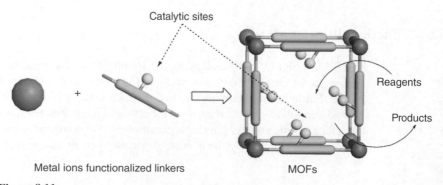

Figure 8.11
Schematic representation showing the use of functional groups in the bridging ligands as active catalysts.
Source: Reproduced with permission of Springer Nature [20].

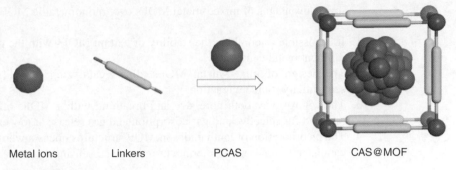

Figure 8.12
Schematic representation of encapsulating catalytic active species inside MOFs. Source: Reproduced with permission of Springer Nature [20].

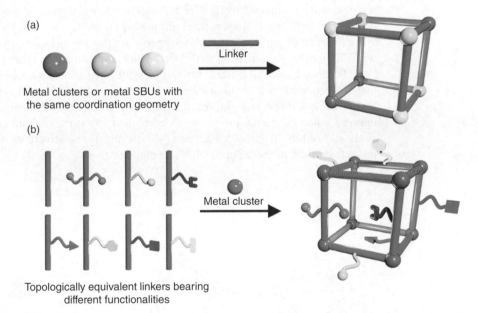

Figure 8.13
(a) Mixed-metal MOFs and (b) mixed linker MOFs. Source: Reproduced with permission of the RSC [21].

8.6.2.1 Mixed Metals

In the majority of reports on mixed-metal MOFs, two (or more) isostructural metal nodes are mixed in the synthesis mixture. It is hoped that the two (or more) isostructural nodes of different metals will be positioned randomly in the MOF. This is shown in Figure 8.13a, taken from the recent review by Roland Fischer [21] .

Mixed-metal MOFs find interesting applications in catalysis, as most enzymes also have a multi-metallic character. We refer here to mixed-metal MOFs only if the metal nodes are part of the crystalline unit. So, grafted metals or embedded metal particles are not considered to be mixed-metal MOFs.

Some advantages of mixed-metal MOFs over monometallic MOFs comprise:

1. It is possible to combine the stability of certain MOFs with the redox activity of different metal centers;
2. The design of mixed-metal MOFs allows catalyzing two different reactions sequentially (tandem reactions);
3. The control over both pore size and breathing within MOFs offers the triggers required for selective and gas adsorption and gas release at low energy cost;
4. The incorporation of lanthanides in MOF structures opens a whole field of applications in luminescence, thermochromism, and sensing.

Sometimes other synthesis routes are possible too. In particular, post-modification techniques are effective. Rather than mixing the components during the synthesis, Cohen and coworkers [22] investigated the metal ion exchange between MIL-53-Br(Al) and MIL-53-Br(Fe). Lillerud [8] studied the replacement of Zr^{4+} with Ti^{4+} and Hf^{4+} ions in the UiO-66. It was found that Zr^{4+} ions in the MOF were partially exchanged by Ti^{4+} or Hf^{4+} ions, and the degree of replacement was dependent on the metal sources used (Figure 8.14).

As a second example, Van Der Voort used a microwave procedure to dope UiO-66 with different transition metals (Ti^{4+}, Ce^{4+}) and lanthanide ions (Nd^{3+}, Eu^{3+}, Yb^{3+}) [24]. Pure Ln-based MOFs are difficult to synthesize and show a limited stability that restricts their use in luminescence and photocatalysis. Doping them into the highly stable UiO-66 material resulted in a new set of Ln-based stable MOFs with unprecedented electronic properties. A combined spectroscopic and computational study allowed to understand the changes in the electronic structures demonstrating the potential use of these doped MOFs in photocatalysis.

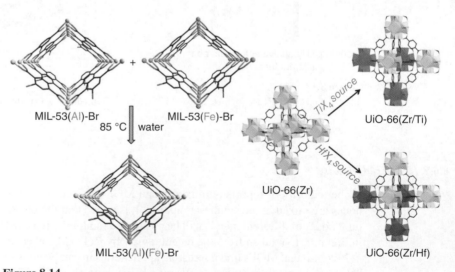

Figure 8.14

Central metal ion substitutions in MIL-53 series (M = Al^{3+} or Fe^{3+}) and UiO-66 series (M = Zr^{4+}, Ti^{4+}, or Hf^{4+}). Source: Reproduced with permission of the RSC [23].

Seth Cohen is currently Professor at the University of California – San Diego. He obtained his bachelor's degrees in Chemistry and in Political Sciences at Stanford and got his Ph.D. at Berkeley. After a postdoc at MIT (Boston), he was tenured at San Diego and became a professor in 2011. He is an important researcher in the field of MOFs, with a focus on the post-synthetic modification methods for the functionalization of Metal–Organic Frameworks.

We refer to the excellent review by Hermenegildo Garcia [25] for more detailed information on the synthesis procedures and application of mixed-metal MOFs.

Herme(negildo) García obtained his Master's and Ph.D. from the University of Valencia (Spain). In 1996, he became full professor at this University and joined the Institute of Chemical Technology (ITQ) at its foundation in 1991. Professor García has focused his research in the field of photochemistry and heterogeneous catalysis where he has studied different materials such as periodic mesoporous organosilicas, carbon nanotubes, diamonds nanoparticles, and metal–organic frameworks, among others.

Photograph: http://www.dicyt.com/viewItem.php?itemId=37044

In some cases, control over the position of the metal nodes is possible to some extent. Van Der Voort has shown that the synthesis method is of high importance [26]. We synthesized two monometallic MOFs, the rigid MIL-47 (V) and the flexible MIL-53 (Cr) and five bimetallic Cr/V structures in two ways: using a solvothermal and microwave assisted synthesis method (see Figure 8.15). To avoid confusion, the MIL-47 (V) has the same topology as the MIL-53 (Cr), the only difference is the flexibility of the framework. The researchers in Versailles (Férey) could have called the MIL-47 also the MIL-53(V), in our opinion. The solvothermal synthesis method produced a uniform dispersion of the metal ions in the lattice. The microwave method, however, produced core-shell compounds having a mixed Cr/V core and a homogeneous Cr shell (Figure 8.15). We showed that vanadium is the most abundant cation in the framework and is preferentially incorporated in the structure, regardless of the synthesis method. The core-shell structures can be explained by a "kinetic quenching" of the reaction. The reaction in the microwave proceeds 50 times faster than the solvothermal route. The nucleation and growth rate of MIL-47(V) are, respectively, 73.6 and 50.6 times faster than those for MIL-53(Cr) during solvothermal synthesis when using V^{3+} and Cr^{3+} salts. So, the microwave synthesis method does not allow enough time for both phases to mix perfectly; the fastest growing phase forms a core with a shell of the slowest growing compound.

Song et al. showed the process of metal exchange nicely in a visual way [27]. They soaked a Co^{2+} based MOF (called ITHD-Co, referring to the net topology) in

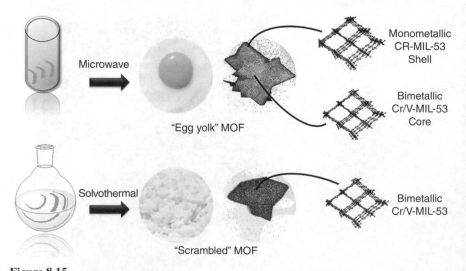

Figure 8.15

Synthesis of bimetallic MOFs using solvothermal and microwave assisted synthesis conditions. Source: Reproduced with permission of the RSC [26].

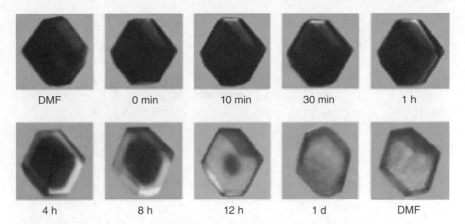

Figure 8.16
Soaking of ITHD-Co in a Zn^{2+} solution, the Zn^{2+} nodes replace the Co^{2+} from the edges to the core, to end up with a ITHD-Zn-MOF. Source: Reproduced with permission of ACS [27].

a solution of $Zn(NO_3)_2$ in DMF. We can see visually how the Zn^{2+} replaces (from outside to inside) the Co^{2+}, ending in a pure ITHD-Zn (Figure 8.16).

8.6.2.2 Mixed Linkers

The principle of mixed ligands is similar to mixed metals. Again, isostructural ligands with different functionalities are required here. The principle is exemplified in Figure 8.13b. Although the principle looks simple, in practice it is often very hard to obtain a nice mixed linker MOF. Typically, the functional groups on the linker will interfere with the coordination to the metal nodes, resulting in an ill-defined, often amorphous material. Sometimes, just as in the case for mixed-metal MOFs, core-shell structures are obtained as well.

One particular example that has been frequently used is the well-known UiO-67 [28], an isorecticular variant of the UiO-66. UiO-67 is composed of biphenyl-4,4'-dicarboxylic acid (H_2BPDC) linkers that are connected to the $Zr_6(\mu_3\text{-}O)_4(\mu_3\text{-}OH)_4(COO)_{12}$ cluster, giving a three-dimensional face-centered cubic (fcu) network. This framework is interesting due to the possibility to replace the BPDC linker with 2,2'-bipyridine-5,5'-dicarboxylic acid (H_2BPy). The structure of mixed linker UiO-67 MOF containing bipyridyl functional groups is shown in Figure 8.17.

This replacement provides an opportunity to the design of the UiO-67(BPy) framework with controlled BPy functional groups per each unit cell. Moreover, the ability to introduce a large variety of active components, including metal complexes, nanoparticles, and organic functional groups into the framework by utilizing the modified ligands directly in the solvothermal synthesis (pre-functionalization) or chemical modification of the framework after synthesis (post-synthetic modification) is an extra advantage to obtain advanced mixed linker MOF materials suitable for more specialized applications. Figure 8.18 shows the different active components that have been applied for the preparation of BPY-functionalized Zr-MOFs.

The presence of these active metal complexes within the UiO-67 framework allows their application in heterogeneous catalysis including C—H activation, aerobic oxidation, hydrogen evolution, and so on.

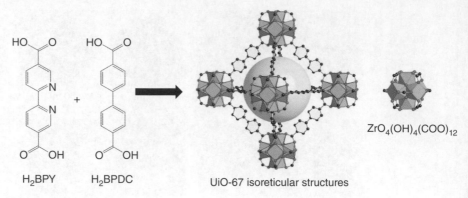

Figure 8.17

Employed linkers and their assembly into Zr-based UiO-67 MOF. Reproduced with permission of Elsevier [28].

There are reports of crystalline solids that have been achieved using linkers with *different relative size*. As an example, biphenyl dicarboxylic acid has been used to form a mixed MOF with a urea-functionalized dicarboxylate that has different size and also different coordination directions with the metal node [29] (Figure 8.19). The obtained mixed linker MOF showed larger pore sizes and enhanced catalytic activity for Henry reactions compared to its pure analog that comprises only the urea-functionalized dicarboxylate linker.

8.6.2.3   *Nanoparticles in MOFs*

A third method to obtain multifunctional frameworks is the preparation of MOFs with well-dispersed nanoparticles (NPs). Several synthesis methods have been reported to introduce nanoparticles in MOFs (Figure 8.20). The most commonly used method is liquid phase impregnation. The MOF is immersed in a solution containing the metal precursor, followed by a reduction to obtain metallic NPs. Although this is a straight-forward method, the most important disadvantage is that it is difficult to solely obtain NPs inside the MOF. A second method of impregnation is the so-called solid phase impregnation as no solvents are used. The embedding of metallic NPs in the MOF is achieved by a simple grinding of the metallic precursors in the presence of the MOFs. During the grinding process, the metallic precursor sublimes and the vapor will diffuse inside the MOF cavities followed by a subsequent reduction. A third method involves the encapsulation of volatile organometallic precursors in the gas phase into the MOF followed by a reduction (chemical vapor deposition).

A very nice example is shown in Figure 8.21. Here, we introduce gold nanoparticles in UiO-66 by simple impregnation, followed by reduction [31]. The resulting tomography pictures show nicely how the gold nanoparticles are truly inside the frameworks, in some cases spanning multiple pores of the UiO-66 framework.

8.6.2.4   *Post-Synthetic Modification*

Just as in the case of the PMOs (see Chapter 7), all kinds of Post-Synthetic Modification (PSM) are possible by modifying the organic linker after the MOF is already synthesized. The possibilities are almost infinite in this case.

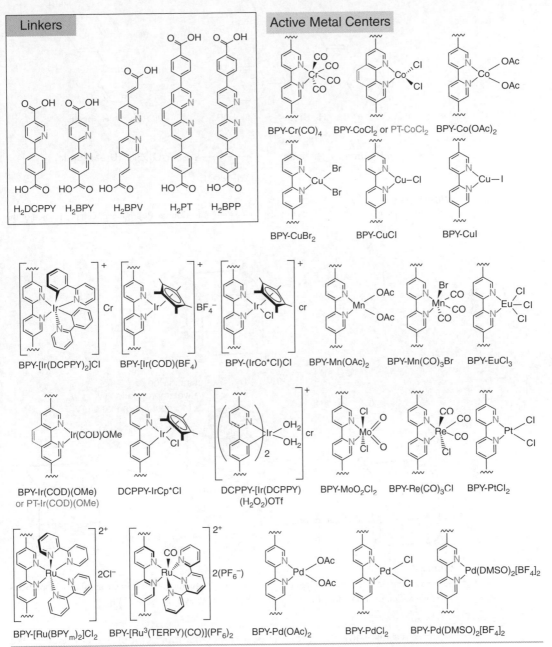

H$_2$DCPPY = 6-(4-carboxyphenyl)nicotinic acid; H$_2$BPY = [2,2′-bipyridine]-5,5′-dicarboxylic acid;
H$_2$BPV = 3,3′-([2,2′-bipyridine]-5,5′-diyl)diacrylic acid; H$_2$BPP = 4,4′-([2,2′-bipyridine]-5,5′-diyl)dibenzoic acid;
H$_2$PT = 4,4′-(1,10-phenanthroline-3,8-diyl)dibenzoic acid; Cp* = Pentamethylcyclopentadientl:
COD = 1,5-Cyclooctadiene; BPY = 2,2′-bipyridine functional group; TERPY = Terpyridine; BPY$_m$ = 2,2′-bipyridine.

Figure 8.18
Applied linkers and active metal complexes for the preparation of BPY-functionalized Zr-MOFs. Source:
Reproduced with permission of Elsevier [28].

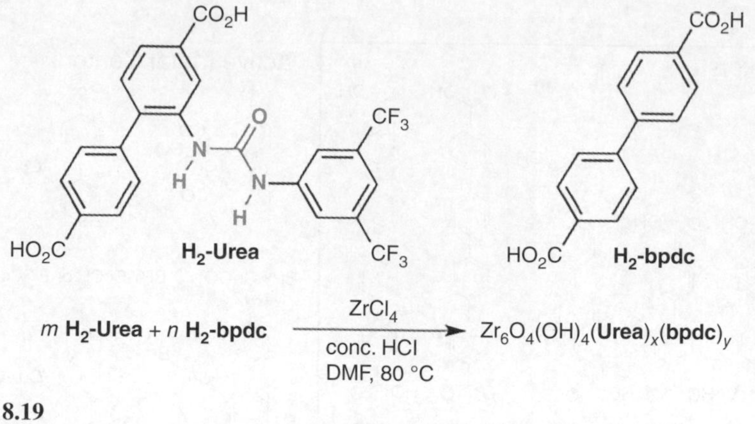

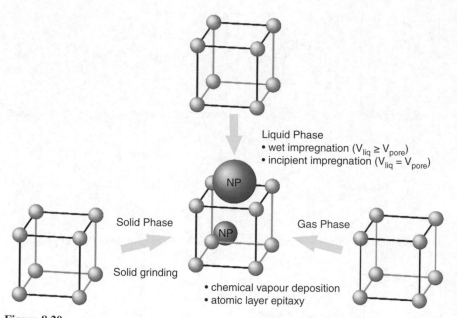

Figure 8.19

Solvothermal synthesis of a "mixed strut" UiO-67-Urea/bpdc. Source: Reproduced with permission of the RSC [29].

Figure 8.20

Schematic representation of the three major types of impregnation methods to encapsulate nanoparticles into MOFs. Source: Reproduced with permission of John Wiley & Sons, Ltd [30].

Here, we just show one example of a somewhat more complex modification. We created a POM@IL-MOF as a highly selective catalyst [32]; POM = Polyoxometallate and IL = Ionic Liquid (Figure 8.22).

As the starting MOF, we used a MIL-101(Cr). The MOF is known for its stability and the typical Cr-node is shown on the top-left of the figure. The Cr-node has two coordinately unsaturated metal sites that are loosely bound to water in ambient conditions and can easily be removed on heating (dehydration) (top right).

At this point, an ionic liquid is introduced (DAIL) that will coordinate through its amine site to the uncoordinated Cr-sites (bottom right). This ionic liquid has an exchangeable Br^- anion that is exchanged with a POM in the last step (bottom left).

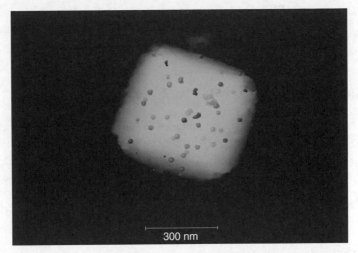

Figure 8.21

Snapshot of the movie of the Au-loaded UiO-66 crystal used for tomographic reconstruction. See supplementary information in Ref. [31] for the full movie. Source: Courtesy of Maria Meledina. Reproduced with permission of the RSC.

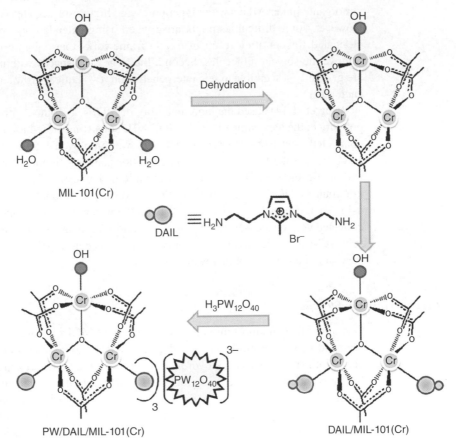

Figure 8.22

Synthetic Scheme for POM@IL-MOF. Source: Reproduced with permission of the RSC [32].

The $[PW_{12}O_{40}]^{3-}$ exchange with the Br^- and attach electrostatically to the cation of the ionic liquid. The obtained material was a very stable and highly selective catalyst for selective alcohol oxidations.

8.7 Applications of MOFs

MOFs are developed for targeted applications, using their optical, magnetic, and electronic properties. MOFs are considered as versatile materials for widespread applications such as, adsorption, separation, catalysis, sensing, luminescence because of their very high surface areas, large-pore volumes, tunable pore sizes, and versatile architectures.

In the following, we will discuss some applications of MOFs in more detail.

8.7.1 MOFs in Gas Storage and Gas Separation

8.7.1.1 *Exceptional High Surface Areas and Porosity*

Some MOFs exhibit an exceptional porosity (up to 90% free volume) and internal surface area extending beyond 6000 $m^2 g^{-1}$ (see Table 8.2). In general, by using a bigger organic linker, MOFs with bigger pores and higher surface areas can be obtained. However, some difficulties might arise when using extended organic linkers. First, elongated linkers often result in fragile frameworks that collapse on activation to remove the solvent molecules. Second, the large void spaces within the framework are generally susceptible to self-interpenetration (two lattices grow and interpenetrate each other).

Wang et al. [41] used the high tunability of MOFs combined with their ultra-high porosity in the harvesting of water. MOF-801, consisting of Zr-based clusters that are connected to each other by fumarate linkers, was employed as an adsorbent to capture and release water from air at ambient conditions and low humidity levels (20%). The authors developed a device that was capable of harvesting 2.81 of water daily per kilogram of MOF material without the use of any additional energy input. Just by using solar energy it was possible to desorb the water from the MOF, which was then collected via a condenser (see Figure 8.23). Particularly in very dry but sunny regions, such as North Africa where the relative humidity is below 20%, this system could be very suitable for overcoming water shortages.

8.7.1.2 *Hydrogen Storage*

One of the first and most important applications of MOFs is gas storage. These very light and very porous materials are ideally suited to capture and store gas. MOFs are very promising materials as hydrogen storage materials. Hydrogen can be used as fuel in combustion engines or in fuel cells. The highly flammable and explosive gas needs to be stored in a safe manner, and when hydrogen gas is adsorbed in micropores it fulfills all safety requirements. The target goals for H_2 storage materials set by the U.S. Department of Energy (DoE) in 2017 are: a gravimetric capacity of 5.5 wt% or 40 $g l^{-1}$ of volumetric capacity at an operating temperature of $-40\,^{\circ}C$ until $60\,^{\circ}C$ at a maximum delivery pressure of 100 bar [42].

Table 8.2 Overview of the reported MOFs with the highest surface area.

MOF	S_{BET}(m^2 g^{-1})	Pore volume (cm^3 g^{-1})	Reference
NU-110^E	7140	4.4	[33]
NU-109^E	7010	3.75	[33]
MOF-210	6240	3.6	[34]
NU-100	6140	2.82	[35]
UMCM-2	5200	2.32	[36]
PCN-68	5110	2.17	[37]
DUT-23-Co	4850	2.03	[38]
MOF-177	4750	1.59	[39]
MOF-205	4460	2.16	[34]
Bio-MOF-100	4300	4.30	[40]
MIL-101	4230	2.15	[7]

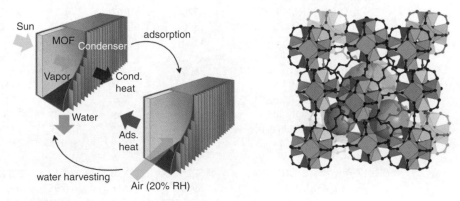

Figure 8.23
Left: A harvesting system for water using a MOF layer and a condenser. Right: structure of MOF-801. Source: Reproduced with permission of AAA [41].

In contrast to many other porous materials where an increase in the H$_2$ adsorption is observed with increasing surface area, there is no such specific correlation between the adsorbed H$_2$ and the MOF surface area (see Table 8.3). It is clear from this table that the hydrogen uptake is especially influenced by the *micropore volume*. This is no surprise (see Chapter 2) as the hydrogen molecules condense at low-pressure in the micropores, showing a typical type I or Langmuir isotherm. The first example was the MOF-5. Yaghi and coworkers [43] obtained a MOF-5 having a Langmuir surface area of 3362 m^2 g^{-1} and a pore volume of 1.19 cm^3 g^{-1}. The material reached a H$_2$ uptake of 1.6 wt% at 77 K and 1 MPa.

Besides the necessity of a large-pore volume, the pore size is also a crucial parameter for H$_2$ adsorption. The optimal MOF material must have pores that closely matches the diameter of H$_2$ to increase the energy of hydrogen binding.

In 2013, Donald Siegel et al. [54]. published a very interesting paper in *Chemistry of Materials*. The authors have checked theoretically more than 60 000 MOF structures from the CSD (Cambridge Structural Database). They published a very interesting figure on the most promising MOFs for hydrogen storage at that time (Figure 8.24).

Table 8.3   Hydrogen uptake by MOFs.

MOF	Surface area BET ($m^2\ g^{-1}$)	Pore volume ($cm^3\ g^{-1}$)	H_2 uptake (wt.%)	Conditions	Refs
$Zn_4O(bdc)_3$, MOF-5	3362	1.19	1.32	77 K, 0.1 MPa	[43]
	2630	0.93	1.0	25 °C, 2 MPa	[44]
	1466	0.52	1.65	25 °C, 4.8 MPa	[44]
	1014	—	1.6	25 °C, 1.0 MPa	[45]
$Zn_4O(Br\text{-}bdc)_3$ IRMOF-2	2544	0.88	1.2	77 K, 0.1 MPa	[2]
$Zn_4O(NH_2\text{-}bdc)_3$ IRMOF-3	3062	1.07	1.4		
$Zn_4O(R^6\text{-}bdc)_3$ IRMOF-6	3263	0.93	1.0		
$Zn_4O(ndc)_3$ IRMOF-8	1466	0.52	1.5	77 K, 0.1 MPa	[46]
$Zn_4O(bpdc)_3$ IRMOF-9	2613	0.9	1.0	77 K, 0.1 MPa	[2]
$Zn_4O(hpdc)_3$ IRMOF-11	1911	0.68	1.62	77 K, 0.1 MPa	[43]
$Zn_4O(pdc)_3$ IRMOF-13	2100	0.73	1.73	77 K, 0.1 MPa	[2]
$Zn_4O(tmbdc)_3$ IRMOF-18	1501	0.53	0.89	77 K, 0.1 MPa	[43]
$Zn_4O(ttdc)_3$ IRMOF-20	4346	1.53	1.31	77 K, 0.1 MPa	[44]
$Zn_4O(btb)_3$ IRMOF-177	4526	1.61	1.25	77 K, 0.1 MPa	[43]
Al(OH)bdc MIL-53 (Al)	1590	—	3.8	77 K, 1.6 MPa	[47]
Cr(OH)bdc MIl-53 (Cr)	1590	—	3.1		
Cu(hfipbb) (H_2hfipbb)$_{0.5}$	—	—	1.0	25 °C, 4.8 MPa	[48]
$Cu_3(tatb)_2$ ($H_2O)_3$	3800	—	1.9	77 K	[46]
$Mn(HCO_2)_2$	240^a	—	0.9	77 K, 0.1 MPa	[49]
$Zn_2(1,4\text{-}(bdc)_2$ (dabco))	1450	—	2.0	77 K, 0.1 MPa	[50]
$Cu_2(bptc)$, MOF-505	1646	0.63	2.48	77 K, 0.1 MPa	[51]
$Cu_2(btc)_{4/3}$, HKUST-1	2175	0.75	2.5	77 K, 0.1 MPa	[44]
$Zn_2(2,5\text{-}(OH)_2\ 1,4\text{-}(bdc)_2)$ MOF-74	1132	0.39	1.7		
$[Zn_4(\mu_4\text{-}O)(L_1)_3(DMF)_2]$	502^a	0.20	1.12	298 K, 4.8 MPa	[52]
$[Zn_4(\mu_4\text{-}O)(L_2)_3]$	396^a	0.13	0.98		

^{a}L_1,6,6′-dichloro-2,2′-diethoxy-1,1′-binaphthyl-4,4′-dibenzoate and L_2, 6,6′-dichloro-2,2′-dibenzyloxy-1,1′-binaphthyl-4,4′-dibenzoate.
Source: Reproduced with permission of Elsevier [53].

Several interesting conclusions can be drawn from this figure. First, it is clear that the vast majority of porous compounds exhibit relatively low H_2 uptake. Still, the authors identified several dozens of MOFs that surpass the targets on a theoretical, materials-only basis. The materials that reach the DoE targets are above and right from the dotted line in the figure. MOF-5, DUT-10, DUT-11, DUT-12 all fulfill the requirement and the best existing and tested MOF is identified as the NU-100.

Farha, Snurr, and Hupp [35] published the NU-100 in *Nature Chemistry* in 2010. They have used *computational modeling* to design and predictively characterize a metal–organic framework (NU-100) with a particularly high surface area. Subsequent experimental synthesis yielded a material, matching the calculated structure, with a high BET (Brunauer, Emmet, Teller) surface area (6143 $m^2\ g^{-1}$). Furthermore, sorption measurements revealed that the material had high storage capacities for hydrogen (164 mg g^{-1}) and carbon dioxide (2315 mg g^{-1}), in excellent agreement with predictions from modeling. Figure 8.25 shows the structure of the NU-100.

Second, Siegel emphasized that the analysis of the MOF mass density and surface area reveals that density decreases with increasing surface area, indicating that

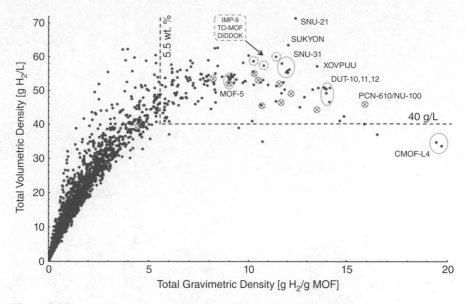

Figure 8.24

Theoretical total (adsorbed + gas phase H_2 at 77 K and 35 bar) volumetric and gravimetric density of stored H_2 in ~4000 MOFs mined from the CSD. The data account only for the mass and volume of the MOF media; mass and volume contributions from the system are neglected. For comparison, the region bounded by the dashed lines represents the DOE 2017 targets for H_2 storage systems. Crossed circles represent common MOFs with incomplete or disordered crystal data in the CSD; structures for these compounds were constructed by hand. Source: Reproduced with permission of ACS [54].

a trade-off exists between gravimetric and volumetric H_2 storage. The concave downward shape of the volumetric versus gravimetric uptake distribution further supports this conclusion: volumetric H_2 density reaches a maximum for surface areas in the range of 3100–4800 $m^2\ g^{-1}$, but then decreases for those compounds having larger surface areas. The data suggests that development of new MOFs should not exclusively target high surface areas, but instead focus on achieving moderate mass densities ($>0.5\ g\ cm^{-3}$) in conjunction with high surface areas.

Third, the discrepancy between these theoretical and the experimental data seems to be mainly due to *incomplete solvent removal* and/or *pore collapse*.

They conclude that [54]

> *looking to the future, we suggest that research efforts targeting MOFs for gas storage emphasize the challenges of structure stability/pore collapse and solvent removal. Many promising compounds in our data set exhibit these deficiencies, and we believe these issues warrant additional effort to quantify the factors that determine whether a given compound can be realized in a robust, solvent-free form.*

8.7.1.3   MOFs in CO_2 Capture and CO_2 Separation

8.7.1.3.1   High Pressure CO_2 Capture

CCU (carbon capture and utilization) is a hot topic, for the obvious reasons of atmospheric pollution and greenhouse effects.

In the early years of their development, mainly MOFs with large surface areas and pore volumes were examined for their potential use in the adsorption of CO_2. Within

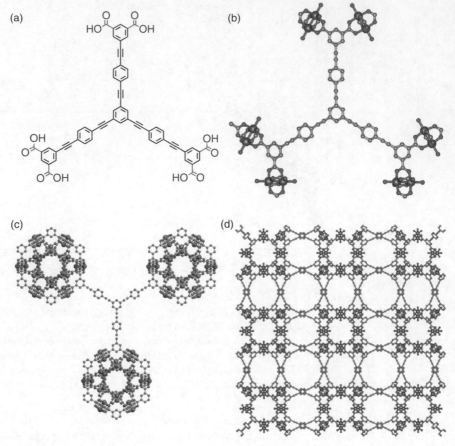

Figure 8.25

Structural features of NU-100. (a) Schematic of the chemical structure of LH_6. (b–d), Representations of the single-crystal X-ray structure of NU-100: LH_6 connecting six paddlewheel units (b), cubaoctahedral building blocks (c) and one of the different cages in NU-100 (d). Hydrogens and disordered solvent molecules are omitted for clarity. Source: Reproduced with permission of Springer Nature [35].

this context, a Zn-based MOF, MOF-177, showed a remarkable CO_2 uptake. A gas cylinder filled with MOF-177, having a BET surface area of $4500\,m^2\,g^{-1}$ was able to store nine times the amount of CO_2 than the same cylinder without the MOF. Besides MOF-177, many other highly porous MOF structures exhibit high gravimetrical CO_2 uptakes. Some of these MOFs are listed in Table 8.4.

8.7.1.3.2 Low-Pressure and Selective CO_2 Capture

Besides the gravimetric uptake and/or volumetric uptake of CO_2, which is a key parameter in evaluating the potential of MOFs as adsorbents, several other criteria also need to be addressed to evaluate the potential of MOFs as adsorbent:

- Selectivity for CO_2 in the presence of other gases such as N_2 and CH_4.
- The chemical and thermal stability of the adsorbent under flue gas conditions, operation conditions, and during the regeneration process.
- Affinity for CO_2 at realistic partial pressures (e.g. 0.15 bar CO_2 in N_2 is a typical exhaust gas concentration).

Table 8.4 High-pressure CO_2 adsorption capacities in selected MOFs between 273 and 313 K.

Chemical formula	Common name	Surface area ($m^2\ g^{-1}$)		Capacity (wt %)	Pressure (bar)	Temp. (K)	Ref.
		BET	Langmuir				
$Zn_4O(BTE)_{4/3}(BPDC)$	MOF-210	6240	10 400	74.2	50	298	[34]
$Zn_4O(BBC)_2(H_2O)_3$	MOF-200	4530	10 400	73.9	50	298	[34]
$Cu_3(TCEPEB)$	NU-100	6143		69.8[d]	40	298	[35]
$Zn_4O(FMA)_3$		1120	1618	69.0[d]	28	300	[55]
$Mg_2(dobdc)$	Mg-MOF-74, CPO-27-Mg	1542		68.9[c]	36	278	[56]
		1800	2060	39.8	40	313	[57]
$Zn_{3.16}Co_{0.84}O(BDC)_3$	Co21-MOF-5		2900	65.0[d]	10	273	[58]
$Be_{12}(OH)_{12}(BTB)_4$	Be-BTB	4030	4400	58.5	40	313	[57]
$Zn_4O(BDC)_3$	MOF-5, IRMOF-1		2900	58.0[d]	10	273	[58]
			3008	45.1[d]	40	298	[59]
$Zn_4O(BTB)_{4/3}(NDC)$	MOF-205	4460	6170	62.6	50	298	[34]
$Ni_5O_2(BTB)_2$	DUT-9			62.1[c]	47	298	[60]
$Zn_4O(BTB)_2$	MOF-177	4500	5340	60.8	50	298	[34]
		4690	5400	60.6	40	313	[57]
		4898	6210	56.8[d]	30	298	[61]
$[Cu_3(H_2O)]_3(ptei)$	PCN-68	5109	6033	57.2[c]	35	298	[37]
$Cr_3O(H_2O)_2F(BDC)_3$	MIL-101(Cr)	4230		56.9[d]	50	304	[62]
		3360	4792	50.2[d]	30	298	[63]
$Ni_2(dobdc)$	Ni-MOF-74, CPO-27-Ni	1218		54.2[c]	22	278	[56]
$[Cu(H_2O)]_3(ntei)$	PCN-66	4000	4600	53.6[c]	35	298	[37]
$Zn_4O(BDC)(BTB)_{4/3}$	UMCM-1	4100	6500	52.7[c]	24	298	[64]
$[Cu(H_2O)]_3(btei)$	PCN-61	3000	3500	50.8[c]	35	298	[37]
$Cu_4(TDCPTM)$	NOTT-140	2620		46.2[d]	20	293	[65]

(Continued)

Table 8.4   *(Continued)*

Chemical formula	Common name	Surface area ($m^2\ g^{-1}$)		Capacity (wt %)	Pressure (bar)	Temp. (K)	Ref.
		Langmuir	BET				
$Tb_{16}(TATB)_{16}(DMA)_{24}$		3855	1783	44.2[d]	43	298	[66]
$Cr_3O(H_2O)_3F(BTC)_2$	MIL-100(Cr)		1900	44.2[d]	50	304	[62]
$Cu_3(BTC)_2$	HKUST-1		1270	42.8	300	313	[67]
			2211	40.1	40	303	[68]
			1571	35.9	15	298	[69, 70]
$H_3[(Cu_4Cl)_3(BTTri)_8]$	Cu-BTTri	2050	1750	42.8[d]	40	313	[57]
$Co(BDP)$	Co-BDP	2780	2030	41.3	40	313	[57]
$Zn_2(BDC)_2(DABCO)$				37.6	15	298	[71]
$Zn(MeIm)_2$	nZif-8, n = nano		1264	35.0[d]	30	298	[72]
$Al(OH)(BDC)$	MIL-53(Al)			30.6[d]	25	304	[73]
$Al(OH)(BDC-NH_2)$	amino-MIL-53(Al)			30.0[d]	30	303	[74]
$Zn_2(BPnDC)_2(bpy)$	SNU-9	1030		29.9[c]	30	298	[75]
$Al(OH)(ndc)$	DUT-4	1996	1308	26.4	10	303	[76]
$Cr_3O(H_2O)_2F(BDC)_3$	MIL-101(Cr)			26.0[d]	5.3	283	[77]
$Zn(F\text{-}pymo)_2$	β-Zn(F-pymo)			26.0[d]	28	273	[78]
$Mn(tpom)(SH)_2$	MSF-2	816	622	25.0[d]	20.3	293	[79]
$Zn_6O_4(OH)_4(BDC)_6$	UiO-66			24.3[d]	18	303	[80]
$Zn(Gly\text{-}Ala)_2$				19.0[d]	15	273	[81]
$Al_{12}O(OH)_{18}(H_2O)_3(Al_2(OH)_4)(BTC)_6$	MIL-96(Al)			18.6[d]	20	303	[82]
$Cr_3O(H_2O)_2F(NTC)_{1.5}$	MIL-102			13.0[d]	30	304	[83]

Source: Reproduced with permission of ACS [84].

- Adsorption/desorption kinetics: for the regeneration of the adsorbent it is important that the CO_2 is adsorbed and desorbed in a reasonable amount of time (preferably as short as possible), to reduce the recycling time and the amount of the required adsorbent.

While the MOFs in Table 8.4 show excellent adsorption capacities, the conditions of the CO_2 sorption are not necessarily realistic. CO_2 often needs to be captured at much lower pressures (in flue gas, typically 0.15 bar) and in the presence of other gases (in the same example of flue gas, 0.85 bar of N_2 and a significant amount of water/steam).

Using the principles of reticular chemistry, MOFs were developed that exhibit a high selectivity for CO_2, particularly *at low pressures* and in the presence of other gases. In Table 8.5 some of the benchmark performing MOFs are presented for low-pressure CO_2 uptake together with the primary contributing adsorption site. The same structural features used for MOFs in catalysis can be employed to design selective MOF-based adsorbents. In the following, these structural features will be described in more detail [109].

8.7.1.3.3 Unsaturated Accessible Metal Sites

Open metal sites or coordinately unsaturated metal sites (CUS) created on removal of solvent molecules represent the first class of adsorption sites that can be employed to selectively adsorb CO_2. Open metal sites possess a partial positive charge due to their Lewis acid character and show an enhanced selectivity for CO_2 because of dipole–quadrupole interactions. The best performing MOF at the moment is Mg-MOF-74. This excellent performance is due to the coordinatively unsaturated metal sites (see Table 8.5 and Figure 8.26) [110]. This MOF is built up from hexagonal channels of 1.2 nm and has a gravimetric uptake of 27.5 wt% at 298 K and a pressure of 1 bar. This high uptake is correlated to the fact that the metal ion is only fivefold coordinated, allowing the coordination of CO_2 to the sixth coordination site. On the downside, in the presence of moisture, the uptake decreases significantly due to the competition of the water molecules for the active binding sites.

A way to solve the competitive adsorption of water is to tether heteroatoms to the adsorption sites using a pre- or post-synthetic functionalization approach. The incorporation of these heteroatoms or ligand functionalities within the backbone or covalently bounded represent the second type of adsorption sites. Different functionalities have been examined, for example aromatic amines, imines, amides, hydroxy, nitro, and cyano groups. The polarizing strength of the functional group as well as the number of functional groups determines the degree of selectivity toward CO_2. In general, *stronger polarizing groups will enhance the adsorption of CO_2*.

8.7.1.3.4 Functional Ligands

It is also possible to embed several functionalities within a single framework by using the mixed ligand approach. This approach might not only result into an enhanced CO_2 adsorption but also into an increased stability of the MOF framework. Wang et al. [111] synthesized two Co based microporous MOF frameworks using two types of linkers: 1,3,5-benzenetricarboxylic acid and 3-amino-1,2,4-triazole (see Figure 8.27). Both frameworks exhibited a good water stability as no structural changes were observed after standing in water at room temperature for one week. Although the surface areas of both frameworks were rather low (BET surface areas of 465 and 356 m^2 g^{-1}) a steep uptake of CO_2 was observed in the low-pressure

Table 8.5 Benchmark MOF adsorbents for low-pressure CO_2 uptake together with the primary adsorption sites.

MOF	Primary adsorption site	Capacity (wt %)	Temperature (K)	Pressure (bar)	Ref
Mg-MOF-74	OMS[a]	27.5	298	1	[85]
[Zn$_2$(tdc)$_2$(MA)]n	Hybrid	27.0	298	1	[86]
Fe-MOF-74	OMS	23.8	298	1	[87]
Mg$_2$(DOBPDC)	OMS	22.0	298	1	[88]
Al-(TCPP)$_{Cu}$	OMS	21.7	298	1	[89]
TEPA-Mg/DOBDC-40	Aliphatic amine	21.1	298	1	[90]
Cu(Me-4py-trz-ia)	Hybrid	21.1	298	1	[91]
Cu-TDPAT	Hybrid	20.6	298	1	[92]
Ni-MOF-74	OMS	20.5	298	1	[93]
rht- MOF-9	Heteroaromatic amine	20.2	298	1	[94]
HKUST-1	OMS	19.8	293	1.1	[95]
NbO-Pd-1	OMS	19.7	298	1	[96]
Co-MOF-74	OMS	19.7	298	1	[93]
[Mg$_2$(dobdc)(N$_2$H$_4$)$_{1.8}$]	Aliphatic amine	19.5	298	1	[97]
Cu-TPBTM	OMS	19.5	298	1	[98]
nbo-Cu$_2$(DBIP)	OMS	19.3	298	0.95	[99]
CPO-27-Mg-c	Aliphatic amine	19.2	298	1	[100]
[Mg$_2$(DHT) (H$_2$O)$_{0.8}$ (en)$_{1.2}$].0.2(en) SIFSIX-2-Cu-i	SBU-based interactions	19.2	298	1.1	[101]
ZJNU-54	Heteroaromatic amine	19.1	298	1	[102]
SIFSIX-1-Cu	SBU-based interactions	19.1	298	1	[103]
JLU-Liu21	Hybrid	18.8	298	1	[104]
ZJNU-44	Heteroaromatic amine	18.6	296	1	[105]
NJU-Bai21, PCN-124	Hybrid	18.4	298	1	[106]
Zn(btz)	Heteroatom	18.0	298	1	[107]
PEI-MIL-101	Aliphatic amine	18.0	298	1	[108]

[a]OMS = coordinatively unsaturated metal site (open metal sites).
Source: Reproduced with permission of Springer Nature [109].

regime and a total uptake of 34.3 and 26.3 wt% at 195 K and a pressure of 1 bar was obtained. Very high selectivities were observed for CO_2 over N_2. Under the typical partial pressure of CO_2 used in industrial flue gas, the ratios of CO_2/N_2 adsorption amounts up to 199. The authors attributed the enhanced CO_2 uptake of both frameworks to the presence of–NH_2 groups, —C=O/—COOH sites, the open Co^{2+} sites and to the high density of uncoordinated triazolate nitrogen atoms within the pore surface.

8.7.1.3.5 *Molecular Sieving Effect*

Next to the affinity factor and the dipolar/quadrupolar interactions, the efficiency of the separation of small molecules is also determined by the pore size of the MOF, acting as a *molecular sieve*. There are clear differences in the kinetic diameter of some of these small molecules; H_2 (2.89 Å), O_2 (3.46 Å), N_2 (3.64 Å), and CO (3.76 Å). If the separation/purification involves larger molecules, for example, aromatics, not only the pore size but also the shape and hierarchy of the porosity need to be taken into account. Table 8.6 summarizes some examples of MOFs that have been explored

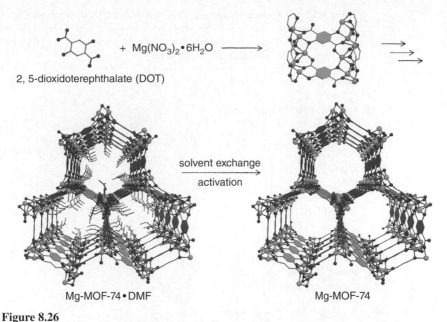

Figure 8.26

Structure of Mg-MOF-74, formed by reaction of the DOT linker with $Mg(NO_3)_2 \cdot 6H_2O$. Source: Taken from https://www.researchgate.net/publication/40042066_Highly_efficient_separation_of_carbon_dioxide_by_a_metal-organic_framework_replete_with_open_metal_sites/figures?lo=1.

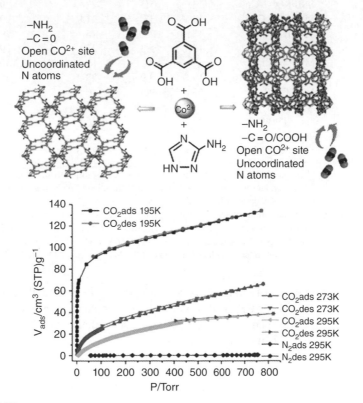

Figure 8.27

Synthesis of mixed ligand based MOFs for enhanced CO_2 adsorption. Source: Reproduced with permission of the RSC [111].

Table 8.6   Selective adsorption of CO_2 over N_2 or CH_4 in some MOFs.

Material	Selectivity[a]		$-Q_{st}$[b] [kJmol^{-1}]	T [K]	Pressure [bar]	Ref.
	CO_2/N_2	CO_2/CH_4				
HKUST-1	101[c]	—	35	293	1	[70]
	—	7.4		296	2	[112]
HKUST-1 (hydrated)	28	7.5	30	298	1	[113]
Bio-MOF-11	75	—	45	298	0.15	[114]
Mg-MOF-74	182.1	105.1[d]	47	296	1	[112]
Ni-MOF-74	30	—	41	298	0.15	[93]
Zn-MOF-74	87.7	5	—	296	1	[112]
MIL-101		9.6[d]	44	296	2	[112]
Cu-TDPAT	34	13.8[d]	42.2	298	0.16	[92]
CU-BTTri	21	—	21	298	1	[115]
	10	—		298	0.15	[115]
en-Cu-BTTri	25	—	90	298	1	[115]
	13	—		298	0.15	[115]
mmen-Cu-BTTri	327	—	96	298	1	[116]
mmen-Mg$_2$(dobpdc)	329		71	298	1	[88]
ZIF-78	50.1[e]	10.6[e]	29	298	1	[117]
ZIF-79	23.2[e]	5.4[e]	—	298	1	[117]
ZIF-81	23.8[e]	5.7[e]	—	298	1	[117]
ZIF-82	35.3[e]	9.6[e]	—	298	1	[117]
UTSA-16	314.7	29.8[d]	34.6	298	1	[112]
SSIFSIX-2-Cu	13.7	5.3	22	298	1	[101]
SIFSIX-2-Cu-i	140	33	31.9	298	1	[101]
SIFSIX-3-Zn	1818	231	45	298	1	[101]
PCN-88	18.1	5.3	27	296	1	[118]
MOF-508b	4[c]	3[c]	19	303	1	[119]
SYSU	19.0	3.9	28.2	298	0.15	[120]
NJU-Bai7	62.8	9.4	40.5	298	0.15	[120]
NJU-Bai8	58.3	15.9	37.7	298	0.15	[120]
Cu$_{24}$(TPBTM)$_s$	22	—	26.3	298	1	[98]
PCN-61	15	—	22	298	1	[98]
SNU-M10	98	—	27.6	298	1	[121]

[a]For calculation of selectivity from ideal absorbed solution theory (IAST), unless otherwise stated.
[b]All Q_{st} values were determined by fitting to the adsorption isotherms, such as the virial or Clausius–Claperyron equation.
[c]For calculations of selectivity from Henry's Law.
[d]Selectivities were calculated from an equimolar mixture of CO_2/CH_4 at 2 bar and 296 K.
[e]From slopes of adsorption isotherms at low pressure.
Source: Reproduced with permission of John Wiley & Sons, Ltd [122].

in the separation of CO_2/N_2 or CO_2/CH_4. In general, the selective adsorption of CO_2 over N_2 or CH_4 is determined by this molecular sieve effect as N_2 and CH_4 cannot be adsorbed within the MOF because of their larger kinetic diameter in comparison to the channel size of the MOF.

For example, Chen and coworkers explored several ultramicroporous MOFs, such as, Mg-MOF-74, MOF-177, ZIF-78, and UTSA-16, for the selective adsorption of CO_2 over CH_4 and N_2 [112]. From all the examined MOFs, UTSA-16

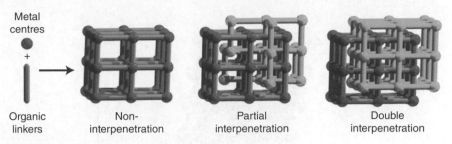

Metal centres

+

Organic linkers → Non-interpenetration

Partial interpenetration

Double interpenetration

Figure 8.28

Example of non-penetrated, partial interpenetrated and double penetrated networks. Source: Reproduced with permission of Springer Nature [123].

(UTSA = University of Texas at San Antonio) exhibited an efficient and reversible CO_2 capture at room temperature and at a low-pressure of 1 bar. UTSA-16 contains a three-dimensional framework and is composed of Co_4O_4 clusters and K^+-polyhedra linkers. The MOF has three-dimensional channels of 3.3×5.4 Å^2 and has small cavities of 4.5 Å exhibiting a CO_2 uptake of 160 cm^3 cm^{-3}. Based on the pure component isotherms fits, the authors calculated the selectivity for CO_2 in a 50/50 CO_2/CH_4 mixture and 15/85 CO_2/N_2 mixture resulting in high values of 29.8 and 314.7, respectively. These values were higher than those of most MOFs used for comparison.

8.7.1.3.6 Interpenetration

Interpenetration is a term to define MOF structures in which two or more structures are intergrown, as shown in Figure 8.28.

Interpenetration creates MOFs with much smaller pores, which is in some cases very beneficial for the CO_2 sorption properties of the materials. So, it is used to tune the pore size and shape of MOFs to allow the separation of CO_2 from other gases; for example, CH_4 and N_2. Within this context, Chen and coworkers reported in 2008 the first example on the use of interpenetrated MOFs to adsorb CO_2 selectively over CH_4 and N_2 in binary and ternary mixtures using fixed-bed breakthrough experiments [119]. The authors used a double interpenetrated MOF, MOF-508b ($Zn(BDC)(4,4'$-Bipy$)_{0.5}$ ($4,4'$-Bipy = $4,4'$-bipyridine)) having one-dimensional pores of about 4.0×4.0 Å in the selective separation of CO_2 (see Figure 8.29). In the first instance, the single-component adsorption isotherms showed that the adsorption capacities of MOF-508b for CO_2, CH_4, and N_2 are 26, 5.5, and 3.2 wt%, respectively, at 303 K and a pressure of 4.5 bar.

In 2012, researchers from the UK published the NOTT-202 (NOTT refers to the University of Nottingham). In this material, the authors optimized the affinity toward CO_2 (playing with the open metal sites and the quadrupolar interactions), flexibility, and interpenetration. The authors made a material with one rigid frame and a second, interpenetrated frame that was only partially occupied. So, they called it *interweaving*. The partially occupied net caused defects throughout the structure and an ideal pore system.

Figure 8.30 shows the low-pressure and high-pressure adsorption isotherms for some small and simple gases. It can be inferred from the Henry regime of these isotherms that particularly the separation efficiency of CO_2 versus the other gases is very high.

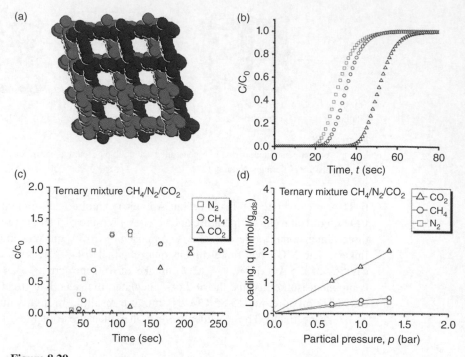

Figure 8.29

(a) One-dimensional pores in MOF-508b, (b) single-component breakthrough curves for N_2 (square), CH_4 (circle), and CO_2 (triangle) at 323 K and 1 bar, (c) multicomponent breakthrough curves for an equimolar ternary mixture at 303 K and 4 bar, (d) adsorption isotherms for an equimolar ternary mixture at 303 K. Source: Reproduced with permission of ACS [119].

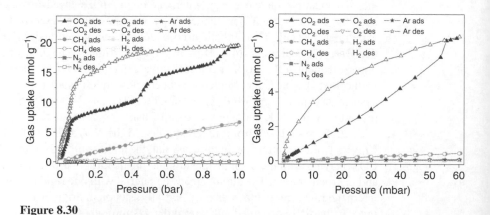

Figure 8.30

Comparisons of low-pressure CO_2, CH_4, N_2, Ar, O_2, and H_2 sorption isotherms at 195 K. The CO_2 isotherms exhibit a marked hysteresis loop, whereas other gas isotherms show good reversibility and low uptake capacities. At 195 K and 60 mbar, the CO_2 uptake was found to be 17–140 times (on a molar basis) higher than uptakes of CH_4, N_2, O_2, Ar, or H_2 under the same conditions. In contrast, at 273 K and 1.0 bar, the CO_2 uptake was only 3.5–22 times higher than uptakes of CH_4, N_2, O_2, Ar, or H_2 under the same conditions. Source: Reproduced with permission of Springer Nature [123].

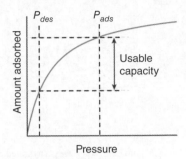 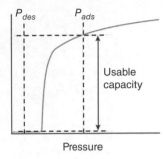

Figure 8.31

Idealized adsorption isotherms comparing the usable capacities for an idealized adsorbent with Langmuir-type adsorption isotherm (left) and a stepped or S-shaped isotherm (right). Source: Reproduced with permission of Springer Nature [124].

8.7.1.4 Breathing MOFs

A very interesting subcategory of MOFs are the so-called breathing MOFs. These flexible structures can open (large-pore state: LP) or close (narrow-pore state) their pore system, as a function of temperature, or pressure on exposure to certain gases. Pore opening leads to a step in the adsorption isotherm. As illustrated in Figure 8.31, when occurring at favorable pressure steps in adsorption isotherms can increase the usable capacity in pressure swing adsorption. Moreover, since such steps occur for certain gases and not for others, and at different pressures for different gases, breathing MOFs also offer perspectives for selective capture.

This flexibility is characterized by a transformation between two or more states on exposing the framework to external physical or chemical stimuli. This results in a structural change of the complete framework. Drastic volume changes may follow from strong host-guest interactions. This phenomenon was already predicted by Kitagawa and Kondo in 1998 [125].

Detailed information on the breathing phenomenon has been obtained by monitoring the physical and chemical properties of MOFs with a variety of analytical techniques. In situ XRD (X-Ray Diffraction) has played a central role in many of the studies. X-ray thermodiffractometry and CO_2 sorption measurements were used to study thermal behavior, and host–guest and guest–guest interactions under exposure to a variety of liquid or gas phase compounds.

8.7.1.4.1 MIL-53

In 2002, MIL-53(Cr) [6] was the first of a series of flexible MOFs. Several other metal nodes followed (Sc, Ga, Fe, all in the +III state, and the famous variant with trivalent Al [5]). They are all BDC (benzenedicarboxylate) linked structures, as shown in Figure 8.32. The MIL-53 framework is constructed from chains of trans-corner-sharing $MO_4(OH)_2$ octahedra, in the other two dimensions connected with BDC. They form a 3D ordered network after synthesis, the MIL-53 as-synthesized (AS) framework is not flexible due to the blocking of the pores by disordered, uncoordinated terephthalic acid molecules, and residual solvent or water molecules. These can be removed by heating or by solvent extraction methods, which are referred to as *activation* of the MOF. The activated structure can reversibly change between a monoclinic hydrated narrow-pore state and an orthorhombic

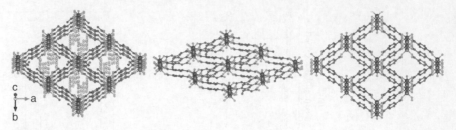

Figure 8.32

The MIL-53(Al) topology as a filled open pore (as synthesized), a hydrated closed pore form (NP-h) and a dehydrated (empty) open configuration (LP). Source: Courtesy of Irena Nevjestic (Ph.D. thesis, U.Gent, 2018).

Table 8.7 Overview of the various forms of MIL-53(Al).

AS (as synthesized) Orthorombic (P*nam*)	NP-h Monoclinic (C*c*)	NP-d Monoclinic (C2/*c*)	LP Orthorombic (I*mcm*)
a = 17.129 Å	a = 19.531 Å	a = 20.756 Å	a = 16.675 Å
b = 12.182 Å	b = 7.612 Å	b = 7.055 Å	b = 12.831 Å
c = 6.628 Å	c = 6.576 Å	c = 6.6067 Å	c = 6.608 Å
$\beta = 90°$	$\beta = 104.24°$	$\beta = 113.58°$	$\beta = 90°$

Source: Courtesy of Irena Nevjestic, Ph.D. thesis, U.Gent, 2018.

large-pore (LP) state as a result of heating or uptake of certain gases. More precise data on the MIL-53(Al) are given in Table 8.7.

8.7.1.4.2   COMOC-2

Another nice example of this breathing behavior was observed in Van Der Voort's vanadium-based MOF, COMOC-2 that consists of $\{VO_6\}$ octahedra that are crosslinked by biphenyl-4,4′-dicarboxylate organic linkers [126]. The CO_2 adsorption isotherm measured at 265 K shows a distinct two-step adsorption process with an initial plateau at approximately 5 mmol g^{-1} and a pressure of 12.5 bar (see Figure 8.33). Above 12.5 bar, a sudden increase in the CO_2 uptake is observed to reach a capacity that is more than the double of the capacity in the first plateau. This corresponds to a structural transition from a "narrow pore" to a "large-pore" (*lp*) form of the material and the pressure at which this transformation takes place is the "gate opening pressure" depending on the adsorbate and temperature. In contrast to the CO_2 adsorption isotherm, the CH_4 adsorption isotherms measured at 265 and 303 K exhibit a classical type I shape, or on the adsorption of CH_4 no breathing behavior is observed allowing a good separation between CO_2 and CH_4. Next to the introduction of guest species, other external stimuli such as light, pressure, and electric field can also result in the reversible change of the channels and pores in soft porous materials [127].

To tailor the breathing behavior Van Der Voort synthesized bimetallic Al/V-COMOC-2 materials using a straightforward solvothermal synthesis approach

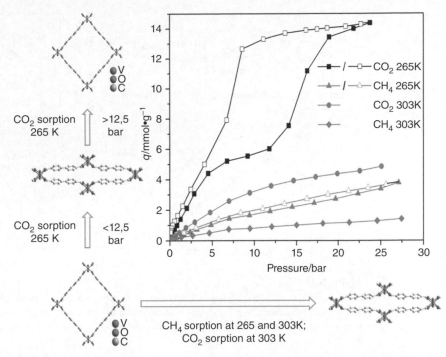

Figure 8.33

Adsorption isotherms of CO_2 and CH_4 at 303 and 265 K. Source: Reproduced with permission of ACS [126].

in which Al is preferentially incorporated [127]. An enhanced thermal stability as well as an increased surface area was obtained with increasing Al content. Additionally, the breathing behavior could be tailored. With increasing the V content, the framework became more flexible and an increase in the hysteresis loop was noted.

8.7.1.4.3 Co-bdp

Jeffrey Long's group has developed a novel family of MOF structures that shows an even richer flexibility than the MIL-53 series and that – very importantly – are also stable in operating humid conditions [128]. They have divalent transition metal nodes (Co^{II}, Fe^{II}, Ni^{II}, and Zn^{II}), connected by 1,4-benzenedipyrazolate (bdp^{2-}) linkers. The structures of the MIL-53 and $M^{II}bdp$ families are very similar. The adsorption isotherms of $Co^{II}bdp$ for CO_2 [57] and CH_4 [124] were published in different papers. When comparing these (Figure 8.34), it becomes immediately clear that this material offers great potential for high pressure CO_2/CH_4 separation. At 10 bar, for example, the uptake of CH_4 is still virtually zero whereas the CO_2 isotherm already starts to saturate. Furthermore, it was shown that the CH_4 adsorption and desorption isotherms can be strongly shifted by exerting mechanical pressure.

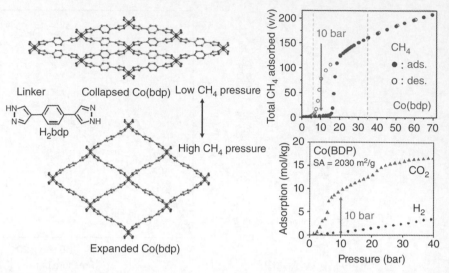

Figure 8.34

Left – Co^{II}bdp in its narrow (collapsed) and large (expanded) pore states [129]; Right – CH$_4$ adsorption and desorption isotherm (top [129]) and adsorption isotherms for CO$_2$ and H$_2$ (bottom [19]) at room temperature. Source: Reproduced with permission of Springer Nature and ACS, adapted from figures in [57, 124].

Jeffrey R. Long is currently Professor of Chemistry and Chemical and Biomolecular Engineering at the University of California, Berkeley (USA), Senior Scientist at the Materials Science Division at the Lawrence Berkeley National Laboratory and the Director of the Center for Gas Sorption. He obtained his bachelor's at Cornell University and his Master's and Ph.D. at Harvard. He is highly active in the field of MOFs, with a special interest in the applications of MOFs in gas sorption and gas separation.

(Photograph: http://alchemy.cchem.berkeley.edu/group-members)

The challenge to reduce the energy required for CO_2 desorption from MOFs in temperature and pressure swing adsorption remains enormous. Moreover, to take advantage of breathing effects in the elimination of CO_2 from post-combustion flue gases from burning fossil fuels ($\sim$15% CO_2 in humid N_2), framework opening should occur near ambient pressure. The quest for more efficient materials and new strategies for adsorbent regeneration is thus still ongoing.

The dynamic behavior of flexible frameworks in the presence of gas mixtures can also be of interest to selectively adsorb one gas over another. A nice example of this was reported by Yaghi's group [130]. In this work, a new ZIF material, having the LTA (Linde Type A) topology, was synthesized solvothermally from $Zn(NO_3)_2.4H_2O$ and an excess of purine in DMF as solvent. This yielded crystalline $Zn(Pur)_2.(DMF)_{0.75}.(H_2O)_{1.5}$ (ZIF-20, pur = purinate) exhibiting a maximum pore aperture of 2.8 Å as determined from the crystal structure. The ZIF-20 material exhibited a five times higher CO_2 uptake in comparison to CH_4 uptake at 273 K and 1 bar suggesting a stronger interaction between the framework and the CO_2 molecules. This encouraged the authors to examine the separation of CO_2 over CH_4 by means of breakthrough experiments using a 50 : 50 v/v gas mixture. Authors noticed that ZIF-20 can completely separate CO_2 from CH_4. Remarkably, the pore aperture of ZIF-20 is smaller than the kinetic diameter of CO_2 and CH_4. The space within ZIF-20 becomes accessible by a *dynamic aperture widening process* in which the "Pur" organic linker swings out to allow the gas molecules to pass. The high CO_2 uptake is probably the result of the presence of uncoordinated N atoms inducing a polar pore wall and for this reason favors the CO_2 over CH_4 adsorption.

8.7.1.5 *Separation of Hydrocarbons*

An important application is the *separation of hydrocarbons*, alkylaromatic and aliphatic isomers, and in the separation of harmful gases [131]. For instance, the separation of acetylene and ethylene is of crucial industrial importance. Ethylene produced nowadays contains about 1% of acetylene. This acetylene heavily poisons the Ziegler–Natta catalysts that are used in the ethylene polymerization, and they form an explosion hazard. The removal of acetylene is at the moment complicated. One route consists of the partial hydrogenation of the acetylene into ethylene over a Pd-catalyst. This catalyst is expensive and, as a side reaction, the olefins also get reduced to alkanes, which is completely undesirable. Another route extracts the olefins using organic solvents. Both routes are costly and energy intensive.

Xiang et al. [132] constructed, following Kittagawa's methods to create mixed-metal MOFs, a chiral and ultramicroporous MOF using the building blocks shown in Figure 8.35 (Kitagawa used the nomenclature M'MOFs for chiral bimetallic MOFs). A very subtle tuning of the pores was possible in these systems. The so-called M'MOF-3 has a separation factor of 25.5 for acetylene/ethylene, which is a promising result for practical application. As a comment, the separation factor is calculated on the individual adsorption isotherms, and not on a realistic mixture of gases.

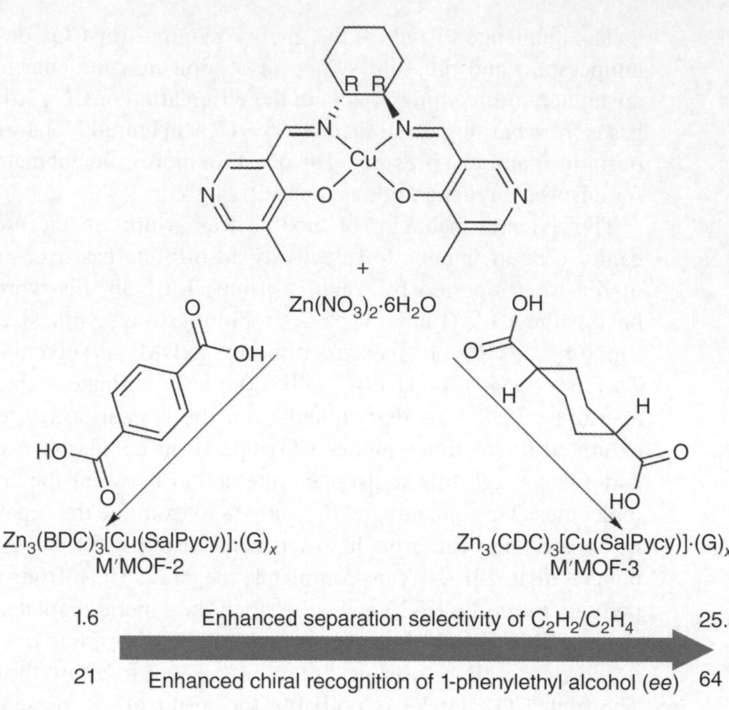

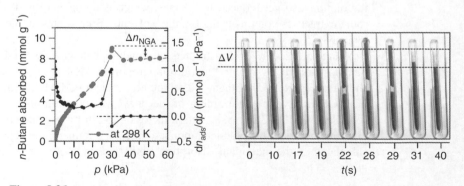

Figure 8.35

M'MOFs developed for the separation of ethylene and acetylene. Source: Reproduced with permission of Springer Nature [132].

8.7.1.6 Negative Breathing

In general, at a constant temperature, the absolute amount of substance in the adsorbed phase increases with increasing pressure of the adsorptive. That is what we see in all the isotherms we have showed in this book so far.

Kaskel et al. [133], showed "negative breathing" on a DUT-49 for the adsorption of *n*-butane at a temperature of 298 K. A distinct negative gas adsorption step was observed at around 30 kPa (see Figure 8.36 left). By using real time adsorption/diffraction/EXAFS measurements, a sudden structural deformation and

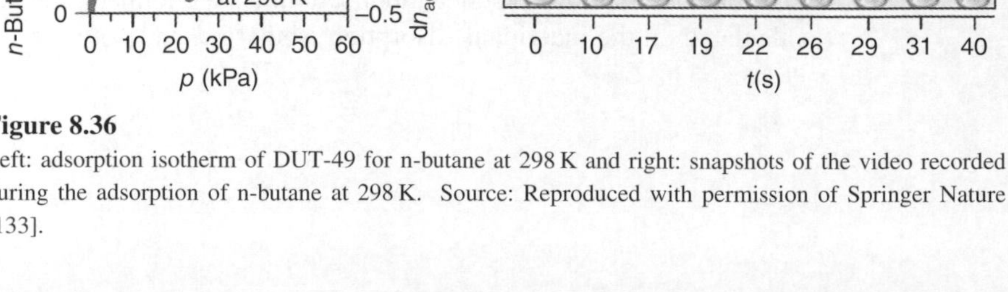

Figure 8.36

Left: adsorption isotherm of DUT-49 for n-butane at 298 K and right: snapshots of the video recorded during the adsorption of n-butane at 298 K. Source: Reproduced with permission of Springer Nature [133].

pore contraction of the MOF was observed that resulted in a release of the guest molecules. These structural changes, which are reversible, could even be detected by visual monitoring the sample during the adsorption experiments. A movement of the particles was caused by the huge gas release during the negative gas adsorption step followed by a pronounced shrinkage of the packed bed volume by approximately 20% (see Figure 8.36, right).

Stefan Kaskel obtained his Ph.D. in 1997 at the University of Tübingen, Germany. He worked at Ames Laboratory (Iowa State University) and at the Max Planck Institut für Kohlenforschung under the directorship of Ferdi Schüth.

In 2004 he became a Professor at the Technical University in Dresden and in 2008 he became (also) manager of the Chemical Surface and Reaction Technology business unit at the Fraunhofer IWS in Dresden.

Kaskel produced a whole series of MOFs that all start with the acronym DUT, in honor of his university.

(Photograph: https://tu-dresden.de/mn/chemie/ac/ac1/die-professur/inhaber-ac1*)*

8.7.2 MOFs in Catalysis

MOFs have already demonstrated potential applications in a huge set of catalytic reactions; for example, oxidation-, reduction-, acid-, and base catalysis. Several reviews have recently appeared on this topic, including one of ours [134]. We will not go into detail here, as the field is very rapidly moving. In general, the confined spaces in MOFs and the almost atomic tunability of the catalytic functions allows for a very highly tunable catalysis as well. Major drawbacks with regard to the zeolites and the oxides always remain the same: lower thermal stability, lower hydrolytical stability, leaching of metal nodes, and loss of crystallinity. However, for specific reactions and applications, to date MOFs offer a valuable solution. As far as we know, MOFs are not yet being used industrially for heterogeneous catalysis at the time of writing (2018).

8.7.2.1 *Chiral Catalysis by Post-Modification*

A special field is the use of chiral MOFs in chiral catalysis or in asymmetric catalysis. Again, there are two major ways to design these catalysts. The first one, constructing an entire MOF out of chiral linkers, has proven to be extremely difficult synthetically

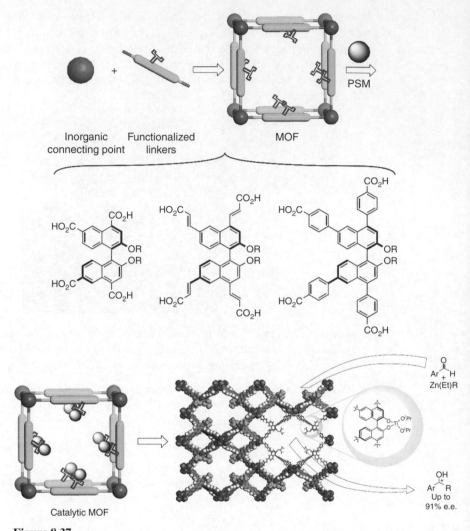

Figure 8.37

Development of isoreticular chiral MOFs as asymmetric catalysts for diethylzinc and alkynylzinc additions. Source: Reproduced with permission of Springer Nature [135].

and often the enantioselectivity of the resulting MOF is somewhat disappointing [134]. The alternative way, the easier way, is to postmodify the MOF with a chiral functionality. We elaborate on one (out of many) example of such a strategy. Lin et al. [135] synthesized a series of isoreticular chiral mesoporous MOFs using various tetracarboxylic acid-based organic linkers that contain orthogonal chiral diethoxy or dihydroxy functional groups in the (R)-enantiomeric form (see Figure 8.37), resulting in a tunable void space that ranges from 73.3 up to 91.9%. In a following step, the chiral dihydroxy groups were postmodified with $Ti(O^iPr)4$ groups to give Lewis acidic $(BINOLate)Ti(O^iPr)_2$ sites.

The asymmetric catalytic activity of the functionalized chiral MOFs was assessed by studying the diethylzinc and alkynylzinc additions that convert aromatic aldehydes into chiral secondary alcohols. They obtained a conversion of 99% for a whole range

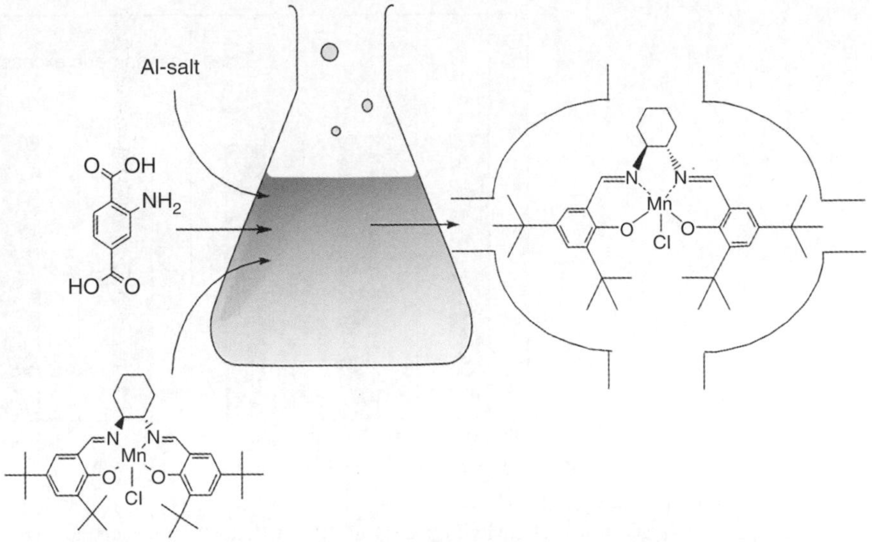

Figure 8.38

"Bottle around the ship" Jacobsen catalyst in MIL-101(Al). Source: Reproduced with permission of the RSC [136].

of aromatic aldehydes with selectivities up to 99% for secondary alcohols. The enantiomeric excess (*ee* up to 91% for 1-naphtaldehyde) was only slightly lower (3%) than that of the homogeneous counterpart. Additionally, the values of the enantiomeric excess changed according to the channel size. With increasing the pore size, a higher *ee* was obtained because of the enhanced diffusion of the reagents throughout the channels.

8.7.2.2 Chiral MOFs as a Ship in a Bottle – A Bottle Around a Ship

Another way to synthesize chiral MOFs is by the ship in a bottle approach. Van Der Voort [136] showed a nice example of the so-called "bottle around the ship" variant, using the Jacobsen catalyst. The Jacobsen catalyst is commercially available and is a chiral Mn complex, as shown in Figure 8.38. It is a famous chiral epoxidation catalyst, very expensive and used homogeneously. We engineered a way in which we simply stirred the commercial Jacobsen catalyst with the ingredients to prepare an NH_2-MIL-101(Al). While the MOF is being formed, some of these large catalyst molecules are trapped inside the pores of the material. So the loading was very low ($0.02\,mmol\,g^{-1}$) and the remaining surface was $2400\,m^2\,g^{-1}$. The material was thoroughly characterized, proving that the Jacobsen catalyst was inside the pores of the MOF and was not degraded.

The catalytic results were very encouraging! We show in Figure 8.39 that the heterogeneous catalyst (same loading of Mn-sites) shows only a minor drop in activity (TON, turn over number) compared to the pure heterogeneous catalyst. This is likely due to the very low loading of the Jacobsen catalyst in the pores. More importantly, the confinement has no effect on the enantioselectivity. The catalyst shows the same *ee* as the homogeneous version. And third, the catalyst can be reused several times without loss in enantioselectivity and minor drops in activity.

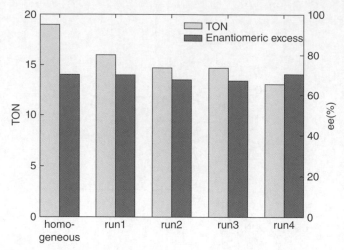

Figure 8.39

TON and enantioselectivity for the salen@NH2-MIL101(Al) compared to homogeneous catalysis with the salen complex for the same conditions. It appears the encapsulation procedure has no effect on the selectivity of the catalyst. Source: Reproduced with permission of the RSC [136].

8.7.2.3   MOFs in Cascade Reactions

The multifunctional MOFs exhibit great potential in cascade reactions. Cascade reactions are multistep chemical transformations that make a multistep reaction go to the desired products in one reactor without the need for separation, purification, and transfer.

$$A \xleftrightarrow{\text{\textit{catalytic function} 1}} B \xleftrightarrow{\text{\textit{catalytic function} 2}} C$$

As just one example, Tang and coworkers made a Pd core IRMOF-3 shell by solvothermal synthesis [137]. IRMOF stands for Isoreticular MOF, from the Yaghi MOF-5 family. They obtained monodispersed spherical core-shell structures having a Pd core of approximately 35 nm surrounded by a MOF shell of 145 nm in thickness (see Figure 8.40). The latter material was examined in a cascade reaction that consists of a Knoevenagel condensation of 4-nitrobenzaldehyde (A) and malononitrile into 2-(4-nitrobenzylidene)malononitrile (B), followed by a hydrogenation of the —NO$_2$ group of the intermediate product B. The reason why the authors studied this cascade reaction is that the target product, 2-(4-aminobenzylidene)malononitrile (C), is a key intermediate in the synthesis of dyes and antihypertensive drugs.

The catalytic results demonstrated that the pure alkaline IRMOF-3 can catalyze the Knoevenagel condensation step. The authors obtained full conversion of substrate A into the intermediate product B, however, no activity was seen for the second step in the cascade reaction. On the contrary, bare Pd nanoparticles exhibited no catalytic activity for the Knoevenagel condensation reaction but demonstrated a good activity in the hydrogenation of B toward C with a selectivity up to 68%. The core-shell-based Pd@IRMOF-3 material could catalyze both steps of the cascade reaction with a selectivity up to 86% toward C. This selectivity could be maintained for at least five successive runs. The conventional supported Pd/IRMOF-3 only showed a selectivity of 71% that decreased to 44% over multiple runs due to sintering and migration of the Pd nanoparticles during the several cycles.

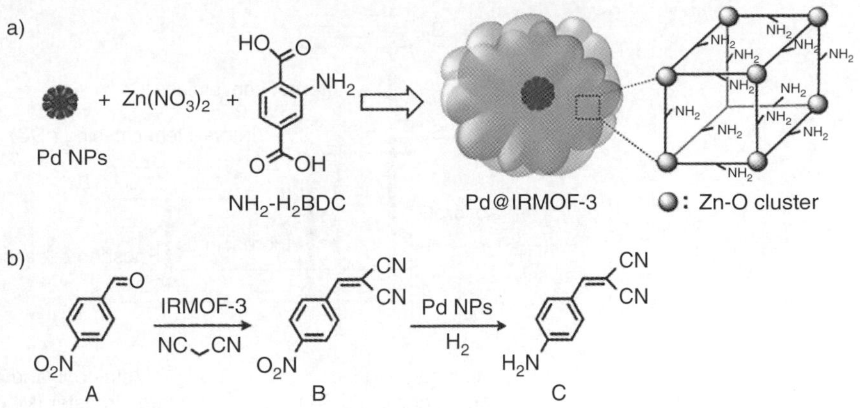

Figure 8.40
Representation of the synthesis route to obtain core-shell Pd@IRMOF-3 for its use in cascade reactions.
Source: Reproduced with permission of ACS [137].

8.7.3   Luminescent MOFs

8.7.3.1   *Luminescence in MOFs*

Luminescence is the process in which *the emission of light occurs following the absorption of radiative excitation energy*. There are generally two types of luminescence pathways (see Figure 8.41). In the first mechanism, the photon emissions are based on a transition from the lowest singlet excited electronic state S_1 to the ground state S_0 (*fluorescence*). In the second mechanism, an *intersystem crossing* occurs from the singlet excited-state S_1 to the tripled excited electronic state T_1 followed by a photon forbidden emission toward the ground state S_0 (*phosphorescence*).

The luminescence in MOFs can be the result of several mechanisms (see Figure 8.41). One of these mechanisms is the so-called "linker centered luminescence" in which the absorption and emissions of the photons occurs through the same ligand. This is the most common mechanism in MOFs, especially when the organic struts have a conjugated π or aromatic system. Next to this mechanism, several other mechanisms can also occur in which the absorption and emission of photons takes place at different parts in the MOF, for example, *ligand to ligand charge transfer* (when there is an intermolecular stacking between adjacent linkers), *ligand to metal charge transfer* (LMCT), *metal to ligand charge transfer* (MLCT), and *metal to metal-based transfers*.

8.7.3.2   *White-Light Emission*

The optimum white-light illumination requires a source with the Commission International de l'Eclairage (CIE) coordinates (0.333, 0.333), with a color rendering index (CRI) above 80 and correlated color temperature (CCT) between 2500 and 6500 K. Currently, most commercially available white LEDs fall into two groups according to their fabrication: (1) those obtained by mixing red, green, and blue LEDs, also called multichip white-light-emitting diodes (MC-WLEDs) and (2) phosphor-converted white-light-emitting diodes (PC-WLEDs) [139].

Ln-MOFs have been widely studied for white-light-emitting phosphors due to the characteristic f–f transitions of lanthanide ions and broad emission bands of organic linkers. Considering that Eu^{3+} and Tb^{3+} ions can display strong red and

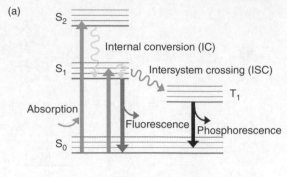

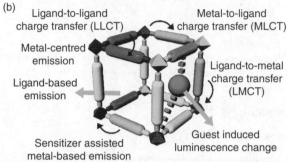

Figure 8.41

(a) Schematic illustration of the two types of pathways for luminescence; (b) Schematic representation of the different mechanisms that can be present within luminescent based MOFs. Source: Reproduced with permission of the RSC [138].

green emissions, respectively, Qian and coworkers [140] reported a two-dimensional Ln-MOF, $La_2(PDC)_3(H2O)_5$ (ZJU-1, PDC = pyridine-2,6-dicarboxylate), which emitted blue (408 nm) light ascribed to the emissive organic PDC linkers. ZJU stands for Zhejiang University. When excited at 312 nm, the co-doped ZJU-1:Tb^{3+}, Eu^{3+} exhibited the characteristic emissions of Tb^{3+}, Eu^{3+}, and PDC ligand simultaneously, resulting in a white-light emission with the CIE coordinate (0.3269, 0.3123).

Although several lanthanide MOFs have been demonstrated for WLEDs, the available luminescence wavelengths of these mixed lanthanide MOFs are almost fixed due to the limitation of types of lanthanide ions. Qian, Cui et al. [141]. developed a new strategy to achieve warm-white LEDs with high CRI by encapsulating cationic dyes DSM (4-(p-dimethylaminostyryl)-1-methylpyridinium) and AF (Acriflavine), which exhibit red-light and green-light emission, respectively, into a blue emitting anionic MOF ZJU-28 to yield the MOF*dye composite. The emission color of the obtained composite ZJU-28*DSM/AF can be easily modulated by simply adjusting the amount and component of dyes. The confinement of organic fluorescence dyes within the pores of MOFs not only provided fine-tuning of emission color, but also efficiently enhanced the fluorescence quantum efficiency by blocking the aggregation-caused quenching process of the dyes.

8.7.3.3   *Luminescent Sensing*

The use of MOFs in sensing, applying the inherent luminescence character of the MOF is a rapidly growing field.

In general, a change in the intensity of the luminescence properties of a certain MOF can be used as a probe in the recognition of the interacting analyte. This change in intensity, depending on the electronic nature of the analyte, is usually employed as the detection method in luminescence-based MOF. A more reliable and efficient pathway is the appearance and/or disappearance of an emission peak.

Ln-MOFs have been widely studied in various sensor applications owing to their inherent porosity and the particular luminescent properties of Ln^{3+} ions. Most of the Ln-MOF sensors show luminescence intensity changes, including luminescence enhancement (turn-on response) and quenching (turn-off response) on recognition of the analytes. Eu^{3+} and Tb^{3+} are commonly used as luminescent centers in Ln-MOF sensors because of their strong, characteristic red emission at around 614 nm and green emission at around 541 nm, respectively. Ln-MOFs succeed in sensing ionic species, small molecules, explosive chemicals, and pH, as well as temperature. In addition, the inherent structural and chemical features of Ln-MOFs make them considerably useful in biosensing and bioimaging applications.

The sensing behavior of functionalized lanthanide MOFs is a complex interplay between the lanthanide nodes, the linkers and the extra functionalities, such as chromophoric (or light harvesting) antennae.

Generally, lanthanide ions (Ln^{3+}) are characterized by successive filling of the $4f$ orbitals, with electronic configurations of $[Xe]4f^n$ ($n = 0$ to 14). These electronic configurations generate a rich variety of electronic levels with the number $14!/n!(14-n)!$, resulting in interesting optical properties. All of the Ln^{3+}, except La^{3+} ($4f^0$) and Lu^{3+} ($4f^{14}$), exhibit luminescent f–f emissions, which almost cover the entire spectrum.

Eu^{3+}, Tb^{3+}, Sm^{3+}, Dy^{3+}, Pr^{3+}, and Tm^{3+} emit in the visible region with the colors red, green, orange, yellow, orange-red, and blue, respectively.

Pr^{3+}, Nd^{3+}, Sm^{3+}, Dy^{3+}, Ho^{3+}, Er^{3+}, Tm^{3+}, and Yb^{3+} show emissions in the near-infrared region, while Ce^{3+} shows a broadband emission from 370 to 410 nm because of the $5d$–$4f$ transition.

Typically, the $4f$–$4f$ transitions of Ln^{3+} are *Laporte forbidden* due to the $4f$ orbitals that are well-shielded by the filled $5s^2 5p^6$ subshells. Consequently, direct photoexcitation of Ln^{3+} ions rarely produces highly luminescent materials due to the low absorption efficiency of the $4f$–$4f$ transitions. This problem can be overcome by the "antenna effect" (Figure 8.42), which commonly uses a strong absorbing chromophore to sensitize Ln^{3+}. The overall process of antenna sensitization involves the following characteristic steps:

(i) the organic ligands can absorb light on excitation;
(ii) the excitation energy is then transferred into Ln^{3+} excited states through intramolecular energy transfer;
(iii) Ln^{3+} ions undergo a radiative process by characteristic luminescence.

This process could effectively increase the luminescence quantum yield of Ln^{3+} in normal conditions at room temperature. Furthermore, the solvent quenching and self-quenching of Ln^{3+} ions are almost nullified in Ln-MOFs due to the separation of Ln^{3+} ions by organic ligands. Consequently, Ln-MOFs exhibit strong luminescence and can be utilized as chemical sensors.

We refer to a very recent review [139] for a more in depth discussion of the use of MOFs in photonic applications in the broadest sense.

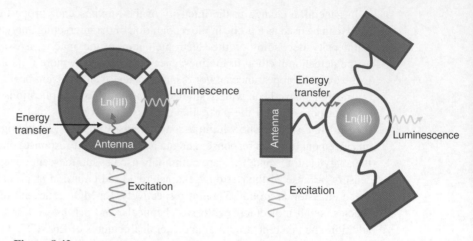

Figure 8.42
The antenna effect for lanthanide[III] (Ln(III)) sensitization, illustrated using the chromophoric chelate (right) and pendant chromophore (left) ligand designs. Source: Reproduced with permission of ACS [142].

Yuanjing Cui was born in Jiangsu, China. He received his BSc and Ph.D. in Materials Science and Engineering from Zhejiang University in 1998 and 2006, respectively. Currently, he is a full professor in the School of Materials Science and Engineering at Zhejiang University. His research interest focuses on organic–inorganic hybrid photonic materials.

Luminescent-based MOFs have several advantages:

- Several MOFs exhibit a low cytotoxicity, making them suitable for sensing in *in vivo* applications.
- The large variety in the inorganic and organic moieties that can be employed allows a tuning of the absorption and emission of photons.

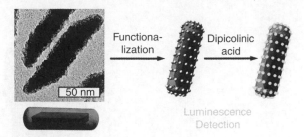

Figure 8.43

Core-shell-based Ln-MOFs postmodified with a Tb complex for sensing dipicolinic acid. Source: Reproduced with permission of ACS [143].

- One can tune the selectivity of the sensing for a specific analyte by changing the pore size or the chemical environment (acidity/basicity, polarizability, hydrophobicity, etc.)
- One can entrap certain guest molecules or attach specific functionalities onto the organic moieties, for example, dye molecules, chromophores, lanthanide ions, contributing to the luminescence properties of the framework.
- The high porosity of the framework allows (i) a good interaction between the MOF and the analyte species and (ii) a pre-concentration of the analyte species in the pores of the MOF due to prior adsorption of the analyte within the pores of the MOF, enhancing the sensing performance.
- Several MOFs have already shown to maintain their fluorescence at higher temperatures. This might be important in applications where high temperatures are required to allow the binding of the analyte.

8.7.3.3.1 *Detection of Dipicolinic Acid*

One of the first examples of the use of MOF based luminescence sensing, was reported by Wenbin Lin and colleagues. They synthesized a nanoscaled MOF toward the sensing of dipicolinic acid (see Figure 8.43) [143]. Dipicolinic acid is the main component of bacterial endospores (counts for 15% of its mass) such as *Bacillus anthracis*, which is used as an agent in the anthrax attacks. The detection of these spores through the sensing of dipicolinic acid is important in the prevention of bioterrorism attacks. Several techniques have already been used to detect the latter acid, such as Raman spectroscopy, potentiometric sensing, and high-performance liquid chromatography. However, these methods suffer from the need to use large instruments as well as the fact that the analysis is time consuming and complicated. In the work of Wenbin Lin, silica coated lanthanide MOFs were prepared having the general formula: $Ln(BDC)_{1.5}(H_2O)_2$ where Ln represents: Eu^{3+}, Gd^{3+}, or Tb^{3+} and BDC stands for 1,4-benzenedicarboxylic acid. The Eu-doped $Gd(BDC)_{1.5}(H_2O)_2@SiO_2$ nanoparticles were modified by anchoring the Tb-ethylenediaminetetraacetic acid monoamide onto the silica shell. Tb^{3+} ions and molecular Tb complexes have been employed as sensing probes for anthrax and other bacterial spores by complexation with dipicolinic acid. In the absence of dipicolinic acid, the core-shell material only gave an Eu^{3+} luminescence on excitation at 278 nm. In the presence of dipicolinic acid, the Tb^{3+} luminescence became visible due to the complexation with the acid. Even in the presence of other interfering molecules such as amino acids, the reported system was able to selectively detect dipicolinic acid with a detection limit of 48 nM.

Wenbin Lin is currently a "James Franck" Professor of Chemistry at the University of Chicago (USA). He obtained his bachelor's degree from the University of Science and Technology of China (USTC) in 1988 and has worked in several American Universities (Illinois, Northwestern, and North Carolina at Chapel Hill).

Wenbin Lin is very active (mostly using MOFs) in the fields of Renewable Energy, Nanomedicine and Sustainable Catalysis.

8.7.3.3.2   Detection of Explosives

Another interesting example of the use of MOFs as sensors is their use in the detection of explosives, not only in the vapor phase but also in the liquid phase. Li and coworkers, reported in 2009 the first report on the use of luminescence-based MOFs for the detection of explosives [92]. In this study block-shaped crystals of $[Zn_2(bpdc)_2\text{-}(bpee)]\cdot 2DMF$ (in which bpdc and bpee stands for 4,4′-biphenyldicarboxylate and 1,2-bis[4-pyridyl]ethylene, respectively) were prepared using a solvothermal synthesis method. The Zn-based MOF was exposed to vapors of the nitroaromatic explosive, 2,4-dinitrotoluene (0.18 ppm at 25 °C), and to vapors of the plastic explosive 2,3-dimethyl-2,3-dinitrobutane (2.7 ppm at 25 °C). The authors observed not only a very fast response for both vapors but also a high selectivity. The fluorescence quenching percentages (defined as $(I_o - I)/I_o * 100\%$ in which I_o = original peak maximum intensity and I = maximum intensity after exposure) were 85% and 84% for 2,4-dinitrotoluene and 2,3-dimethyl-2,3-dinitrobutane, respectively, after only 10 seconds of exposure time. Additionally, the authors observed that this process was fully reversible. On heating the material for only 1 minute at 150 °C the photoluminescence could be recovered.

8.7.3.3.3   Ion, Small Molecule, and Gas Sensing

The number of papers that have appeared on both anion and cation sensing is tremendous. Van Der Voort and Zhao recently published a review in *Materials* (open access) discussing the latest developments in this field [144]. Most MOFs in this field are tailor-made, tuning in on the selective detection of one species in the presence of many other species. We cannot cover this topic in all its richness here. We give a few examples.

Formaldehyde (HCHO) is widespread in construction, furniture, and particle board, posing an impact on human health, such as watery eyes, asthma, and respiratory irritation. Yu and coworkers [145] developed a ratiometric luminescence HCHO probe through incorporation of Eu^{3+} ions into NH_2-UiO-66 under microwave irradiation conditions. The dual-emitting luminescence originated from the characteristic red emission of Eu^{3+} ions (615 nm) and linker-to-cluster (Eu-oxo or Zr-oxo) charge transfer transition-related emission (465 nm). The interaction of the free amino groups with HCHO can drastically enhance emission around 465 nm due to the added electron transfer from the amino group with lone pair electrons to the positively charged HCHO. This is in contrast to the emission of Eu^{3+} at 615 nm that was only slightly enhanced. Then, a ratiometric luminescence HCHO probe was formed based on the intensity ratio of two emission bands at 465 and 615 nm. The results indicated that the fabrication of a ratiometric luminescence probe based on multiband luminescent MOFs can serve as a common sensing method for organic molecules.

Chromium is extensively used in various industrial processes causing Cr(VI) anions (CrO_4^{2-}) and $Cr_2O_7^{2-}$ to often be present in all kinds of industrial wastewater. It is one of the most prevalent, toxic heavy-metal ions of which excess intake can cause serious protein and DNA disruption, as well as damage to the human enzyme system. The detection of Cr(VI) anions was realized through a cationic Eu-MOF $[Eu_7(mtb)_5(H_2O)_{16}].NO_3.8DMA.18H_2O$ (H_4mtb = 4-[tris(4-carboxyphenyl)methyl]benzoic acid) with a luminescence turn-off response [146]. The Cr(VI) anions can absorb the excitation light and hinder the energy absorption of the Eu-MOF, resulting in luminescence quenching (Figure 8.44a,b). This highly stable Eu-MOF sensor with excellent sensitivity and selectivity can also be utilized in real environmental conditions, such as lake water and sea water, suggesting the possible application of MOF chemical sensors in environmental fields. The performance of this sensor for chromate compared to several other anions and cations is shown in Figure 8.44.

8.7.3.3.4 pH Sensing

The need for fast pH sensing in industry, biomedicine, and many other environmental fields up to biological systems and living cells has become a top priority. The advantages of luminescence-based pH probes include quick response times, high sensitivity, and easy operation. As an example, Chen and coworkers designed a pH-sensitive MOF nanoparticle using DMF and 1,10-phenanthroline (Phen) as ligands with Tb^{3+} ions based on the intramolecular-charge-transfer (ICT) effect [147]. A DMF molecule contains both an electron-donor and -acceptor part, allowing it to generate ICT. Furthermore, it can change the Tb^{3+}-based luminescence through the antenna effect. Consequently, the protonation of H^+ could change the charge transfer of DMF and further change the antenna effect for Tb^{3+}, in turn resulting in a change of Tb^{3+}-based luminescence. The Phen molecule in the nanoparticle was used to improve such a change and reduce the luminescence quenching effect of Tb^{3+} by replacing the coordinated water molecules. The emission intensity of DMF–Tb was improved approximately four times, while the emission intensity of DMF–Tb–Phen was improved 10 times due to a decrease of the ICT effect and increase of the antenna effect on the Tb^{3+} ions on adding H^+. This MOF nanoparticle pH sensor with high specificity and sensitivity was used in strong acidic conditions, indicating its potential applications in biological systems.

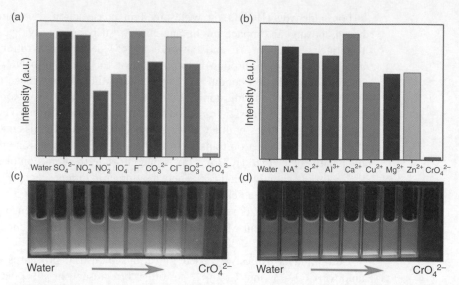

Figure 8.44

(a) and (b): Luminescence intensity of the $^5D_0 \rightarrow {}^7F_2$ transition of Eu^{3+} at 616 nm dispersion in different aqueous solutions of various anions and cations. (c) and (d) Actual photo of the luminescence (excited at 365 nm) for these solutions. Only the chromate containing solution gets quenched. Source: Reproduced with permission of ACS Ref. [146].

As a second example, Qian and coworkers fabricated a fluorescence pH sensor by encapsulating Eu^{3+} ions into the pores of the nanoscale *UiO-67-bpydc* (bpydc = 2,20-bipyridine-5,50-dicarboxylic acid) [148]. The luminescence intensity of *Eu^{3+}@UiO-67-bpydc* shows a significant luminescence turn-off response in acidic solutions while exhibiting fluorescence enhancement in basic solutions. This is because protonation and deprotonation of the ligands first change the excited-state energy level of the ligands followed by a change in ligand-to-Eu energy transfer efficiency, explaining the different changes in the Eu^{3+}-based luminescence. This *Eu^{3+}@UiO-67-bpydc* pH sensor is stable within a wide pH range of 1.06–10.99 and can thus be used in physiological environments (pH = 6.80–7.60). The bio-compatibility of Eu^{3+}@UiO-67-bpydc was further confirmed by an MTT (MTT = 3-(4,5-dimethylthiazol-2-yl)-2,5-diphenyl-tetrazolium bromide) assay. Cell imaging results demonstrate that the Eu^{3+}@UiO-67-bpydc pH probe could be a promising candidate for monitoring pH both *in vitro* and *in vivo*.

Very recently, the same group reported another luminescence pH sensor based on a nanoscale mixed Ln-MOF *Eu$_{0.034}$Tb$_{0.966}$(fum)$_2$(ox)(H$_2$O)$_4$* (fum = fumarate, ox = oxalate) [149]. The *Eu$_{0.034}$Tb$_{0.966}$(fum)$_2$(ox)(H$_2$O)$_4$* pH sensor shows high stability in aqueous solutions. Moreover, its morphology and size can easily be adjusted by changing the amount of CTAB (cetyl trimethyl ammonium bromide) surfactant. The mixed Ln-MOF exhibits both Tb^{3+} (545 nm) and Eu^{3+} (618 nm) emissions, which can be used for sensing pH values ranging from 3.00 to 7.00 in a ratiometric manner.

8.7.3.3.5 *Temperature Sensing*

Temperature is an important parameter in human life and scientific investigations. Among the approaches to measure the temperature, luminescence-based measurements have recently achieved attention because of their non-invasiveness, fast response, accuracy, high spatial resolution, and ability to work in strong electro or magnetic fields There are several different parameters for assessing the temperature change of a material: band intensity, spectral position, band-shape, polarization, lifetime, and bandwidth. It is most common to select the steady state intensity. However, the luminescent thermometers depending on a single emission are susceptible to errors because of sample concentration changes and drifts of the optoelectronic system.

Qian and coworkers fabricated the first *self-calibrated luminescent temperature* (ratiometric) sensor using a mixed Ln-MOF $Eu_{0.0069}Tb_{0.9931}$ *DMBDC* (DMBDC = 2,5-dimethoxy-1,4-benzenedicarboxylate) [150].

For Tb-DMBDC and Eu-DMBDC, the characteristic luminescence gradually decreases because of thermal activation of nonradiative decay pathways. However, $Eu_{0.0069}Tb_{0.9931}$ *DMBDC* exhibits a significant temperature dependent luminescent behavior as the temperature increases from 10 to 300 K. The Tb^{3+}-based emission in $Eu_{0.0069}Tb_{0.9931}$ *DMBDC* decreases as the temperature increases, while that of the Eu^{3+} ions increases. The good linear relationship between the I_{Tb}/I_{Eu} ratio and temperature in the range of 50–200 K suggests that $Eu_{0.0069}Tb_{0.9931}$ *DMBDC* is an excellent temperature thermometer within this temperature range. These results suggest that mixed Ln-MOFs featuring temperature-dependent luminescence can be ideal candidates for self-referencing temperature sensing. For further examples of MOF thermometers, the reader is referred to the review by Luis D. Carlos [151].

Luís António Ferreira Martins Dias Carlos (born in Coimbra, Portugal), got his Ph.D. in physics from the University of Évora in 1995 working on photoluminescence of polymer electrolytes incorporating lanthanide salts. In 1996 he joined the Department of Physics at the University of Aveiro. Currently, he is full professor there. Since 2009 he has been the Vice-Director of the Centre for Research in Ceramics and Composite Materials (CICECO) at Aveiro, Portugal. He is member of the Lisbon Academy of Sciences (Physics section) since 2011 and was visiting professor of S. Paulo State University (UNESP), Brazil, and of University of Montpellier 2, France. He is widely recognized for this contribution in the field of luminescent thermometers.

8.7.3.3.6   *Anti-Counterfeit Technology*

Van Der Voort and Kaczmarek demonstrated that MOFs exhibit potential in multistage security control showing temperature and wavelength dependence luminescence properties [152]. This was achieved by anchoring a Ln^{3+}-β-diketonate complex onto an Al-based bipyridine MOF, MOF-253. This MOF was chosen as bipyridine groups show (i) good luminescence harvesting properties and (ii) are ideal chelating groups for the Ln^{3+} complex. The MOF-253, grafted with $Ln(acac)_3$ and $Ln(tfac)_3$ (Ln = 10%Sm, 90%Tb, acac = acetylacetone, tfac = trifluoroacetylacetone) exhibited a yellow or orange emission color when excited at 302 nm that could be assigned to the Tb^{3+} and Sm^{3+} transitions. When the material is excited at 365 nm, emissions from the host itself takes place and a green emission color is obtained. Additionally, when the temperature was changed from 280 K up until room temperature, a change from green to yellow/orange emission color is obtained on 302 nm excitation whereas at 365 nm excitation, the color remains unchanged (green) in the same temperature range. We observed that not only the β-diketonate ligand was critical in tuning the properties of the material, but also the Ln^{3+} ratio was crucial. Higher doping percentages of Sm^{3+} (15–20%) showed temperature-dependent emission colors more shifted toward red that would make it more difficult to distinguish the colors by human eye. Lowering the Sm^{3+} doping percentage to 5–10% shows yellow/orange and green emission when placed under a UV lamp at 302 nm excitation and 365 nm excitation, respectively.

8.8   Industrial Applications of MOFs

Since the discovery of the first MOF, they have found their way into industry. So far, more than a thousand patents are reported on MOFs that range from new MOF components and synthesis methods to applications and improvements thereof [153]. Most of these patents are filled by North American companies while most of the patents from Germany originates from the chemical company BASF. Since 2016, some of these MOF products are available on the open market.

One of these commercial MOF products is ION-X (see Figure 8.45), which was launched in the autumn of 2016 by NuMat Technologies, a spin-out company of which Omar Farha is a co-founder [154]. ION-X is used for the storage of toxic gases such as arsine, phosphine, and boron trifluoride that are used as dopants in the electronic industry. Typically, these gases are stored in gas cylinders under high pressures. In contrast to the conventional methods, the gases are stored below atmospheric pressure in the MOF based ION-X or, in other words, a vacuum needs to be applied to flow them out.

A second commercial MOF product is TruPick, which was developed by the spin-out company MOF technologies in collaboration with Decco, a company holding expertise in fruit and vegetable storage [154]. The TruPick's MOF is able to store the gas 1-methylcyclopropene, which retards the ripening process because it can bind onto the ethylene binding site in the enzyme of fruit and for this reason can block the effect of ethylene. TruPick is commercially available as disposable sachets (similar to the silica gel packs put in shoe boxes) and releases the 1-methylcyclopropene gas when placed in water. At this moment, TruPick is sold in the U.S. and in Turkey but this product will be soon available on the European markets as well.

Figure 8.45
The first commercialized product, ION-X by NuMat, cofounded by Hernandez (left) and Farha
(right). Source: Reproduced with permission of ACS. Photograph: https://cen.acs.org/articles/95/i32/
Semiconductor-industry-begin-using-MOFs.html

Also, the German chemical company, BASF, who received the French Pierre Potier
Prize in 2012 because of their sustainable development and green chemistry of MOFs,
is selling MOFs in small volumes via Merck and Sigma Aldrich [153]. Moreover,
BASF have already sponsored several MOF projects, such as the EcoFuel Asia tour
in 2007 (see Figure 8.46) [155]. In this project, the goal was to drive a car using a
MOF enhanced fuel tank for a tour throughout 14 countries. The basolite C300, a
commercialized Cu-MOF from BASF was used because of its high storage capacity
for natural gas. Pellets of the latter MOF were placed in the traditional tank that was
filled afterward with compressed natural gas. An increase of approximately 20% was
observed in the distance that the car could drive before refueling.

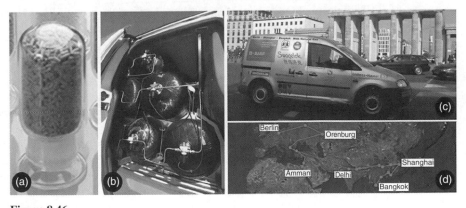

Figure 8.46
The EcoFuel Asia tour 2007 from Berlin to Bangkok using a tank filled with Basolite C300. (a) Basolite
C300; (b) MOF fuel tank with CH_4; (c) Volkswagen Caddy EcoFuel car; and (d) journey map. Source:
Reproduced with permission of the RSC [153].

A third application of MOFs has been launched by the spin-off company, MOFgen, of the Russel Morris Laboratory (University of St. Andrews, UK). For this application, MOFgen worked together with several medical device companies to develop MOF containing coatings for indwelling devices, such as urinary catheters, and textiles, such as hospital gowns [154]. In these devices the MOFs act as reservoirs for antimicrobial agents as they can be loaded with several biologically active gases, therapeutic agents, and metal ions.

Russel Morris is Professor at St Andrews University (Scotland). He performed his Ph.D. under the supervision of Anthony Cheetham at Oxford. He is an expert in zeolite synthesis (Assembly-Disassembly-Organization-Reassembly) methods. He is very active in the field of MOFs. He is the driving force behind MOFgen Ltd.

Besides the previously presented commercial MOF products, many other industrial applications of MOFs are in the pipeline, for example, MOFs to separate carbon dioxide from various mixtures (developed by the spin-off company of Jeffrey Long, University of California, Mosaic Materials) and the use of MOFs for water harvesting via BASF in collaboration with Yaghi and coworkers [154].

8.9 Transmission Electron Microscopy

The Authors thank Maria Meledina for her contribution to this paragraph.

Modern transmission electron microscopy (TEM) is an extremely powerful technique providing a set of tools for precise and versatile MOF characterization. Its history starts in the 1930s when two German physicists Ernst Ruska and Max Knoll constructed the first transmission electron microscope. Later, in 1986, Ernst Ruska was awarded with the Nobel Prize in Physics for "fundamental work in electron optics, and for the design of the first electron microscope." In principle, TEM is highly similar to optical microscopy, but in the case of TEM electromagnetic lenses are used, whereas the glass lenses in optical microscopy serve the purpose of object image magnification. The main benefit of TEM arises from using electrons instead of light photons: accelerated electrons of several hundred kV have a shorter wavelength leading to a significant improvement in resolution.

However, for a rather long period of time, lens aberrations have been the main cause of impaired resolution: again, similar to the artifacts of optical microscopy, the aberrations of electromagnetic lenses in TEM prevented the achievement of sub-Å resolution. The development of the aberration-corrected microscope is a recent revolutionary achievement in electron microscopy finally allowing the collection of information at an extremely local, sub-Å level. Another powerful unit in TEM is a monochromator that also allows chromatic aberration minimization in TEM mode and is often used to achieve optimized energy resolution in spectroscopy. Nowadays, the most up-to-date microscopes are usually equipped with aberration correctors placed at different locations within the microscope that serve as resolution improvement for (scanning) transmission electron microscopy ((S)TEM) modes and a monochromator. Alternatively, to the monochromator, the so-called cold field emission electron gun (cold FEG) can be exploited for improving the spatial resolution in TEM mode and energy resolution ramp-up in spectroscopy.

MOFs are extremely obstinate materials for TEM characterization. Built of metal ions or clusters connected by organic linkers into a 3D structure, MOFs rapidly lose their crystallinity on electron-beam illumination, mainly due to ionization damage. Nevertheless, in many cases, there is no alternative way to investigate certain features (e.g. the defects or the loaded particles' precise locations) of materials expect TEM. All TEM techniques in principle can be applied for MOF characterization. However, to collect reliable information about the original material, one has to constantly keep in mind the extreme electron-beam sensitivity of materials and minimize the electron dose as much as possible.

In this part, the most common TEM techniques are briefly described and some examples of their exploitation for MOF characterization are given.

8.9.1 Electron Diffraction and Bright Field Imaging

A diffraction pattern arises when an electron beam generated by the electron source of the TEM is diffracted by the crystal lattice of the investigated material. Electron diffraction (ED) is displayed in the back focal plane of the objective lens. It is closely related to the one of the most common TEM modes – bright field (BF) imaging. Both ED and BF TEM modes use parallel electron-beam illumination. The back focal plane of the objective lens where the diffraction pattern is displayed (Fourier space) is the Fourier Transform of the sample's image in real space. During usual TEM operation, there is a constant switch between Fourier space and real space by just pressing a button. If the studied specimen is in a well-defined orientation, the diffraction pattern is highly symmetrical. The BF TEM image or the ED pattern is finally magnified onto the camera with intermediate and projector lenses.

To choose the area of interest for ED investigations in the specimen, a selected area aperture can be used. In this case, the technique is called selected area electron diffraction (SAED) (see Figure 8.47). Tilting one crystal into several highly symmetrical orientations (zone axis), recording ED patterns, and noting the angles between these orientations, one can determine the crystal structure of the material as well as cell parameters. Bragg's law is used to index the diffraction patterns allowing correlation of the distances between the bright-contrast spots with interplanar spacings. It is also possible to define the spatial symmetry of the studied crystal

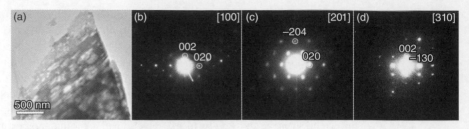

Figure 8.47
(a) BF TEM image showing a COK-18 particle and SAED patterns of the COK-18 material recorded along the (b) , (c) [201] and (d) [310] zone axis. Source: Reproduced with permission of ACS [156].

by determining the present and absent reflections in the diffraction patterns. The polycrystalline specimens consisting of nanosized partials typically produce "ring patterns": randomly oriented nanocrystals giving rise to a rotation averages with specific radii matching materials' interplanar distances, but not carrying angular information. The ring patterns are often used for phase identification.

In the BF TEM imaging mode, the electrostatic potential of the sample scatters the initial beam with a change of both wave amplitude and phase of the electrons that passed through the investigated specimen. These effects cause the contrast, allowing image formation. For crystalline specimens, the diffraction contrast originating from coherent elastic scattering is typical. It strongly depends on the orientation of the specimen with respect to the incoming beam and can be changed by tilting the sample. For amorphous materials, the mass-thickness contrast serves for image formation. It arises from incoherent elastic scattering. The thicker areas of the specimen or those areas with heavier elements scatter more electrons compared to thinner areas. By inserting an objective aperture (placed in the back focal plane of the objective lens) it is possible to cut off the axis-scattered electrons coming from the thicker specimen and high-mass density parts. These areas will appear dark in the image. An objective aperture in general can be used to cut off the undesired electrons to ramp up the image contrast or, for example, to select only the specific reflections in the ED pattern that will serve for specific image formation.

8.9.2   High-Resolution Transmission Electron Microscopy

In the case of high resolution transmission electron microscopy (HR TEM) the phase-contrast serves for image formation (see Figure 8.48). An interference (fringe) pattern is formed by several diffracted beams and the transmitted beam interfering with each other. As a result, the contrast in the image plane depends on the phase of various beams. In HR TEM mode, an objective aperture is often inserted.

The exit-wave is the electron wave at the exit plane of the investigated specimen appearing as a result of the incident electron wave interaction with the sample. The exit-wave carries both thickness and atomic position information about the specimen. This information still needs to be "delivered" by the lenses onto the camera. The final image relation to the exit-wave is described by the contrast transfer function (CTF). In the ideal case, all the frequencies in the exit-wave function are transported equally. However, lenses in the electron microscope suffer from aberrations causing the distortions. Some aberrations can be played down by working at the so-called Scherzer defocus: in this case the defocus balances with the spherical aberration and

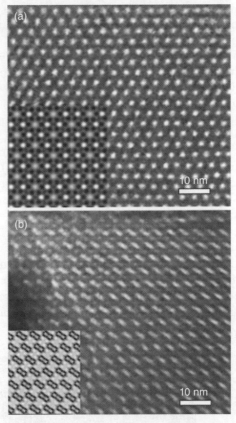

Figure 8.48

HR TEM images of MIL-101 structure taken along the (a) [111] and (b) [011] zone axes. The inserts contain the calculated images coming in good agreement with the experimental data. Source: Reproduced with permission of ACS [157].

the CTF attains the ideal situation. Scherzer defocus corresponds to the highest spatial frequency without the phase reversal and the limit of the direct interpretation (point resolution) is reached at the Scherzer defocus.

For accurate HR TEM image interpretation, it is often necessary to run an image simulation. Such parameters as specimen thickness, defocus, and lens aberrations need to be taken into account and a slight variation of each of them is possible to simulate the image in conditions as close to the experimental ones as possible.

Nowadays, aberration-corrected microscopes have a unit allowing correction of the spherical aberration – the so-called Cs-corrector. The most advanced instruments are even more enhanced by a chromatic aberration corrector (Cc-corrector) allowing outstanding improvement of point resolution down to 0.5 Å at 300 kV acceleration voltage.

8.9.3 Scanning Transmission Electron Microscopy

Scanning transmission electron microscopy (STEM) modes exploit the focused electron beam to study the area of interest in the specimen. The position of the beam on the specimen is correlated with, the amount of the detected electrons (intensity) and an image is formed. In STEM the diameter of the focused beam (often called the "probe") is coupled to the resolution. Magnification can be changed by changing the scanned area size.

A set of detectors is placed under the sample. An important parameter in STEM is the scattering angle collected by the detector (typically referred to as the "collection semi-angle"). It can be changed by camera length variation resulting in collection of different information. Low angle scatters are recorded by a disk detector on the microscope optical axis. This set-up corresponds to the bright field scanning transmission electron microscopy (BF STEM) regime that results in similar contrast. Annular dark field scanning transmission electron microscopy (ADF-STEM) mode corresponds to the next ring of detectors. In ADF-STEM, both mass-thickness and diffraction contrasts contribute to image formation (see Figure 8.49). Using the outermost detector ring leads to the elimination of the diffraction contrast and the so-called Z-contrast high angle annular dark field scanning transmission electron microscopy (HAADF-STEM) images can be taken. In this mode, the intensity is correlated to the atomic number Z of the scattering atoms: the higher atomic numbers lead to higher intensities in the image. However, the lighter elements, such as oxygen, cannot be detected in HAADF-STEM mode. For lighter element position imaging, annular bright field scanning transmission electron microscopy (ABF STEM) can be applied. Some of the microscopes also provide opportunities to use several detectors at the

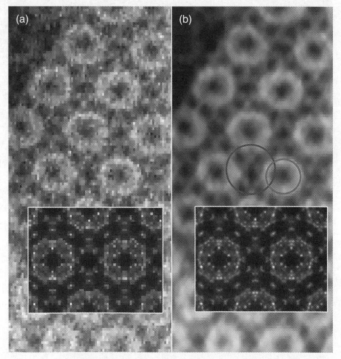

Figure 8.49

(a) HR ADF-STEM image of the MIL-101 structure viewed along the [011] zone axis, together with an ADF-STEM image simulation as inset. (b) Low-pass filtered HR ADF-STEM image with the ADF-STEM image simulation as inset. The circles show two different pore types: the smaller one corresponds to a small pore with a diameter of 29 Å and the larger one to a large-pore with a diameter of 34 Å. Source: Reproduced with permission of John Wiley & Sons, Ltd [158].

same moment, which is extremely handy in some cases for simultaneous collection of different information.

8.9.4 Energy Dispersive X-Ray Spectroscopy

Fast incoming electrons of the beam transfer energy to the inner-shell electrons. These inner-shell electrons can leave the shell creating holes. The outer-shell electrons lower the energy and drop to the created holes together with the emitting X-rays to balance the energy. Every element and shell combination possesses a characteristic energy. This effect is exploited in energy dispersive X-ray spectroscopy (EDX) to determine the elements present in the investigated specimen. A combination of EDX with STEM imaging allows the recording of maps to visualize the elemental distribution within the mapped area.

8.9.5 Electron Energy Loss Spectroscopy

After the interaction with the sample, electrons carry the chemical information about it (see Figure 8.50). Electron energy loss spectroscopy (EELS) analyses the energy of the electrons after going through the investigated sample: a spectrometer is situated under the sample and is used to settle the electrons into a spectrum with the kinetic energy loss versus the number of the electrons (intensity). The EELS spectrum contains several areas. First of all, there is the so-called zero-loss peak (ZLP) at $0\,eV$ resulting from the electrons that passed the specimen without detectable energy loss. The low-loss region arises after the ZLP with typical energy from zero to $\sim 50\,eV$ corresponding to the outer-shell electron excitations. Starting from $\sim 50\,eV$, the core-loss region shows up. The information on the inner-shells of the atoms can be ejected from the core loss. The onset position and the shape of the EELS edge is influenced by the electronic state and the bonding of atoms. Careful inspection of the EELS edge creates opportunities to investigate the "fingerprints" of the oxidation states of the elements, and to distinguish the possible allotropes. To record maps, EELS can also be combined with STEM and in every image pixel an EELS spectrum can be recorded.

8.9.6 Electron Tomography

The results of the imaging and spectroscopy techniques only correspond to 2D projection images and maps of 3D objects. However, in many cases it is not possible to make an absolutely reliable conclusion on the exact position of certain features (e.g. the location of the loading nanoparticles within the MOF) based on the 2D data (see Figure 8.51). Electron tomography serves for direct determination of featured positions. To record the so-called tilt-series, one needs to acquire a set of images of investigated objects at different tilt-angles. To avoid the contribution of the tilt-angle depending on diffraction contrast for crystalline materials, STEM is often used rather than TEM. The tilt-series needs to go through alignment: a cross-correlation procedure needs to be applied to minimize the position difference. The aligned data set finally goes into the reconstruction step. For the moment, there are several reconstruction techniques available. However, one of the most popular is the Simultaneous Iterative Remonstration Technique (SIRT) starting from the weighed back projection

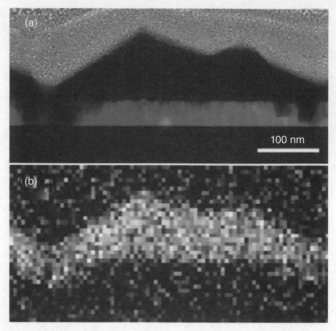

Figure 8.50

(a) ADF-STEM image of ZIF-8 film grown from ZnO and (b) EELS elemental map of N. Source: Reproduced with permission of John Wiley & Sons, Ltd [159].

reconstruction and going into iterations to compare and optimize the estimations of every iteration with the projection data.

Exercises

1 MIL-47 $V^{IV}O(CO_2\text{-}C_6H_4\text{-}CO_2)$, is a fully saturated three-dimensional vanadium-based MOF in which the vanadium center is coordinated to four oxygen atoms from the terephthalate moieties and to two oxygen atoms on the O-V-O axis. This MOF was examined as a catalyst in the oxidation of cyclohexene to cyclohexene oxide. In a typical catalytic test, a flask was charged with 30 ml of chloroform (used as solvent), 5 ml of cyclohexene (density = 0.811 g ml^{-1}), and 2.9343 g of 1,2,4-trichlorobenzene (used as internal standard). As an oxidant *tert*-butylhydroperoxide (5.5 M in decane) was used. The molar ratio cyclohexene/oxidant was ½. The conversion of the substrate was determined by means of GC and 0.0488 g of catalyst was used. In the table, the peak areas of each detected component are given. Extra information: the response factors for cyclohexene and cyclohexene oxide were determined, giving the following equations:

$$\frac{mol\ cyclohexene}{mol\ internal\ standard} = 0.978\ \frac{peak\ area\ cyclohexene}{peak\ area\ internal\ standard}$$

$$\frac{mol\ cyclohexene\ oxide}{mol\ internal\ standard} = 1.179\ \frac{peak\ area\ cyclohexene\ oxide}{peak\ area\ internal\ standard}$$

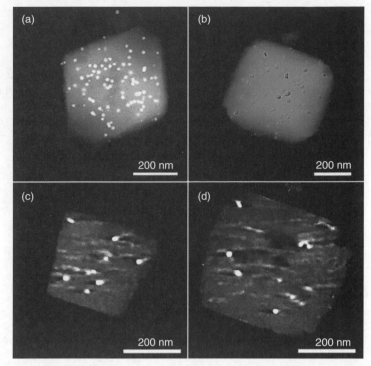

Figure 8.51
(a) HAADF-STEM image of the Au-loaded UiO-66 crystal used for tomographic reconstruction. (b) Reconstructed volume, with the UiO-66 matrix, and the Au nanoparticles. (c and d) Orthoslices through the reconstructed volume, evidencing the bright-contrast Au nanoparticles are spread throughout the UiO-66 matrix. Source: Reproduced with permission of the RSC [31].

Peak areas of each component as a function of reaction time

Time (hour)	Peak area cyclohexene	Peak area epoxide	Peak area int. stand.
0	5 570 631	8094	5 665 661
0.26	10 418 570	791 989	12 163 880
0.51	9 216 289	1 583 225	11 996 810
0.78	8 303 621	2 751 276	12 440 740
1.03	7 069 347	2 865 072	11 359 010
1.28	7 383 883	3 743 521	12 871 720
1.61	7 767 293	5 045 173	15 041 710
1.86	7 009 639	5 835 951	16 037 570
2.11	6 041 736	5 091 334	13 490 720
3.33	4 722 039	5 773 944	12 899 400
3.95	3 900 131	5 928 189	12 307 730
4.3	4 307 299	7 315 926	14 787 520

a. Determine the vanadium content (expressed in mmol g^{-1}) in MIL-47.
b. How many mililiters of the oxidant need to be added to have a substrate/oxidant ratio of ½?

c. How much catalyst needs to be employed in the catalytic reaction to have a vanadium loading of 0.42 mmol?

d. Determine the TON and TOF (turn over frequency) based on the GC peak areas shown in the table.

2 Gold nanoparticles were embedded in the UiO-66 structure using a wet impregnation approach. To determine the gold content in the UiO-66, a calibration curve was made by means of XRF using PbO as an internal standard (see Figure 8.52).

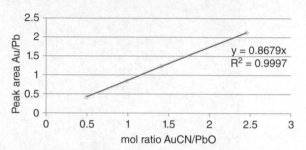

Figure 8.52
Calibration curve to determine the Au loading in UiO-66.

Determine the Au loading in the UiO-66 material based on the following data. For the XRF measurements, 0.022048 g of Au@UiO-66 material and 2.156 mg of PbO was used. The peak area of Au and Pb was measured four times to minimize experimental errors.

Peak area Au	Peak area Pb
1003.798	4876.647
972.75	4726.982
952.277	4644.349
963.159	4656.653

Answers to the Problems

1 a. The MIL-47 has a unit cell of $V^{IV}O(CO_2\text{-}C_6H_4\text{-}CO_2)$ that corresponds to a molecular weight of $231 \, mol \, g^{-1}$. So, 1 g of MOF contains: $\frac{1}{231} mol \, V = 0.00433 \, mol$

b. $5 \, mL \times 0.811 \, g \, mL^{-1} = 4.055 \, g$; Molecular weight of cyclohexene $= 82.143 \, g \, mol^{-1}$

  a. $\dfrac{g}{82.143 \, g \, mol^{-1}} = 0.049 \, mol; V_{oxidant} = \dfrac{0.0987 \, mol}{5.5 \, L^{-1}} = 0.0179 L$

c. 1 g of MOF contains 4.33 mmol V; to have a loading of 0.42 mmol, one needs to add 0.097 g of catalyst;

d. Step 1: take the ratio of the peak area of cyclohexene (CH)/internal standard (IS) and the ratio of the peak area of cyclohexene oxide (CO)/IS for each measured sample:

Time (hour)	Peak area CH/ peak area IS	Peak area CO/peak area IS
0	0.983 227 023	0.001428606
0.26	0.856 517	0.065109899
0.51	0.768 228 304	0.131 970 499
0.78	0.667 453 946	0.22 115 051
1.03	0.62 235 591	0.252 229 023
1.28	0.573 651 618	0.290 833 004
1.61	0.516 383 643	0.335 412 197
1.86	0.437 076 128	0.363 892 472
2.11	0.447 843 851	0.377 395 276
3.33	0.366 066 561	0.447 613 377
3.95	0.316 884 673	0.481 663 881
4.3	0.291 279 336	0.494 736 508

Step 2: Determine the molar ratio of CH/IS and the molar ratio of CO/IS using the information about the response factors:

Time (hour)	mol CH/mol IS	mol CO/mol IS
0	0.961 596 029	0.001684327
0.26	0.837 673 626	0.076764571
0.51	0.751 327 281	0.155 593 218
0.78	0.652 769 959	0.260 736 452
1.03	0.60 866 408	0.297 378 019
1.28	0.561 031 282	0.342 892 112
1.61	0.505 023 202	0.39 545 098
1.86	0.427 460 453	0.429 029 225
2.11	0.437 991 286	0.444 949 031
3.33	0.358 013 097	0.527 736 172
3.95	0.30 991 321	0.567 881 716
4.30	0.28 487 119	0.583 294 342

Step 3: determine the mmol internal standard

$$\frac{2.9343 \text{ g}}{181.45 \text{ g mol}^{-1}} * 1000 = 16.17 \text{mmol}$$

Step 4: Determine the mmol of cyclohexene and the mmol of cyclohexene that is converted taking into account the mmol of internal standard that is used:

Time (hour)	mmol CH	mmol CH converted
0	1 555 035 121	0
0.26	1 354 635 282	200 399 839
0.51	121 500 118	3 400 339 408
0.78	1 055 620 221	4 994 149 005
1.03	9 842 948 518	5 707 402 693
1.28	9 072 659 636	6 477 691 575
1.61	816 693 074	7 383 420 471
1.86	6 912 632 726	8 637 718 485
2.11	708 293 101	8 467 420 201
3.33	5 789 571 949	9 760 779 262
3.95	5 011 729 587	1 053 862 162
4.30	4 606 765 133	1 094 358 608

Step 5: Determine the mmol of V that is used during the reaction:
(0.0488 g)/(4.3 mmol g^{-1}) = 0.01135 mmol V = 0.00001135 mol V
Step 6: determine the TON at the end of the reaction and TOF at the beginning of the reaction

$$\text{TON:} \quad \frac{mol\ cyclohexene\ converted}{mol\ V} = \frac{10.9435\ \text{mol}}{0.00001135\ \text{mol}} = 964$$

$$\text{TOF:} \quad \frac{TON}{time} = \frac{3.40033}{\frac{0.00001135}{0.5166}} = 579\text{h}^{-1}(\text{TON was determined after}$$

0.5166 hours of reaction)

2 In first instance, the ratio of the peak area of Au/Pb needs to be determined for each measurement and the average of these ratios needs to be determined:

Peak area (Au)	Peak area (Pb)	Ratio Au/Pb
1003.798	4876.647	0.205837741
972.75	4726.982	0.205786694
952.277	4644.349	0.205039931
963.159	4656.653	0.206835038

Average ratio Au/Pb = 0.2058

$$\text{Mol PbO} = \frac{0.02156\ \text{g}}{223\ \text{g mol}^{-1}} = 9.66 \times 10^{-6}\text{mol}$$

Based on the XRF calibration curve we can calculate the mmol g^{-1} Au in the sample.

$$\frac{0.205874851}{0.8679} \times 9.66 \times 10^{-6} = 2.29 \times 10^{-6} \text{ mol}$$

Or in other words: the sample contains: $\frac{2.29 \times 10^{-6} \text{ mol}}{0.022048 \text{ g}} = 0.104$ mmol g^{-1} Au

References

1. (a) Eddaoudi, M., Li, H., and Yaghi, O.M. (1999). *J. Am. Chem. Soc.* 122: 1391–1397; (b) Li, H., Eddaoudi, M., O'Keeffe, M., and Yaghi, O.M. (1999). *Nature* 402: 276–279.
2. Eddaoudi, M., Kim, J., Rosi, N. et al. (2002). *Science* 295: 469–472.
3. Chui, S.S.Y., Lo, S.M.F., Charmant, J.P.H. et al. (1999). *Science* 283: 1148–1150.
4. Worrall, S.D., Bissett, M.A., Hirunpinyopas, W. et al. (2016). *J. Mater. Chem. C* 4: 8687–8695.
5. Loiseau, T., Serre, C., Huguenard, C. et al. (2004). *Chem. Eur. J.* 10: 1373–1382.
6. Serre, C., Millange, F., Thouvenot, C. et al. (2002). *J. Am. Chem. Soc.* 124: 13519–13526.
7. Ferey, G., Mellot-Draznieks, C., Serre, C. et al. (2005). *Science* 309: 2040–2042.
8. Cavka, J.H., Jakobsen, S., Olsbye, U. et al. (2008). *J. Am. Chem. Soc.* 130: 13850–13851.
9. Biswas, S., Zhang, J., Li, Z.B. et al. (2013). *Dalton Trans.* 42: 4730–4737.
10. Mondloch, J.E., Bury, W., Fairen-Jimenez, D. et al. (2013). *J. Am. Chem. Soc.* 135: 10294–10297.
11. Wang, T.C., Vermeulen, N.A., Kim, I.S. et al. (2016). *Nat. Protoc.* 11: 149–162.
12. Howarth, A.J., Katz, M.J., Wang, T.C. et al. (2015). *J. Am. Chem. Soc.* 137: 7488–7494.
13. Huang, X.C., Lin, Y.Y., Zhang, J.P., and Chen, X.M. (2006). *Angew. Chem. Int. Ed.* 45: 1557–1559.
14. Lee, Y.R., Jang, M.S., Cho, B.Y. et al. (2015). *Chem. Eng. J.* 271: 276–280.
15. Leus, K., Bogaerts, T., De Decker, J. et al. (2016). *Microporous Mesoporous Mater.* 226: 110–116.
16. Czaja, A.U., Trukhan, N., and Muller, U. (2009). *Chem. Soc. Rev.* 38: 1284–1293.
17. Klinowski, J., Paz, F.A.A., Silva, P., and Rocha, J. (2011). *Dalton Trans.* 40: 321–330.
18. Mueller, U., Schubert, M., Teich, F. et al. (2006). *J. Mater. Chem.* 16: 626–636.
19. Ameloot, R., Stappers, L., Fransaer, J. et al. (2009). *Chem. Mater.* 21: 2580–2582.
20. Ma, L. (2010). *Functional Metal-Organic Frameworks: Gas Storage, Separation and Catalysis*. Springer.
21. Schneemann, A., Bon, V., Schwedler, I. et al. (2014). *Chem. Soc. Rev.* 43: 6062–6096.
22. Kim, M., Cahill, J.F., Fei, H.H. et al. (2012). *J. Am. Chem. Soc.* 134: 18082–18088.
23. Han, Y., Li, J.R., Xie, Y.B., and Guo, G.S. (2014). *Chem. Soc. Rev.* 43: 5952–5981.
24. Hendrickx, K., Joos, J.J., De Vos, A. et al. (2018). *Inorg. Chem.* 57: 5463–5474.
25. Dhakshinamoorthy, A., Asiri, A.M., and Garcia, H. (2016). *Catal. Sci. Technol.* 6: 5238–5261.
26. Depauw, H., Nevjestic, I., De Winne, J. et al. (2017). *Chem. Commun.* 53: 8478–8481.
27. Song, X., Kim, T.K., Kim, H. et al. (2012). *Chem. Mater.* 24: 3065–3073.
28. Tu, T.N., Nguyen, M.V., Nguyen, H.L. et al. (2018). *Coord. Chem. Rev.* 364: 33–50.
29. Siu, P.W., Brown, Z.J., Farha, O.K. et al. (2013). *Chem. Commun.* 49: 10920–10922.
30. Leus, K., Jena, H.S., and Van der Voort, P. (2018). *Metal-Organic Frameworks: Applications in Separations and Catalysis*. Wiley-VCH Verlag GmbH & Co. KGaA.
31. Leus, K., Concepcion, P., Vandichel, M. et al. (2015). *RSC Adv.* 5: 22334–22342.
32. Abednatanzi, S., Leus, K., Derakhshandeh, P.G. et al. (2017). *Catal. Sci. Technol.* 7: 1478–1487.
33. Farha, O.K., Eryazici, I., Jeong, N.C. et al. (2012). *J. Am. Chem. Soc.* 134: 15016–15021.
34. Furukawa, H., Ko, N., Go, Y.B. et al. (2010). *Science* 329: 424–428.
35. Farha, O.K., Yazaydin, A.O., Eryazici, I. et al. (2010). *Nat. Chem.* 2: 944–948.

36. Koh, K., Wong-Foy, A.G., and Matzger, A.J. (2009). *J. Am. Chem. Soc.* 131: 4184–4185.
37. Yuan, D.Q., Zhao, D., Sun, D.F., and Zhou, H.C. (2010). *Angew. Chem. Int. Ed.* 49: 5357–5361.
38. Klein, N., Senkovska, I., Baburin, I.A. et al. (2011). *Chem. Eur. J.* 17: 13007–13016.
39. Furukawa, H., Miller, M.A., and Yaghi, O.M. (2007). *J. Mater. Chem.* 17: 3197–3204.
40. An, J., Farha, O.K., Hupp, J.T. et al. (2012). *Nat. Commun.* 3.
41. Kim, H., Yang, S., Rao, S.R. et al. (2017). *Science* 356: 430–432.
42. Suh, M.P., Park, H.J., Prasad, T.K., and Lim, D.W. (2012). *Chem. Rev.* 112: 782–835.
43. Rowsell, J.L.C., Millward, A.R., Park, K.S., and Yaghi, O.M. (2004). *J. Am. Chem. Soc.* 126: 5666–5667.
44. Rowsell, J.L.C. and Yaghi, O.M. (2006). *J. Am. Chem. Soc.* 128: 1304–1315.
45. Hirscher, M. and Panella, B. (2005). *J. Alloys Compd.* 404: 399–401.
46. Sun, D.F., Ma, S.Q., Ke, Y.X. et al. (2006). *J. Am. Chem. Soc.* 128: 3896–3897.
47. Ferey, G., Latroche, M., Serre, C. et al. (2003). *Chem. Commun.* 2976–2977.
48. Pan, L., Sander, M.B., Huang, X.Y. et al. (2004). *J. Am. Chem. Soc.* 126: 1308–1309.
49. Dybtsev, D.N., Chun, H., Yoon, S.H. et al. (2004). *J. Am. Chem. Soc.* 126: 32–33.
50. Dybtsev, D.N., Chun, H., and Kim, K. (2004). *Angew. Chem. Int. Ed.* 43: 5033–5036.
51. Chen, B.L., Ockwig, N.W., Millward, A.R. et al. (2005). *Angew. Chem. Int. Ed.* 44: 4745–4749.
52. Ferey, G. (2007). *Stud. Surf. Sci. Catal.* 170: 66–84.
53. Sels, B. and Kustov, L. (2016). *Zeolites and Zeolite-Like Materials*, 1e. Elsevier.
54. Goldsmith, J., Wong-Foy, A.G., Cafarella, M.J., and Siegel, D.J. (2013). *Chem. Mater.* 25: 3373–3382.
55. Xue, M., Liu, Y., Schaffino, R.M. et al. (2009). *Inorg. Chem.* 48: 4649–4651.
56. Dietzel, P.D.C., Besikiotis, V., and Blom, R. (2009). *J. Mater. Chem.* 19: 7362–7370.
57. Herm, Z.R., Swisher, J.A., Smit, B. et al. (2011). *J. Am. Chem. Soc.* 133: 5664–5667.
58. Botas, J.A., Calleja, G., Sanchez-Sanchez, M., and Orcajo, M.G. (2010). *Langmuir* 26: 5300–5303.
59. Choi, J.S., Son, W.J., Kim, J., and Ahn, W.S. (2008). *Microporous Mesoporous Mater.* 116: 727–731.
60. Gedrich, K., Senkovska, I., Klein, N. et al. (2010). *Angew. Chem. Int. Ed.* 49: 8489–8492.
61. Jung, D.W., Yang, D.A., Kim, J. et al. (2010). *Dalton Trans.* 39: 2883–2887.
62. Llewellyn, P.L., Bourrelly, S., Serre, C. et al. (2008). *Langmuir* 24: 7245–7250.
63. Zhang, Z.J., Huang, S.S., Xian, S.K. et al. (2011). *Energy Fuels* 25: 835–842.
64. Mu, B., Schoenecker, P.M., and Walton, K.S. (2010). *J. Phys. Chem. C* 114: 6464–6471.
65. Tan, C.R., Yang, S.H., Champness, N.R. et al. (2011). *Chem. Commun.* 47: 4487–4489.
66. Park, Y.K., Choi, S.B., Kim, H. et al. (2007). *Angew. Chem. Int. Ed.* 46: 8230–8233.
67. Moellmer, J., Moeller, A., Dreisbach, F. et al. (2011). *Microporous Mesoporous Mater.* 138: 140–148.
68. Hamon, L., Jolimaitre, E., and Pirngruber, G.D. (2010). *Ind. Eng. Chem. Res.* 49: 7497–7503.
69. Liang, Z.J., Marshall, M., and Chaffee, A.L. (2009). *Energy Procedia* 1: 1265–1271.
70. Liang, Z.J., Marshall, M., and Chaffee, A.L. (2009). *Energy Fuels* 23: 2785–2789.
71. Liang, Z.J., Marshall, M., and Chaffee, A.L. (2010). *Microporous Mesoporous Mater.* 132: 305–310.
72. Nune, S.K., Thallapally, P.K., Dohnalkova, A. et al. (2010). *Chem. Commun.* 46: 4878–4880.
73. Bourrelly, S., Llewellyn, P.L., Serre, C. et al. (2005). *J. Am. Chem. Soc.* 127: 13519–13521.
74. Couck, S., Denayer, J.F.M., Baron, G.V. et al. (2009). *J. Am. Chem. Soc.* 131: 6326–6327.
75. Park, H.J. and Suh, M.P. (2010). *Chem. Commun.* 46: 610–612.
76. Senkovska, I., Hoffmann, F., Froba, M. et al. (2009). *Microporous Mesoporous Mater.* 122: 93–98.
77. Chowdhury, P., Bikkina, C., and Gumma, S. (2009). *J. Phys. Chem. C* 113: 6616–6621.
78. Galli, S., Masciocchi, N., Tagliabue, G. et al. (2008). *Chem. Eur. J.* 14: 9890–9901.
79. Zhang, J., Xue, Y.S., Liang, L.L. et al. (2010). *Inorg. Chem.* 49: 7685–7691.
80. Wiersum, A.D., Soubeyrand-Lenoir, E., Yang, Q.Y. et al. (2011). *Chem. Asian J.* 6: 3270–3280.
81. Rabone, J., Yue, Y.F., Chong, S.Y. et al. (2010). *Science* 329: 1053–1057.

82. Loiseau, T., Lecroq, L., Volkringer, C. et al. (2006). *J. Am. Chem. Soc.* 128: 10223–10230.

83. Surble, S., Millange, F., Serre, C. et al. (2006). *J. Am. Chem. Soc.* 128: 14889–14896.

84. Sumida, K., Rogow, D.L., Mason, J.A. et al. (2012). *Chem. Rev.* 112: 724–781.

85. Bao, Z.B., Yu, L.A., Ren, Q.L. et al. (2011). *J. Colloid Interface Sci.* 353: 549–556.

86. Lu, Y.P., Dong, Y.L., and Qin, J. (2016). *J. Mol. Struct.* 1107: 66–69.

87. Marcz, M., Johnsen, R.E., Dietzel, P.D.C., and Fjellvag, H. (2012). *Microporous Mesoporous Mater.* 157: 62–74.

88. McDonald, T.M., Lee, W.R., Mason, J.A. et al. (2012). *J. Am. Chem. Soc.* 134: 7056–7065.

89. Liu, Y.Y., Yang, Y.M., Sun, Q.L. et al. (2013). *ACS Appl. Mater. Interfaces* 5: 7654–7658.

90. Cao, Y., Song, F.J., Zhao, Y.X., and Zhong, Q. (2013). *J. Environ. Sci. China* 25: 2081–2087.

91. Forrest, K.A., Pham, T., McLaughlin, K. et al. (2014). *Chem. Commun.* 50: 7283–7286.

92. Li, B.Y., Zhang, Z.J., Li, Y. et al. (2009). *Angew. Chem. Int. Ed.* 48: https://doi.org/10.1002/anie.200804853.

93. Yazaydin, A.O., Snurr, R.Q., Park, T.H. et al. (2009). *J. Am. Chem. Soc.* 131: 18198–18199.

94. Luebke, R., Weselinski, L.J., Belmabkhout, Y. et al. (2014). *Cryst. Growth Des.* 14: 414–418.

95. Aprea, P., Caputo, D., Gargiulo, N. et al. (2010). *J. Chem. Eng. Data* 55: 3655–3661.

96. Spanopoulos, I., Bratsos, I., Tampaxis, C. et al. (2016). *Chem. Commun.* 52: 10559–10562.

97. Liao, P.Q., Chen, X.W., Liu, S.Y. et al. (2016). *Chem. Sci.* 7: 6528–6533.

98. Zheng, B.S., Bai, J.F., Duan, J.G. et al. (2011). *J. Am. Chem. Soc.* 133: 748–751.

99. Liang, Z.Q., Du, J.J., Sun, L.B. et al. (2013). *Inorg. Chem.* 52: 10720–10722.

100. Bernini, M.C., Blanco, A.A.G., Villarroel-Rocha, J. et al. (2015). *Dalton Trans.* 44: 18970–18982.

101. Nugent, P., Belmabkhout, Y., Burd, S.D. et al. (2013). *Nature* 495: 80–84.

102. Jiao, J.J., Dou, L., Liu, H.M. et al. (2016). *Dalton Trans.* 45: 13373–13382.

103. Burd, S.D., Ma, S.Q., Perman, J.A. et al. (2012). *J. Am. Chem. Soc.* 134: 3663–3666.

104. Liu, B., Yao, S., Shi, C. et al. (2016). *Chem. Commun.* 52: 3223–3226.

105. Song, C.L., Hu, J.Y., Ling, Y.J. et al. (2015). *J. Mater. Chem. A* 3: 19417–19426.

106. Lu, Z.Y., Bai, J.F., Hang, C. et al. (2016). *Chem. Eur. J.* 22: 6277–6285.

107. Cui, P., Ma, Y.G., Li, H.H. et al. (2012). *J. Am. Chem. Soc.* 134: 18892–18895.

108. Lin, Y.C., Yan, Q.J., Kong, C.L., and Chen, L. (2013). *Sci. Rep.* 3.

109. Trickett, C.A., Helal, A., Al-Maythalony, B.A. et al. (2017). *Nat. Rev. Mater.* 2: 17045.

110. Britt, D., Furukawa, H., Wang, B. et al. (2009). *Proc. Natl Acad. Sci. U.S.A.* 106: 20637–20640.

111. Liu, B., Zhao, R.L., Yuc, K.F. et al. (2013). *Dalton Trans.* 42: 13990–13996.

112. Xiang, S.C., He, Y.B., Zhang, Z.J. et al. (2012). *Nat. Commun.* 3.

113. Yazaydin, A.O., Benin, A.I., Faheem, S.A. et al. (2009). *Chem. Mater.* 21: 1425–1430.

114. An, J., Geib, S.J., and Rosi, N.L. (2010). *J. Am. Chem. Soc.* 132: 38–39.

115. Demessence, A., D'Alessandro, D.M., Foo, M.L., and Long, J.R. (2009). *J. Am. Chem. Soc.* 131: 8784–8786.

116. McDonald, T.M., D'Alessandro, D.M., Krishna, R., and Long, J.R. (2011). *Chem. Sci.* 2: 2022–2028.

117. Phan, A., Doonan, C.J., Uribe-Romo, F.J. et al. (2010). *Acc. Chem. Res.* 43: 58–67.

118. Li, J.R., Yu, J.M., Lu, W.G. et al. (2013). *Nat. Commun.* 4.

119. Bastin, L., Barcia, P.S., Hurtado, E.J. et al. (2008). *J. Phys. Chem. C* 112: 1575–1581.

120. Du, L.T., Lu, Z.Y., Zheng, K.Y. et al. (2013). *J. Am. Chem. Soc.* 135: 562–565.

121. Choi, H.S. and Suh, M.P. (2009). *Angew. Chem. Int. Ed.* 48: 6865–6869.

122. Li, B., Wang, H.L., and Chen, B.L. (2014). *Chem. Asian J.* 9: 1474–1498.

123. Yang, S.H., Lin, X., Lewis, W. et al. (2012). *Nat. Mater.* 11: 710–716.

124. Mason, J.A., Oktawiec, J., Taylor, M.K. et al. (2015). *Nature* 527: 357–361.

125. Kitagawa, S. and Kondo, M. (1998). *B. Chem. Soc. Jpn.* 71: 1739–1753.

126. Liu, Y.Y., Couck, S., Vandichel, M. et al. (2013). *Inorg. Chem.* 52: 113–120.

127. Depauw, H., Nevjestic, I., Wang, G.B. et al. (2017). *J. Mater. Chem. A* 5: 24580–24584.

128. (a) Choi, H.J., Dinca, M., and Long, J.R. (2008). *J. Am. Chem. Soc.* 130: 7848; (b) Salles, F., Maurin, G., Serre, C. et al. (2010). *J. Am. Chem. Soc.* 132: 13782–13788.

129. Llewellyn, P.L., Bourrelly, S., Serre, C. et al. (2006). *Angew. Chem. Int. Ed.* 45: 7751–7754.
130. Hayashi, H., Cote, A.P., Furukawa, H. et al. (2007). *Nat. Mater.* 6: 501–506.
131. (a) Li, H., Wang, K.C., Sun, Y.J. et al. (2018). *Mater. Today* 21: 108–121; (b) Herm, Z.R., Bloch, E.D., and Long, J.R. (2014). *Chem. Mater.* 26: 323–338.
132. Xiang, S.C., Zhang, Z.J., Zhao, C.G. et al. (2011). *Nat. Commun.* 2.
133. Krause, S., Bon, V., Senkovska, I. et al. (2016). *Nature* 532: 348–352.
134. (a) Zhu, L., Liu, X.Q., Jiang, H.L., and Sun, L.B. (2017). *Chem. Rev.* 117: 8129–8176; (b) Leus, K., Liu, Y.Y., and Van Der Voort, P. (2014). *Catal. Rev.* 56: 1–56; (c) Huang, Y.B., Liang, J., Wang, X.S., and Cao, R. (2017). *Chem. Soc. Rev.* 46: 126–157.
135. Ma, L.Q., Falkowski, J.M., Abney, C., and Lin, W.B. (2010). *Nat. Chem.* 2: 838–846.
136. Bogaerts, T., Van Yperen-De Deyne, A., Liu, Y.Y. et al. (2013). *Chem. Commun.* 49: 8021–8023.
137. Zhao, M.T., Deng, K., He, L.C. et al. (2014). *J. Am. Chem. Soc.* 136: 1738–1741.
138. Lustig, W.P., Mukherjee, S., Rudd, N.D. et al. (2017). *Chem. Soc. Rev.* 46: 3242–3285.
139. Cui, Y.J., Zhang, J., He, H.J., and Qian, G.D. (2018). *Chem. Soc. Rev.* 47: 5740–5785.
140. Rao, X.T., Huang, Q., Yang, X.L. et al. (2012). *J. Mater. Chem.* 22: 3210–3214.
141. Cui, Y.J., Song, T., Yu, J.C. et al. (2015). *Adv. Funct. Mater.* 25: 4796–4802.
142. Moore, E.G., Samuel, A.P.S., and Raymond, K.N. (2009). *Acc. Chem. Res.* 42: 542–552.
143. Rieter, W.J., Taylor, K.M.L., and Lin, W.B. (2007). *J. Am. Chem. Soc.* 129: 9852–9853.
144. Zhao, S.N., Wang, G.B., Poelman, D., and Van der Voort, P. (2018). *Materials* 11: 1–26.
145. Li, C.M., Huang, J.P., Zhu, H.L. et al. (2017). *Sensor Actuat. B Chem.* 253: 275–282.
146. Liu, W., Wang, Y.L., Bai, Z.L. et al. (2017). *ACS Appl. Mater. Interfaces* 9: 16448–16457.
147. Qi, Z.W. and Chen, Y. (2017). *Biosens. Bioelectron.* 87: 236–241.
148. Zhang, X., Jiang, K., He, H.J. et al. (2018). *Sensor Actuat. B Chem.* 254: 1069–1077.
149. Xia, T.F., Zhu, F.L., Jiang, K. et al. (2017). *Dalton Trans.* 46: 7549–7555.
150. Cui, Y.J., Xu, H., Yue, Y.F. et al. (2012). *J. Am. Chem. Soc.* 134: 3979–3982.
151. Rocha, J., Brites, C.D.S., and Carlos, L.D. (2016). *Chem. Eur. J.* 22: 14782–14795.
152. Kaczmarek, A.M., Liu, Y.Y., Wang, C. et al. (2017). *Adv. Funct. Mater.* 27: 1700258–1700262.
153. Silva, P., Vilela, S.M.F., Tome, J.P.C., and Paz, F.A.A. (2015). *Chem. Soc. Rev.* 44: 6774–6803.
154. Notman, N. (2017). *Chem. World* 14: 44–47.
155. Jacoby, M. (2008). *Chem. Eng. News* 86: 13–16.
156. Wee, L.H., Meledina, M., Turner, S. et al. (2017). *J. Am. Chem. Soc.* 139: 819–828.
157. Lebedev, O.I., Millange, F., Serre, C. et al. (2005). *Chem. Mater.* 17: 6525–6527.
158. Meledina, M., Turner, S., Filippousi, M. et al. (2016). *Part. Part. Syst. Char.* 33: 382–387.
159. Khaletskaya, K., Turner, S., Tu, M. et al. (2014). *Adv. Funct. Mater.* 24: 4804–4811.

9 Beyond the Hybrids – Covalent Organic Frameworks

In 2005, Omar Yaghi [1] reported the synthesis of a Covalent Organic Framework (COF) for the first time. It was the start of the development of many new compounds and structures.

COFs are defined as "crystalline" porous polymers, built up entirely by light elements such as boron, carbon, nitrogen, oxygen, and silicon. These are connected to each other by strong covalent bonds. Except for the Si-variants, COFs are metal-free components, closely related to polymers. So "porous polymers" or "porous ordered polymers" are terms that are also sometimes used.

COFs have some similar characteristics to metal organic frameworks (MOFs), the first one being that they were both developed by Omar Yaghi, following the principles of isoreticular chemistry. In general, COFs have very low densities (down to $0.17 \, \mathrm{g \, cm^{-3}}$), high surface areas (up to $4210 \, \mathrm{m^2 \, g^{-1}}$) and an inherent porosity and pore aperture that can be tuned by using bigger/smaller building units. COFs have a higher thermal and chemical stability than MOFs allowing pre- or postsynthetic modifications for specific applications without introducing a significant change in the porosity and crystallinity.

9.1 Classification and Nature of COFs

Since the discovery of the first COF-based frameworks in 2005 by Yaghi and co-workers several COFs linkages have been introduced (Figure 9.1). The first and most commonly used types of linkage are the B—O-based linkages that include:

1) the boroxine-based linkages formed by a self-condensation of boronic acids such as 1,4-benzenediboronic acid, applied in the synthesis of COF-1.
2) boronate ester linkages and borosilicate linkages constructed through the reaction of boronic acids with catechols and silanols, respectively.
3) Spiroborate linked ionic COFs assembled via the condensation of diols and trialkylborate in the presence of a basic catalyst.

COFs-based on B—O-based linkages are *highly crystalline* due to *high reversibility* of the reactions to allow the corrections of defects. This high reversibility, however, results in a *low hydrolytic stability* and low chemical stability, leading to the decomposition of the framework in the presence of water and acids.

Introduction to Porous Materials, First Edition.
Pascal Van Der Voort, Karen Leus and Els De Canck.
© 2019 John Wiley & Sons Ltd. Published 2019 by John Wiley & Sons Ltd.

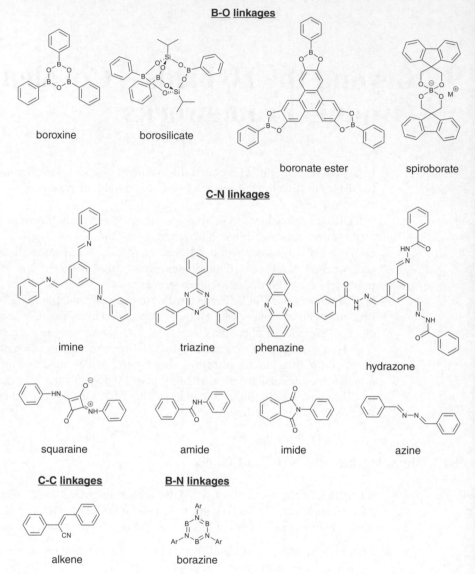

Figure 9.1
Linkages for the synthesis of COFs.

A *second type* of linkages are the C—N-based linkages including:

1) *Imine*-based COFs, formed through imine condensation of an aldehyde and an amine.
2) *Hydrazone* linkages produced through the condensation of aldehydes and hydrazide linkers.
3) *Squaraine* COFs which are composed of squaric acid and amines
4) Aromatic-based C=N-based COFs including the *triazine* and *phenazine* frameworks
5) *Imide* and *amide*-based linkages

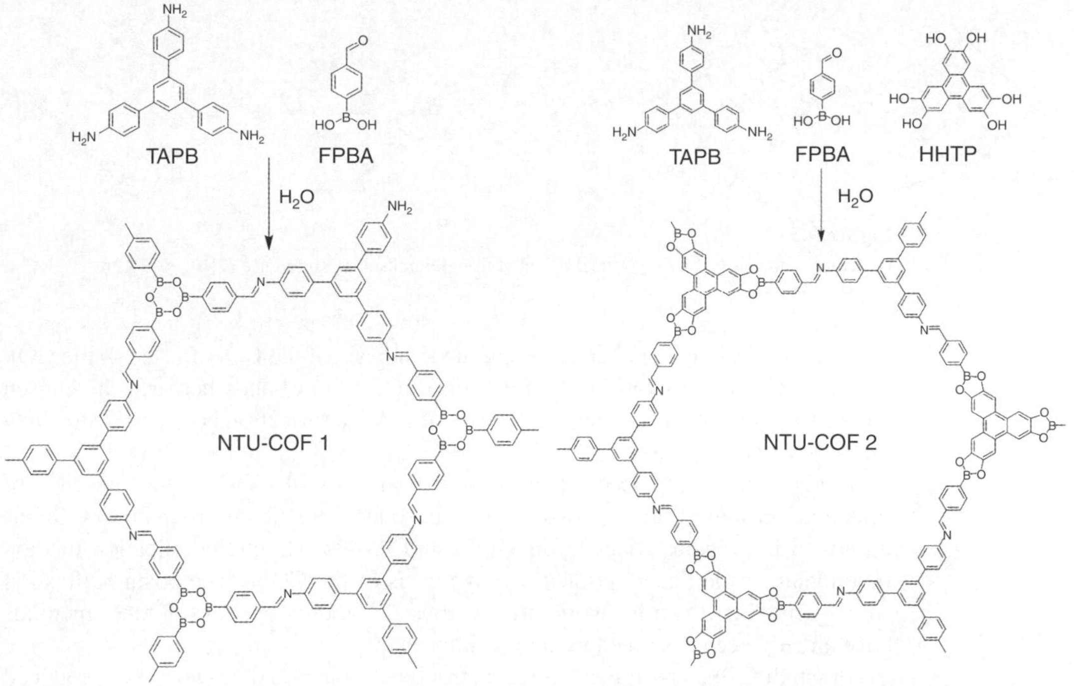

Figure 9.2
Synthesis of NTU-COF 1 and NTU-COF-2 by two types of covalent bonds.

The C—N linked COFs are much more stable in water, basic, and acid media but have in general a *lower crystallinity*. These condensation reactions are less reversible, resulting in more stable chemical bonds, but less error correction during the synthesis.

The C—C-based linkages and borazine (B—N) linkages produced through the decomposition of amineborane functionalized building units form the *third* and *fourth* type of linkages, respectively. They are less commonly used.

Zhao et al. reported the first multiple component COF materials very recently, involving the formation of two types of covalent bonds [2] (see Figure 9.2). The materials were referred to as NTU-COF 1 and 2 (NTU = Nanyang Technical University).

9.2 Design of COFs

COF materials contain two components: the organic monomers or building units and the linkages (the bonds formed between the organic monomers). This means that one needs a library of building units. These building units designed using the principles of reticular chemistry (similar to the synthesis of MOFs). As recently described by Yaghi et al. [3], there are four steps in the reticular synthesis of COFs (Figure 9.3).

In step 1 and 2 the target network topology is defined, followed by a deconstruction in its fundamental geometric units. In step 3, one looks into the library of building units to select monomers corresponding to the geometric units defined in step 2. The chosen building units need to be rigid and well-defined to obtain a porous structure.

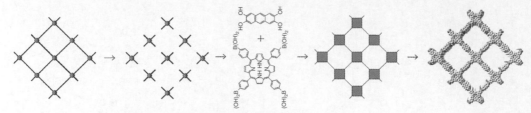

Figure 9.3
Reticular synthesis of COFs. Source: Adapted from Diercks et al. (reference [3]).

The monomers may not change during the synthesis of the COF. In step 4, the COF framework is synthesized by the formation of covalent bonds between the chosen monomers. It is important that this process of linkage formation is *reversible* to allow the self-correction of defects.

Another important aspect in the design principles of COFs is the presence of functional groups. There are two ways to introduce functional groups into COF, as described in previous chapters on MOFs and PMOs. The premodification method functionalizes the linkers prior to the synthesis of the COFs. The postmodification strategy modifies the framework after synthesis. The pros and cons of these methods have already been described in the previous chapters.

Although COFs are in general rigid structures, Yaghi and co-workers introduced in 2016 the so-called woven COFs exhibiting an unusual behavior in elasticity [4]. In this study, the authors synthesized a 3D-based COF by imine condensation of Cu(I)-bis[4,4′(1,10-phenanthroline-2,9-diyl)dibenzaldehyde]tetrafluoroborate, $Cu(PDB)_2(BF_4)$, with benzidine in a mixture of THF and aqueous acetic acid that resulted in a dark-brown crystalline solid, denoted as COF-505 (see Figure 9.4) [4].

The 1D threads are designed to intersect at regular intervals by means of metal templates. The Cu centers, used to template the formation of the framework are topologically independent of the weaving within the resulting framework allowing the reversible removal of the Cu centers without loss of the COF structure. Additionally, the authors observed a 10-fold increase in elasticity upon demetallation. This was assigned to the loose interaction between the threads after removal of copper allowing the threads to carry out large motions without unzipping the structure.

Another recent study conducted by Wang et al. [5] showed that COFs can also exhibit a distinct *breathing behavior*, just like the same MOFs (see previous chapter). They prepared a 3D interpenetrated imine COF by mixing (3,3′-bipyridine)-6,6′-dicarbaldehyde and tetra(4-anilyl)methane in the presence of 1,4-dioxane and aqueous acetic acid in a sealed glass ampule (see Figure 9.5). The resulting LZU-301 in THF showed a crystal structure transformation, compared to the dry LZU-301. The authors noted that the original 200 peak for the activated sample split into two peaks after solvation, accompanied by a shift to two lower theta values. Upon the removal of THF from the pores, the original desolvated XRPD pattern was retained demonstrating that this process was fully reversible. After solvation, an increase in the pore size from 5.8×10.4 Å^2 up to a pore size of 9.6×10.4 Å^2 was observed, corresponding to a unit cell volume expansion of 35%. In comparison to MOFs, a clear gate opening effect was also observed when collecting the THF vapor adsorption isotherm of LZU-301.

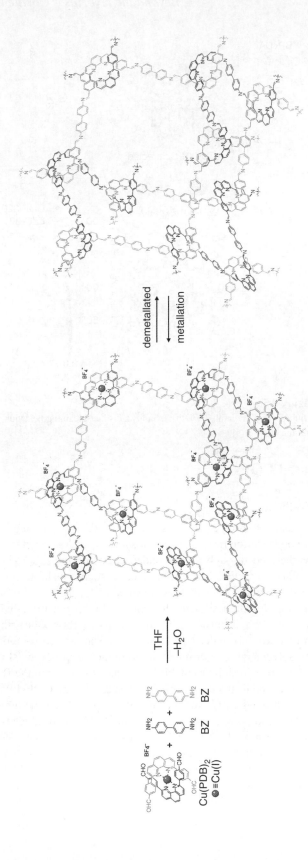

Figure 9.4

Strategy for the design and synthesis of weaving frameworks. Source: Reproduced with permission of AAA [4].

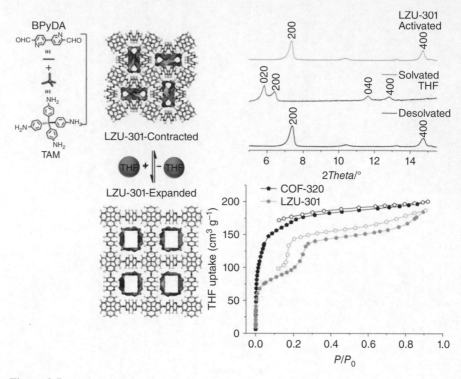

Figure 9.5
Synthesis of a 3D imine COF that exhibits a distinct breathing behavior. Source: Reproduced with permission of ACS [5].

9.3 Boron-Based COFs

9.3.1 Introduction

Typically, boron-based COFs are classified into two categories depending on the type of condensation reaction used for synthesis. The first class of boron-based COFs are synthesized through a self-condensation of *single* boronate SBUs. The first COF, denoted as COF-1, was synthesized in such a fashion. For the synthesis of COF-1, 1,4-benzenediboronic acid was heated at 120 °C for 72 hours in a sealed Pyrex tube in a solution of mesitylene-dioxane [1]. In this reaction, three boronic acid molecules were converted into a six-membered B_3O_3 (boroxine) ring through the release of three water molecules (see Figure 9.6). The limited solubility of the building unit in the solvent mixture ensures that the reaction proceeds slowly. The closed reaction system provides the availability of water to maintain the reversibility of the process to allow the formation of crystals. The crystalline COF-1 was isolated at a 71% yield and exhibited a BET surface area of 711 m^2 g^{-1} on the removal of guest molecules.

The second class of boron containing COFs are produced through a co-condensation of *two or more* building units. An example of this class of COF materials is COF-5 that was obtained through a dehydration reaction of a boronic acid, phenylboronic acid, and a trigonal catechol building unit, 2,3,6,7,10,11-hexahydroxytriphenylene, generating a five membered BO_2C_2 boronate ester ring (see Figure 9.6). Several boronic acids and diols can be used, which results in the huge diversity of boronate ester-based COFs reported to date (see Figure 9.7).

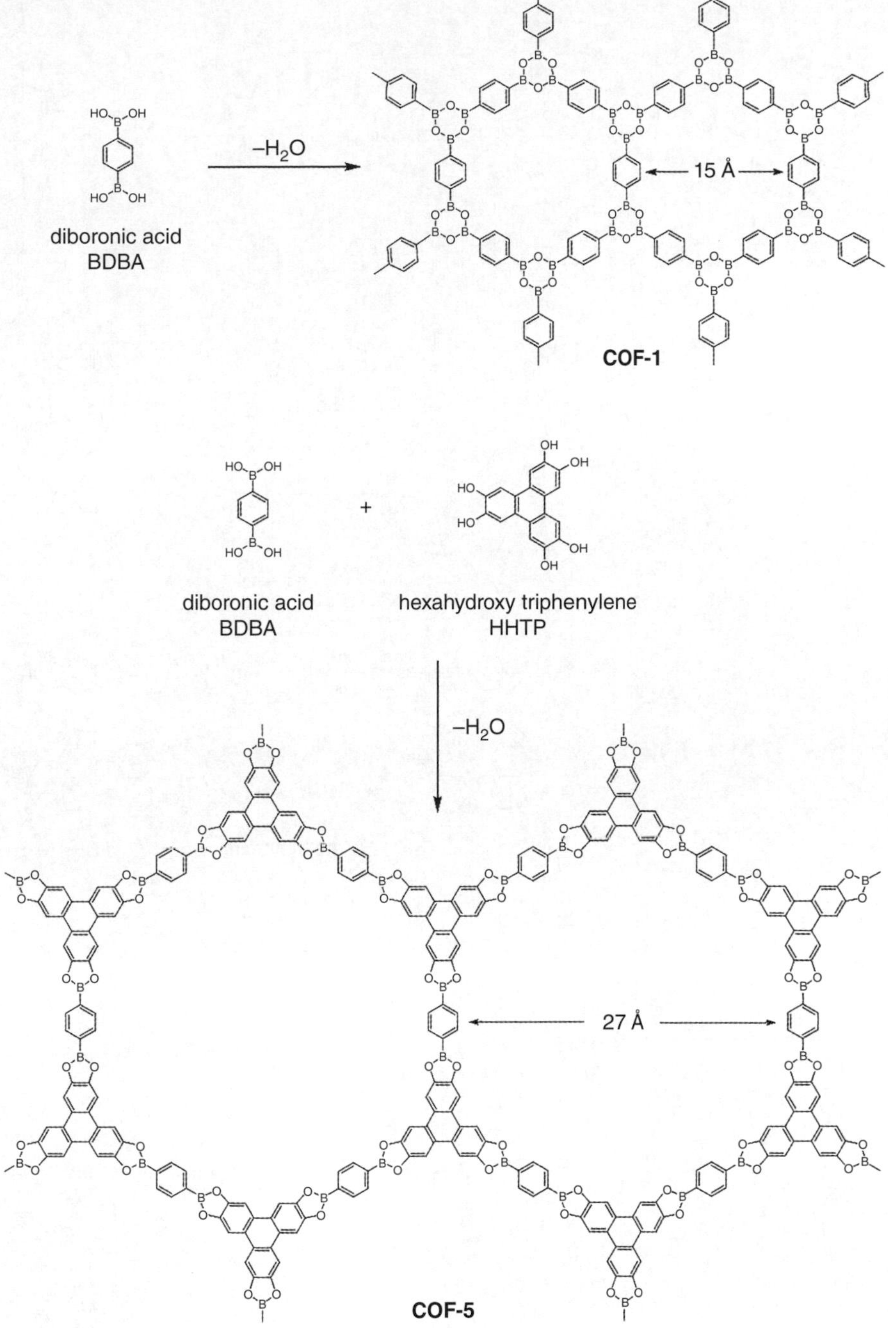

Figure 9.6

Solvothermal synthesis of the first boron containing COFs, denoted as COF-1, and COF-5. Source: Reproduced with permission of AAA [1].

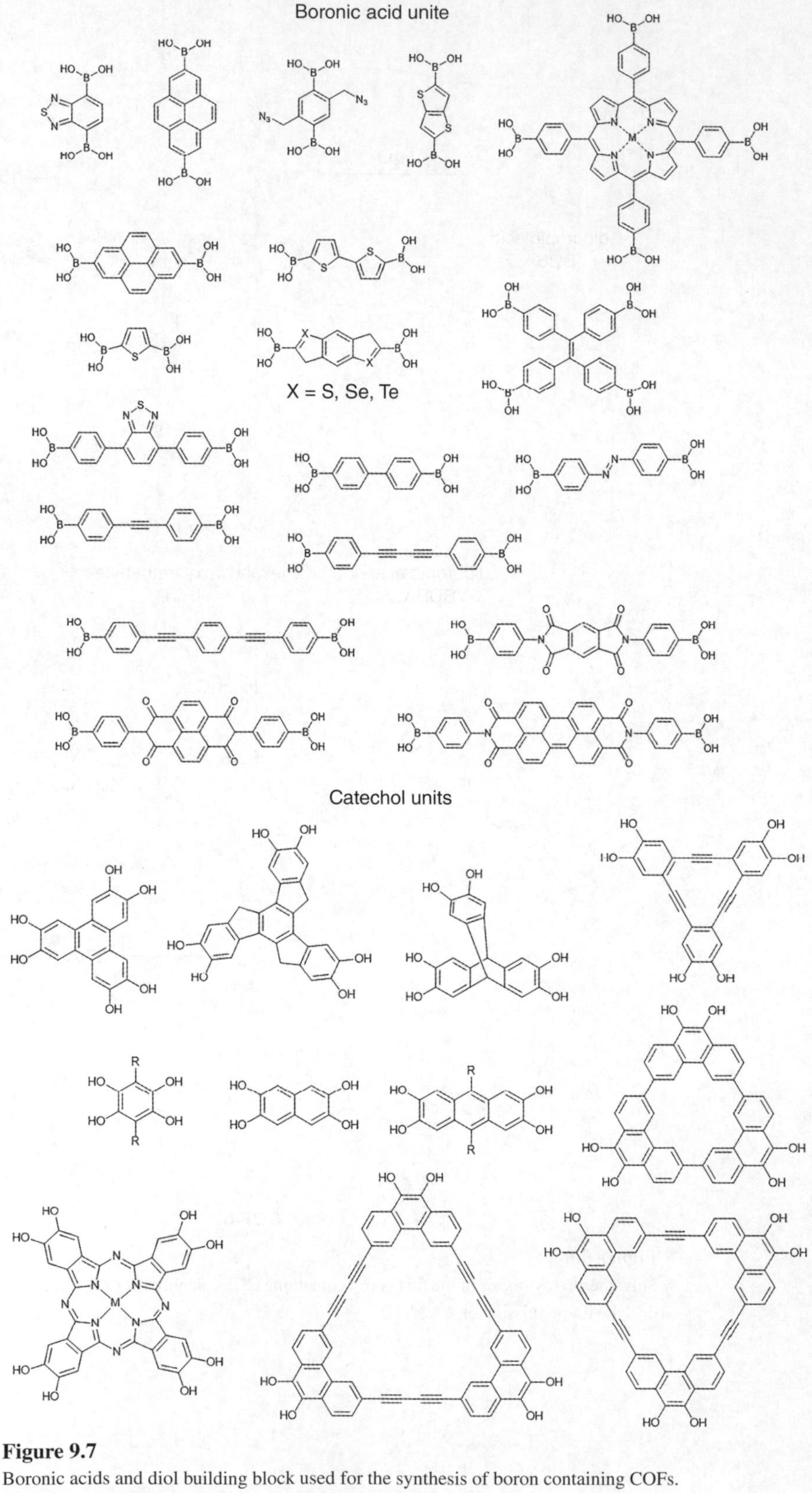

Boronic acid unite

X = S, Se, Te

Catechol units

Figure 9.7
Boronic acids and diol building block used for the synthesis of boron containing COFs.

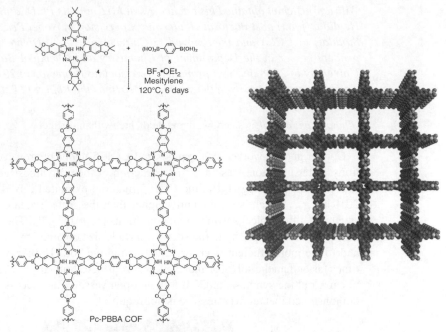

Figure 9.8
Lewis acid assisted synthesis of an eclipsed 2D phthalocyanine linked COF. Source: Reproduced with permission of Springer Nature [6].

9.3.2 Other Synthetic Routes to Obtain Boron-Based COFs

Other synthesis routes have been reported too. Dichtel reported on a Lewis acid assisted synthesis of boronate ester linked COFs. As such, they avoided the direct use of unstable and insoluble polyfunctional catechol monomers because these are prone to oxidation [6]. They used protected catechols to form boronate ester linked COFs, as the protecting groups increase the solubility (due to a decrease of the polarity of the compound) and prevent auto-oxidation. A new phthalocyanine-based COF was synthesized using the tetrafunctional catechol, denoted as phthalocyanine tetra(acetonide) (see Figure 9.8). The isolated phthalocyanine-based COF, had an excellent thermal stability (stable up to 500 °C) and a moderate Langmuir surface area of 506 m^2 g^{-1}.

Will(iam) Dichtel obtained his bachelor's at MIT and his Ph.D. at Berkeley in 2005. He did a (joint) post-doctorate under the supervision of Nobel Prize winner Frasier Stoddart at UCLA and then moved to Caltech (all USA). Dichtel began his independent career in the Department of Chemistry and Chemical Biology at Cornell University in 2008 and was promoted to Associate Professor in 2014. He moved to Northwestern University in the summer of 2016 as the Robert L. Letsinger Professor of Chemistry.

(Photograph: https://sites.northwestern.edu/dichtel/william-dichtel*)*

Cooper and co-workers [7] synthesized COF-5 under microwave heating conditions. After filtration, washing, and extraction of the as synthesized material under nitrogen, the resulting crystalline COF-5 material exhibited a BET surface area of $2019\,m^2\,g^{-1}$, which is significantly higher than the surface area of COF-5 synthesized under the typical solvothermal conditions ($1590\,m^2\,g^{-1}$). The authors assigned this difference in porosity to the smaller particle size obtained by microwave heating and to the more efficient microwave extraction method. The reaction was also 200 times faster than the solvothermal synthesis route. Additionally, the authors were able to carry out the synthesis of COF-5 in an open vessel. The success of this approach demonstrated that an overpressure is not required.

Andy Cooper obtained his Ph.D. in 1994 at Nottingham under the supervision of Professor Martyn Poliakoff. He held an 1851 Fellowship and a Royal Society NATO Fellowship at the University of North Carolina at Chapel Hill, USA, working with Professor Joseph M. DeSimone. In 1997, he obtained a Ramsay Memorial Research Fellowship at the Melville Laboratory for Polymer Synthesis in Cambridge, working with Professor Andrew B. Holmes on polymerization in supercritical CO_2. In 1998, he was awarded a Royal Society University Research Fellowship and joined Liverpool in January 1999, where he now holds a Personal Chair. He is the founding Director of the Center for Materials Discovery and the Materials Innovation Factory

Photograph: www.liverpool.ac.uk/chemistry/staff/andrew-cooper

9.3.3 Methods to Increase the Stability of Boron-Based COFs

As mentioned earlier, boron-based COFs are highly crystalline, but can be easily hydrolyzed. Lavigne and co-workers proposed a method to enhance the hydrolytic stability by introducing hydrophobic alkyl groups in the pores of the COF [8]. The authors showed that the hydrolytic stability improves with increasing length of the alkyl chain.

Zhang et al. [9] reported on the first *spiroborate*-based COFs (see Figure 9.9). Spiroborates are ionic derivatives of boronic acids exhibiting a good stability in water, methanol, and in basic conditions. They form an attractive alternative for the instable boroxine and boronate ester-based COFs [10]. The materials showed a high stability in water for at least two days. This enhanced hydrolytic stability is due to the additional Lewis base coordination to the boron centers in anionic spiroborate esters.

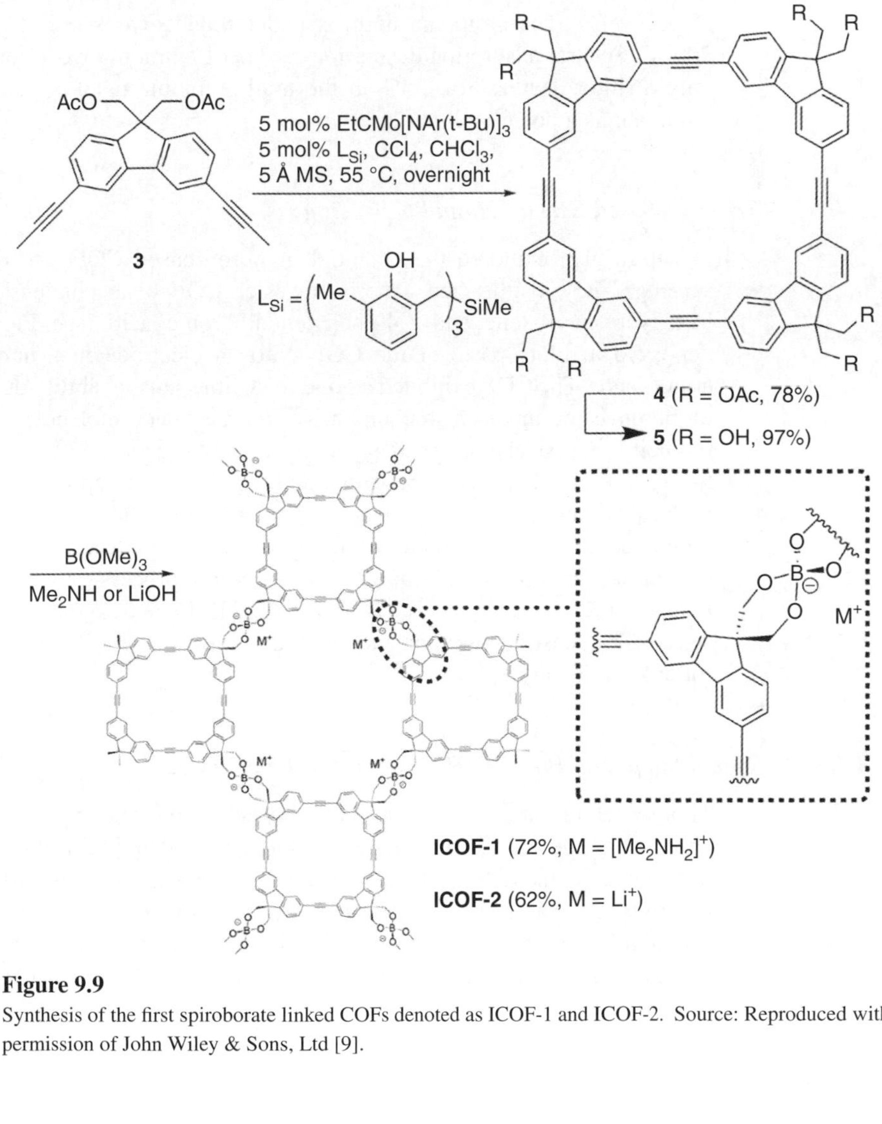

Figure 9.9
Synthesis of the first spiroborate linked COFs denoted as ICOF-1 and ICOF-2. Source: Reproduced with permission of John Wiley & Sons, Ltd [9].

The ICOF-2 also exhibited a good stability in basic conditions (1 M LiOH, 2 days). Due to the acid instability of the spiroborate linkages, both ICOFs were unstable in aqueous acidic conditions.

9.3.4   Applications of Boron-Containing COFs

9.3.4.1   *The Use of Boron COFs in the Adsorption of Ammonia*

The presence of a high density of Lewis acid boron atoms in boroxine and boronate ester-based COFs allows interaction with Lewis basic gases such as ammonia; Yaghi and co-workers screened several 2D and 3D COFs for the uptake of ammonia [11]. COF-10 (see Figure 9.10), consisting of hexahydroxytriphenylene and biphenyldiboronic acid as building units exhibits an exceptional uptake of ammonia of 15 mol kg^{-1} at 25 °C. This value is much higher than state of the art adsorbents such as Amberlyst 15, Zeolite 13X and MCM-41 (11, 9$^-$, and 7.9 mol kg^{-1}, respectively). The ammonia desorbs under reduced pressure at a temperature of 200 °C. Several adsorption/desorption cycles of ammonia were conducted showing only a slight decrease of 4.5% in the total ammonia uptake after the third cycle. Ammonia is a potential hydrogen carrier.

9.3.4.2   *A Truxene-Based COF in Humidity Sensing*

Pal et al. [12] explored the potential of boron-based COFs for relative humidity sensing. They synthesized a boronate ester COF by a condensation reaction of hexahydroxy truxene and 1,4-phenylenediboronic acid (see Figure 9.11). They deposited a dilute paste of the COF onto an electrode in a humidity-controlled atmosphere. The COF exhibited a good reversible sensing ability due to the presence of boron ester linkages that interact with the water molecules. At low relative humidity, a first chemisorbed layer of water molecules forms on the surface of the sensor via Lewis acid–base interactions. This low amount of adsorbed water molecules hinders the free movement of water molecules and therefore results in a high impedance. At high relative humidity, multilayer adsorption of water molecules occurs, which favors the charge transportation process by the *Grotthuss chain reaction* ($H_2O + H_3O^+ \rightarrow H_3O^+ + H_2O$). This increases the conductivity of the sensor. The sensor showed an excellent long-term stability at each relative humidity for at least 70 days.

9.3.4.3   *A Thienothiophene-Based COF as a Photovoltaic Device*

Thomas Bein and co-workers [13] investigated the potential of COFs as a photovoltaic device. They prepared a COF using thieno[2,3-b]thiophene-2,5 diyldiboronic acid and 2,3,6,7,10,11 hexahydroxytriphenylene as building units (see Figure 9.12). The resulting boronate ester-based COF (TT-COF) exhibited a high crystallinity, a BET surface area of 1810 m^2 g^{-1}, and a pore size of 3 nm. Bein loaded a fullerene electron acceptor ([6,6]-phenyl-C$_{61}$-butyric acid methyl ester) into

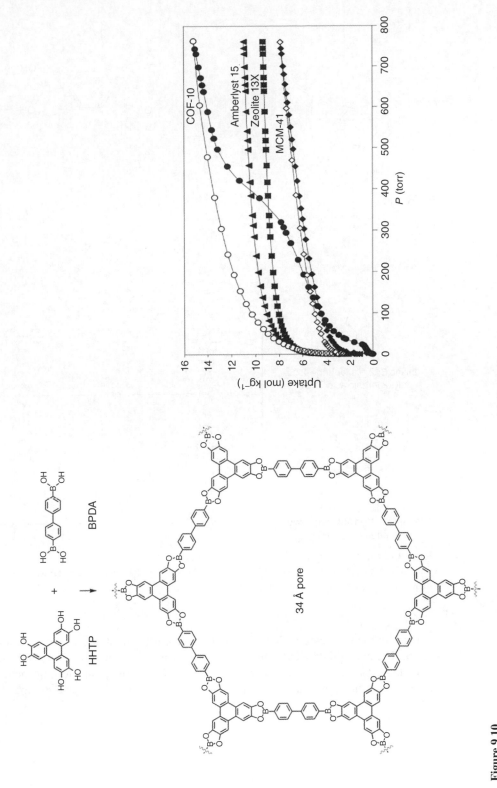

Figure 9.10

The potential use of COF-10 as adsorbent of ammonia. Source: Reproduced with permission of Springer Nature [11].

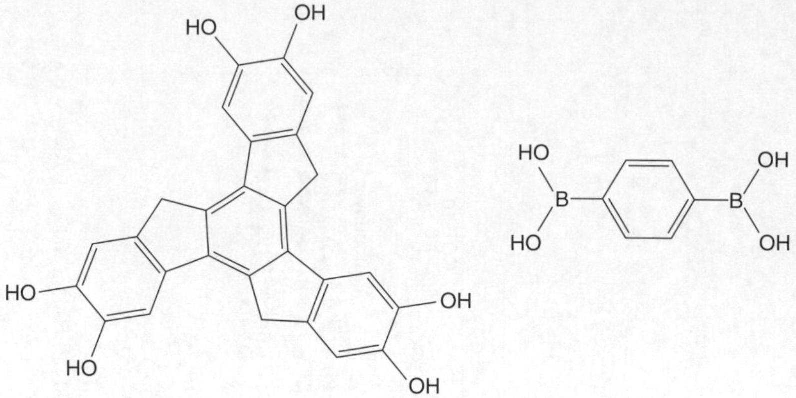

Figure 9.11

Hexahydroxy truxene and 1,4-phenylenediboronic acid.

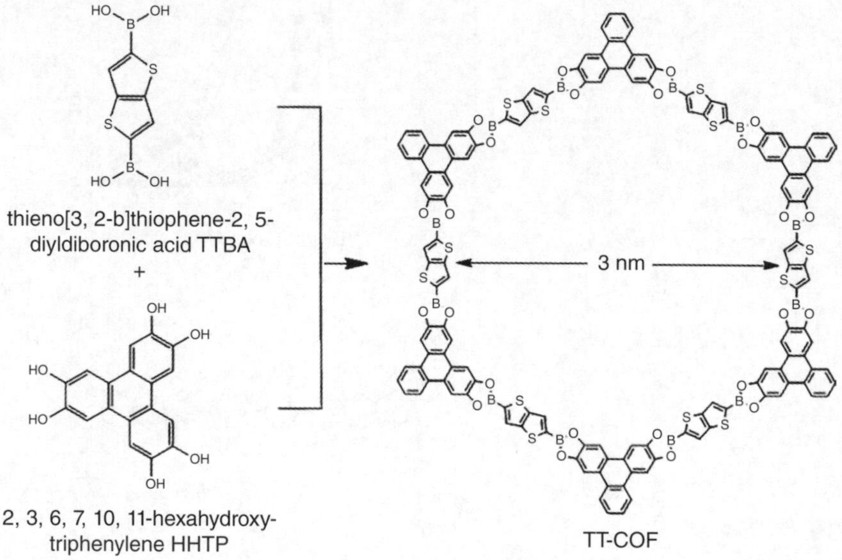

thieno[3, 2-b]thiophene-2, 5-
diyldiboronic acid TTBA

+

2, 3, 6, 7, 10, 11-hexahydroxy-
triphenylene HHTP

TT-COF

3 nm

Figure 9.12

Synthesis of a thienothiophene-based COF toward its use as a photovoltaic device. Source: Reproduced
with permission of John Wiley & Sons, Ltd [13].

the pores of the synthesized COF. The authors measured the fluorescence emission
spectra. The unmodified COF exhibits a blue emission on excitation at $\lambda = 380$ nm.
In the loaded COF, the emission band at $\lambda = 487$ nm was completely quenched
demonstrating the light induced interactions between the COF host and the loaded
fullerene.

Thomas Bein obtained his Ph.D. in Chemistry from the University of Hamburg (Germany) and the Catholic University Leuven (Belgium) in 1984. Afterward, he was a visiting research scientist at the DuPont Central Research and Development Department in Wilmington (USA). From 1986 to 1991 he was Assistant Professor of Chemistry at the University of New Mexico in Albuquerque (USA). In 1991, he joined the Purdue University (Indiana) as Associate Professor and was promoted to Full Professor of Chemistry in 1995. In 1999 he was appointed as the Chair of Physical Chemistry at the University of Munich (LMU), where he also served as Director of the Department of Chemistry. His current research interests cover the synthesis and physical properties of functional nanostructures, with an emphasis on porous materials for targeted drug delivery and nanostructured materials for solar energy conversion.

Photograph: http://www.cup.lmu.de/en/departments/chemistry/people/prof-dr-thomas-bein

9.4   Covalent Triazine Frameworks

9.4.1   Ionothermal Synthesis of Covalent Triazine Frameworks

Antonietti, Thomas, and co-workers reported an alternative to the boric acid-based COFs [14]. They introduced a new dynamic covalent reaction that allows the formation of porous crystalline covalent triazine-based frameworks (CTF) through the *reversible ionothermal trimerization of simple and cheap aromatic polynitriles* (see Figure 9.13). This trimerization reaction relies on the use of high temperatures, so molten salt metals (Lewis acid salts) at temperatures above 400 °C are needed.

Figure 9.13
Schematic reaction scheme for the synthesis of CTF-1. Source: Reproduced with permission of ACS [14].

Generally, $ZnCl_2$ is used as a metal salt but also binary mixtures of $ZnCl_2$ with various alkali chlorides, for example, NaCl, KCl, or LiCl can be used. CTFs have a low crystallinity but have an exceptionally high mechanical, chemical, and thermal stability, withstanding harsh conditions up to 600 °C. Again, this is the consequence of the reversibility of the condensation reaction. As the triazine formation is less reversible, more stable bonds are formed, but there is much less "error correction" in the formation of the framework. Furthermore, their structural and textural properties are different to those of the typical carbon blacks as well as the mesoporous carbons of the CMK family [14].

Markus Antonietti was born in 1960 in Mainz (Germany) and studied Physics and Chemistry. He obtained his doctoral degree in Chemistry in 1985 at Mainz University. He obtained his German Habilitation in physical chemistry at the same university in 1990. In 1991 he became lecturer (in Mainz) and in the same year Professor in Marburg University. Since 1993 and up to now, Antonietti has been director and scientific member of the Max Planck Institute of Colloids and Interfaces in Potsdam (Berlin, Germany). Additionally, since 1995 he has been Full Professor at Potsdam University. Antonietti has more than 700 publications, more than 90 patents, and has won several prestigious awards. He is one of the founding fathers of the Covalent Triazine Frameworks.

Photograph: https://www.mpg.de/389855/kolloid_grenzflaechen_wissM

Arne Thomas, born in 1975, studied chemistry at Justus-Liebig-Universität in Gießen, Philipps-Universität in Marburg, and at Heriot-Watt University in Edinburgh. He obtained his Ph.D. under the supervision of Markus Antonietti. He obtained an Alexander von Humboldt grant to work as a post-doctoral researcher at the University of Santa Barbara (USA) in the group of professor Galen D. Stucky. In 2005 he returned to the Max Planck Institute as a group leader where he invented the first CTF material, CTF-1, in 2008. In 2009 he was appointed as Full Professor at the Technische Universität Berlin at the Department of Inorganic Chemistry for Functional Materials.

Photograph: http://polymat-spotlight.eu/team/arne-thomas/

CTF-1 was synthesized through a trimerization of 1,4-dicyanobenzene in a vacuum sealed quartz ampule at 400 °C for 40 hours in the presence of $ZnCl_2$ (see Figure 9.13). Hereafter, the obtained black-colored solid was stirred in H_2O and HCl to remove most of the $ZnCl_2$ [14]. The resulting CTF-1 has a 2D structure with a hexagonal packing of the pores (Figure 9.13), exhibiting the hcb-topology.

The surface area and pore volume of the CTFs is tunable by changing the synthesis temperature and $ZnCl_2$/monomer ratio. In general, by increasing the temperature or increasing the $ZnCl_2$/monomer ratio, a higher surface area, and pore volume is obtained, together with a shift in the pore size distribution toward the mesopore range without affecting the micropores. At higher temperatures, irreversible reaction pathways take place including the thermal decomposition of the cyano-alkyl moieties via C—C coupling of neighboring aromatic nitriles accompanied by nitrogen loss ("carbonization").

Besides 1,4-dicyanobenzene, several other nitrile-based monomers can be used to synthesize CTF frameworks. We give an overview in Figure 9.14 [15]. Most of the obtained CTF structures show little to no crystallinity, except for a few monomers such as 1,4 dicyanobenzene (CTF-1), 1,3,5- tricyanobenzene (CTF-0), and 2,6-dicyanonaphthalene (CTF-2) [16].

Although the ionothermal synthesis method is a very simple and straightforward one, the harsh conditions limit their practical applications as many monomers tend to be unstable and exhibit carbonization at these high temperatures. For this reason, milder synthesis routes are being explored.

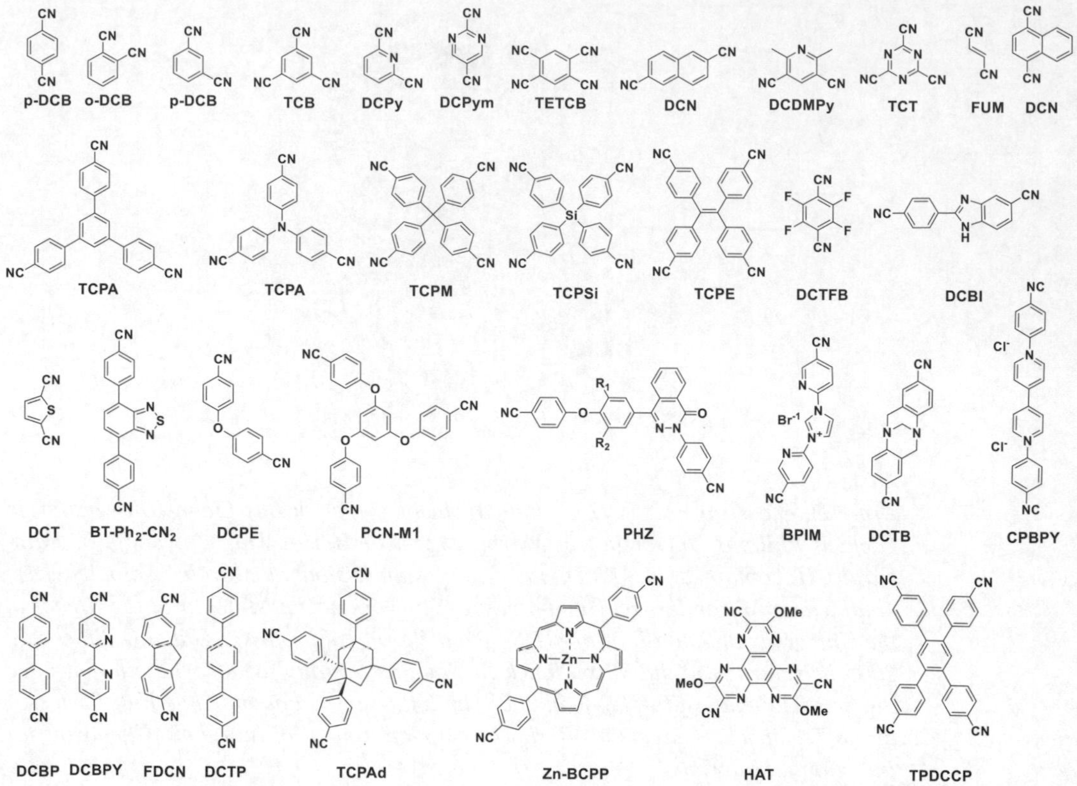

Figure 9.14

Nitrile building blocks used for the synthesis of CTF materials.

9.4.2 Acid Assisted Synthesis Route

In 2012, Andrew (Andy) Cooper's group used strong Brønsted acids (as replacements for the molten ZnCl$_2$) to catalyze the trimerization of nitriles [17]. They tested the very strong trifluoromethanesulfonic acid (TFMSA, triflic acid) as a catalyst under both room temperature and microwave-assisted conditions. They obtained pale yellow to brown colored, fluorescent powders. The "non-black" character of the materials becomes very important in photo-catalytic and luminescence applications.

Another big advantage of this method is that irreversible pathways, such as CH cleavage and carbonization, do not occur.

9.4.3 Mechanochemical Synthesis

Borchardt et al. reported the *mechanochemical* synthesis of CTF materials by *Friedel-Crafts alkylation* [18]. In first instance, carbazole was used as a model monomer in the presence of AlCl$_3$ and ZnCl$_2$ as activating reagent and the bulking agent, respectively. After 1 h of ball milling, using tungsten carbide balls, a porous yet amorphous CTF of 570 m^2 g^{-1} could be obtained. Figure 9.15 shows that other monomers were also investigated to evaluate the generality of the latter approach. All the monomers bearing a large conjugated π-system led to a high yield in the Friedel–Crafts alkylation.

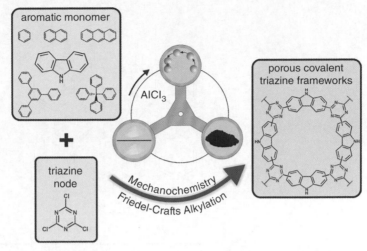

Figure 9.15
Mechanochemical synthesis of CTFs by Friedel–Crafts alkylation of different monomers with cyanuric chloride. Source: Reproduced with permission of John Wiley & Sons, Ltd [18].

This Friedel–Crafts alkylation approach to synthesize CTFs significantly broadens the variety of possible monomers that can be used, as no cyano groups are required.

9.4.4 Applications of CTFs

In the following, we describe four case studies highlighting the potential of CTFs toward their use in catalysis and liquid phase adsorption.

9.4.4.1 Oxidation of Methane to Methanol

The high chemical stability of CTFs in both acidic and basic media and their extreme thermal and mechanical stability render CTFs excellent catalytic support materials for the anchoring of homogeneous metal complexes and for the embedding of nanoparticles. This was demonstrated by Palkovits and Schüth and co-workers in the low temperature oxidation of methane to methanol [19]. Nowadays, methanol is produced via a two-step process involving the conversion of methane into CO and H_2 (syngas) at 1123 K. In a second step, methanol is produced over a $Cu/ZnO/Al_2O_3$ catalyst that operates at 513–533 K and pressurized syngas of 50–100 bar [20]. As this industrial process is energy intensive, the direct conversion of methane to methanol at low temperatures is a very attractive but challenging reaction due to the high binding energy of CH_3-H (435 kJ mol^{-1}) and the risk of over-oxidation to CO_2. In 1998, Periana's group developed a dichlorobipyrimidyl platinum complex with a remarkable catalytic performance in the low temperature oxidation of methane. They obtained a methane conversion above 90% and a selectivity of 81% toward methylbisulfate [21]. So, Schüth selected a pyridine-based CTF with numerous accessible bipyridyl units that would resemble the coordination sites for Pt coordination in the molecular Periana catalyst (see Figure 9.16) [19]. The obtained Pt-CTF was evaluated as a catalyst in the low temperature oxidation of methane to methanol using similar reaction conditions as was applied for the homogeneous variant. This heterogeneous catalyst was

$$CH_4 \xrightarrow[\substack{\text{40 bar} \\ \text{concentrated sulphuric acid} \\ \text{215 °C} \\ \text{2.5 hours}}]{} CH_3OH$$

Figure 9.16
Anchoring of the homogeneous Periana's platinum bipyrimidine complex onto the pyridine CTF material toward its use in the low temperature oxidation of methane. Source: Reproduced with permission of John Wiley & Sons, Ltd [19].

just as active and just as selective as the molecular Periana catalyst. Recyclability and stability studies demonstrated that the catalyst was stable during at least five cycles with TONs up to 300.

Regina Palkovits is currently Professor for Heterogeneous Catalysis and Chemical Technology at the RWTH Aachen University (Germany). She did her Master's thesis under the supervision of Stephan Kaskel (Dresden, Germany), Ph.D. in the group of Ferdi Schüth (2003–2006) at the Max Planck Instute for Kohlenforschung and her post-doctorate in Utrecht University with Bert Weckhuysen (2007). She then became group leader in the Max Planck Institute for Kohlenforschung and was appointed as a professor in Aachen in 2013.

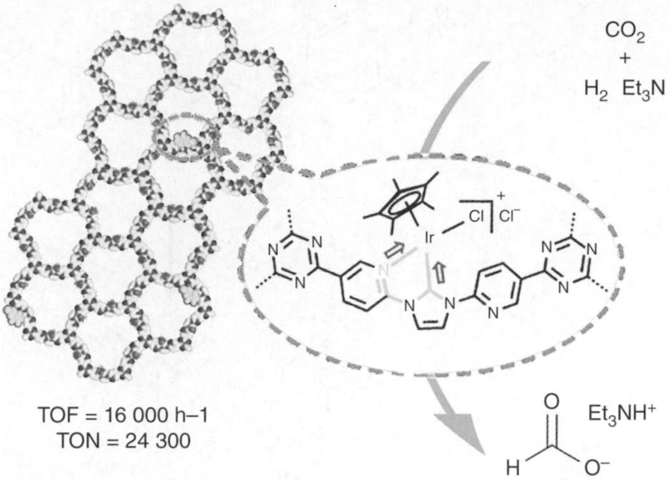

CO$_2$
+
H$_2$ Et$_3$N

TOF = 16 000 h–1
TON = 24 300

Et$_3$NH$^+$

Figure 9.17
Ir(III) anchored NHC-based CTF in the conversion of CO$_2$. Source: Reproduced with permission of ACS [22].

9.4.4.2 CTFs in the Conversion of CO$_2$ into Formic Acid

Yoon et al. [22] anchored an Ir(III) complex onto a N-heterocyclic carbene- (NHC)-based CTF using 1,3-bis(5cyanopyridyl)-imidazolium bromide as the monomer toward its use in the hydrogenation of CO$_2$ to formic acid (see Figure 9.17). During the last 20 years, excellent progress on homogeneous catalysts for this atom-economic conversion of CO$_2$ into formate/formic acid has been reported. Especially homogeneous Ir- (III)-based complexes have shown tremendous catalytic activity and selectivities exhibiting TONs up to 222 000 and TOFs of approximately 53 800 h^{-1} [23]. The big disadvantage, however, of these homogeneous Ir-based complexes is that the reverse decomposition of formic acid back into CO$_2$ and H$_2$ can occur during the separation steps. To overcome this disadvantage, Yoon used an electron donating NHC-based CTF for the first time as catalytic support for the CO$_2$ hydrogenation. After a careful evaluation of the reaction conditions, it was observed that at a temperature of 120 °C and a total pressure of 8.0 MPa, a TOF of 16 000 h^{-1} could be obtained, which is the highest value reported to date for any heterogeneous catalytic system. Moreover, after 15 hours of reaction time, the formate concentration amounts 0.282 M, corresponding to a TON of 24 300, which is one of the highest TONs ever published for this reaction using a heterogeneous catalyst.

9.4.4.3 Pyridinic CTFs as Supports for Formaldehyde Decomposition

The opposite reaction was studied by Bavykina and Gascon [24]. They prepared a mixed linker COF using two starting nitriles, these being 2,6-pyridinedicarbonitrile and 4,4′-biphenyldicarbonitrile. The idealized structure of the resulting COF is shown in Figure 9.18. This CTF was loaded with the same Ir(I) complex, after which the Cl$^-$ counterion was exchanged for the triflate counterion, a much weak coordinating anion. The obtained heterogeneous catalyst was highly active for the production of hydrogen while being recyclable. While Ir(I) complexes are very sensitive to

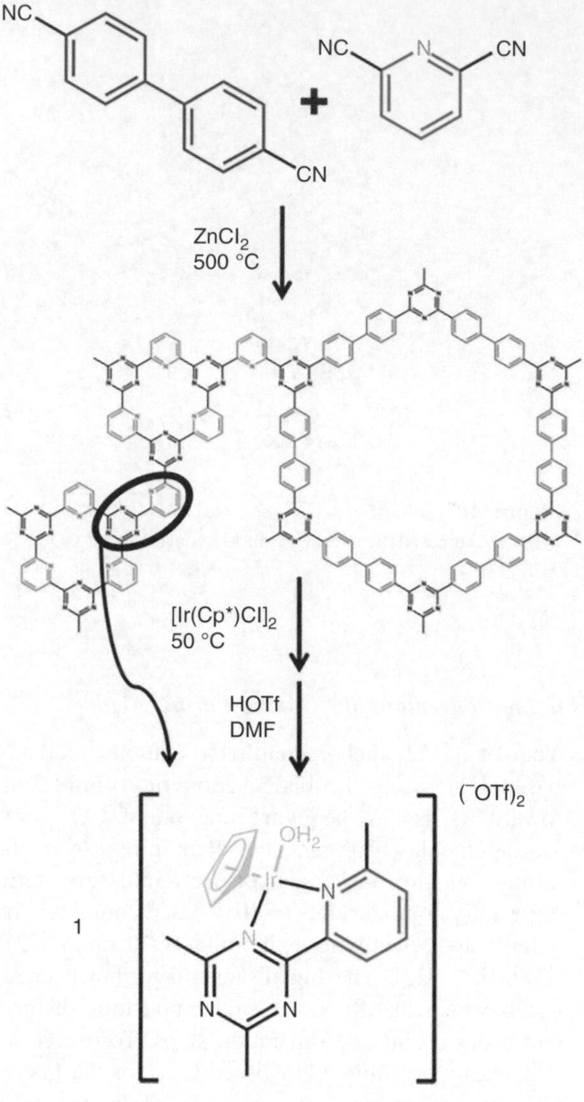

Figure 9.18
Synthesis of mixed linker CTF. Source: Reproduced with permission of ACS [24].

air, as they immediately oxidize toward Ir(III), this catalyst becomes oxidized during the catalytic reaction and is air stable in between catalytic runs. The catalyst showed a very high turnover frequency up to $27\,000\,h^{-1}$ for a 3 M formic acid solution at 80 °C.

Figure 9.19 shows the mechanism for the catalytic reaction, consisting of three steps: in step 1, the formic acid is deprotonated on the basic N-sites. The second step, the β-hydride elimination with removal of CO_2, is probably the rate-limiting step. Thanks to the pyridinic sites that are able to deprotonate the formic acid, no auxiliary base is required. In the final step, hydrogen is released.

Figure 9.19
Simplified scheme displaying the catalytic cycle within the CTF polymer backbone (sketched gray). Methyl groups of the Cp* ligand and coordination of the labile aqua ligand are not depicted. (1) formic acid activation, (2) b-hydride elimination, and (3) hydrogen release. Source: Reproduced with permission of John Wiley & Sons, Ltd [24].

Jorge Gascon is currently Professor and director of the Advanced Catalytic Materials section of KAUST, the King Abdullah University of Science and Technology, Saudi Arabia. He obtained his Ph.D. at the University of Zaragoza (Spain) and moved to the TUDelft (the Netherlands), where he worked as postdoc, assistant, and associate professor in the team of Professor Freek Kapteijn. He became Full Professor in 2016. A year later, he moved to KAUST.

9.4.4.4 CTFs in the Adsorption of Arsenic Species from Waste Water

CTFs are also promising adsorbents, especially in liquid (aqueous) media, because of their high solvolytic/thermolytic stability [26]. Within this context, Van Der Voort and Leus used CTF-1 as host material to embed Fe_2O_3 nanoparticles toward the removal of As^{III} and As^V species from waste water. A combination of ADF-STEM imaging together with EDX, EELS, and ^{57}Fe Mössbauer spectroscopy showed that most of the nanoparticles are distributed throughout the framework. Optimal removal efficiencies (>99%) were obtained for both arsenate and arsenite species. At pH 7, an adsorption capacity of 198 and 102 mg g^{-1} for As^{III} and As^V was obtained. This is significantly higher than most other iron-based adsorbents such as Fe_xO_y and zero-valent iron-based adsorbents [27]. There was no decrease in the adsorption efficiency in the presence of other competing ions such as calcium, magnesium, or natural organic matter. In order to study the regeneration potential four adsorption-desorption cycles were carried out, using a 0.1 M NaOH and 5% H_2O_2 solution to remove the adsorbed As species. There were no significant changes in the removal performance during the successive runs.

9.5 Imine COFs

9.5.1 Solvothermal Synthesis: COF-300

Imine-based COFs have a high crystallinity, as the condensation of an amine with an aldehyde is a very reversible process. The first crystalline 3D imine-based COF, denoted as COF-300, was (again) reported by Yaghi and co-workers in 2009 (see Figure 9.20) [25]. They used the typical solvothermal assisted Schiff base condensation reaction in which the aldehyde, terephthaldehyde, and the amine, tetra-(4-anilyl)-methane, were mixed in a Pyrex tube using 1,4-dioxane as a solvent and 3 M aqueous acetic acid as a catalyst. In the following, the tube was evacuated, flame sealed and heated to 120 °C for 72 h to obtain a yellow solid. Again, this assures that the water does not leave the reaction medium and the reversibility of the reaction is maintained. Since tetrahedral building blocks were used, a 3D diamond structure was obtained having a BET surface area and a total pore volume of 1360 m^2 g^{-1} and 0.72 cm^3 g^{-1}, respectively.

The imine formation is a well-known condensation reaction in which an imine bond is formed accompanied with the release of water. It is a highly reversible reaction, which allows the correction of errors and for this reason results in highly ordered crystalline materials. Imine formation does not require the use of expensive catalysts.

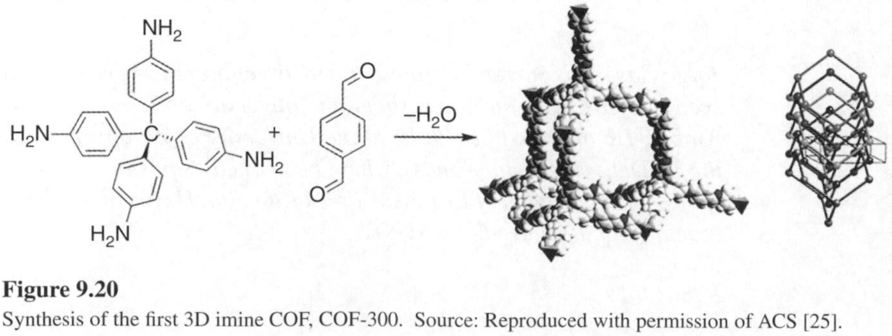

Figure 9.20
Synthesis of the first 3D imine COF, COF-300. Source: Reproduced with permission of ACS [25].

Since the first report on imine COFs, many aldehydes and amines have been synthesized to construct 2D and 3D imine COFs having high surface areas and pore
volumes.

Figure 9.21 shows all the different building blocks (amines and aldehydes, respectively) that have been used to date.

Still, the solvothermal synthesis route has some inherent disadvantages. First, the
harsh synthesis conditions, for example high pressures, and temperatures, makes it
difficult to control the size and shape of the resulting imine COF. Second, long reaction times (up to several days) are required to obtain the crystalline solid. Milder
synthesis routes are reported that show a higher control over the size and shape of the
resulting imine COFs.

aldehyde monomers used for the synthesis of imine COFs

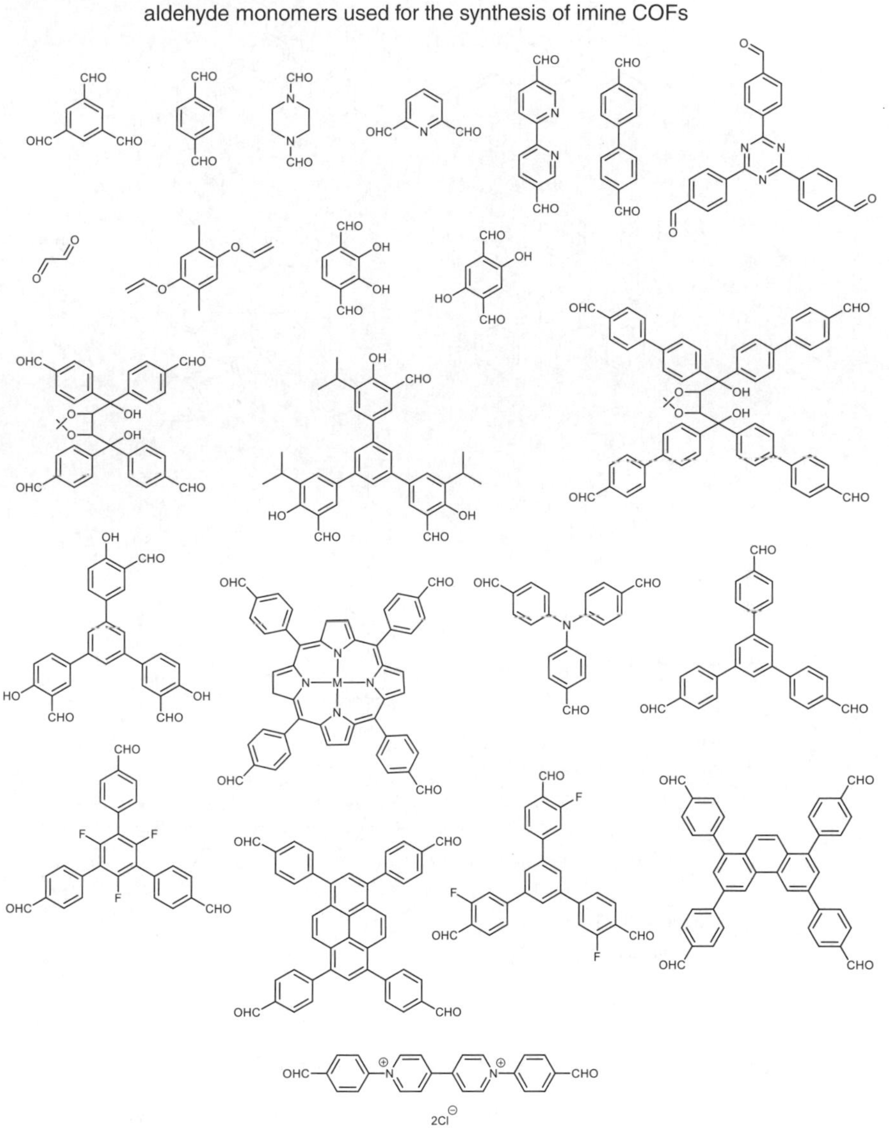

Figure 9.21
Aldehyde and amine building blocks used for the synthesis of imine-based COFs.

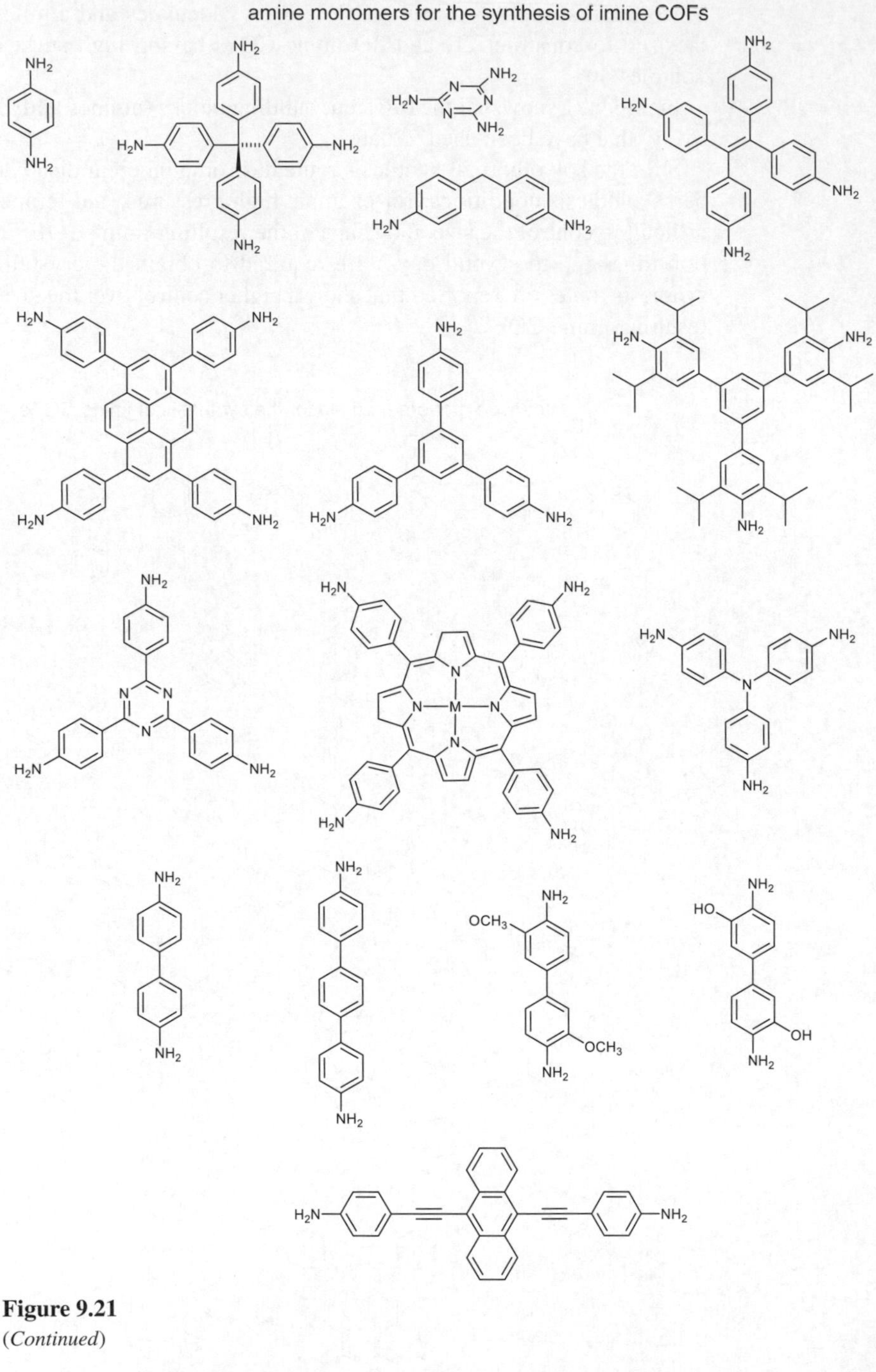

amine monomers for the synthesis of imine COFs

Figure 9.21
(*Continued*)

9.5.2 Room Temperature Synthesis of Imine COFs

Zhao et al. [28] reported the room temperature synthesis of imine COFs. The authors synthesized the imine COF, LZU-1 (LZU = Lanzhou University), by mixing 1,3,5-triformylbenzene (TFB) with *p*-phenylenediamine (PDA) in the presence of

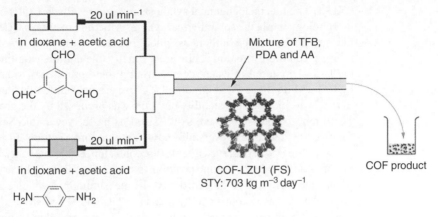

in dioxane + acetic acid

20 ul min⁻¹

Mixture of TFB, PDA and AA

in dioxane + acetic acid

20 ul min⁻¹

COF-LZU1 (FS)
STY: 703 kg m⁻³ day⁻¹

COF product

Figure 9.22

Schematic overview of the continuous flow synthesis of COF-LZU-1. Source: Reproduced with permission of ACS [28].

dioxane and acetic acid in a glass vial. The resulting mixture was sonicated until a homogeneous suspension was obtained and was left undisturbed for three days at room temperature. The resulting LZU-1 had a good crystallinity and even a higher BET surface area than the solvothermal analog. The authors attributed this to the fact that the milder synthesis conditions allowed a better crystallization process by the reversibility of the reaction to allow error corrections. The good solubility of the monomers having a rigid conformation and strong π interactions were assumed to be the driving force to allow the reversible error correction process. Zhao also demonstrated the first example of continuous flow synthesis of an imine COF (see Figure 9.22). By using a continuous laminar flow system consisting of two syringes acting as reservoirs, the authors were able to synthesize 41 mg of LZU-1 in a reaction time of 1 h. The calculated space–time yield, defined as kg of product per m^3 reaction tube, was $703 \, \text{kg} \, m^{-3} \, day^{-1}$. To their surprise, the BET surface area of the LZU-1 synthesized under continuous flow conditions was even higher than the LZU-1 obtained at room temperature under batch conditions. The authors assigned this observation to the better crystallization process under flow conditions due to the local super-saturation at the interface between the two flow streams.

9.5.3 Liquid Assisted Grinding

Banerjee and co-workers reported in 2014 the use of liquid assisted grinding as a method to synthesize imine COFs [29]. The imine COF, LZU-1, was used as a prototype to demonstrate the applicability of this grinding method. A catalytic amount of liquid is added to bring the reactants in closer proximity leading to an increased reaction rate. For the synthesis of LZU-1, the aldehyde and amine building block were placed in a 5 ml stainless steel jar, with one stainless steel ball in the presence of 1–2 drops of mesitylene : dioxane (1 : 1) and one drop of 3 M acetic acid. The resulting mixture was milled at room temperature to obtain a dark yellow powder. The resulting material showed a similar FT-IR, ^{13}C solid state NMR, PXRD, and TGA profile in comparison to the solvothermally synthesized LZU-1. Only a much lower BET surface area was obtained in comparison to the traditionally synthesized

LZU-1. The mechanochemical synthesis method results in COFs having a thin layer morphology while the solvothermal synthesis method results in COFs having a flower like morphology. The authors assumed that due to this thin layer morphology, the long-range pore formation is hindered. Finally, the authors examined the stability of the LZU-1 material in water, demonstrating that both the solvothermal synthesized LZU-1 and liquid assisted grinding LZU-1 were not stable in water.

A way to increase the stability of COFs was proposed by the group of Heine [30]. Two imine-based COFs were synthesized using the reversible Schiff base reaction followed by a second irreversible enol to keto tautomerization (Figure 9.23). No loss in crystallinity was observed after this second step as this step only involves the shifting of bonds while the atomic positions are kept the same. Evidence for the presence of the keto form was obtained from FT-IR measurements, which showed the absence of –OH and imine groups and by means of ^{13}C CP-MAS NMR measurements. The resulting COFs exhibited very high stabilities in water and in acidic conditions for at least seven days due to the irreversible enol to keto tautomerization.

9.5.4 Applications of Imine COFs

9.5.4.1 *Bimetallic Chiral Imine COFs in Cascade Reactions*

A very nice example on the use of imine-based COFs as a chiral catalyst in cascade reactions has been presented by Cui et al. [31] A chiral Zn(salen)-based imine COF was synthesized employing the typical solvothermal synthesis method by mixing the chiral 1,2-diaminocyclohexane and 1,3,5-tris(3′-tert-butyl-4′-hydroxy-5′-formylphenyl)benzene building blocks in DMF/EtOH for three days at 120° to obtain a yellow crystalline solid. The imine COF showed very high stabilities in the presence of boiling water and in acidic (1 M HCl) and alkaline (9 M NaOH) solutions. This is due to the *t*-butyl groups, as the unalkylated variants dissolved completely in base. The authors exchanged the Zn^{2+} by several mono and bimetallic metals such as Cr^{2+}, Co^{2+}, Mn^{2+}, Fe^{2+}, and V^{4+} (see Figure 9.24). For instance, the resulting bimetallic *Cr—Mn chiral salen-based COF* was evaluated in the sequential epoxidation of alkenes followed by the ring opening reaction to afford the corresponding amino alcohol. In this cascade reaction, Mn^{2+} promotes the epoxidation reaction whereas Cr^{2+} is responsible for the ring opening of the epoxides. The catalytic tests were performed at 0 °C in the presence of 1 mol% of catalyst using 2,2-dimethyl benzopyran as substrate and 2-(*tert*-butylsulfonyl)iodosylbenzene (sPhIO) as an oxidant. After a reaction time of 24 h, a conversion of 84% was achieved with an *ee* value up to 91%. Moreover, the chiral COF could be recycled up to five runs without a significant loss in activity and enantioselectivity. After five runs a conversion of 76% was still noted and an *ee* value of 85%.

9.5.4.2 *Porphyrin COF in the Electrochemical Reduction of CO₂ to CO*

Chang et al. [32] developed a Co-porphyrin COF in which the struts were connected through imine bonds (see Figure 9.25). The resulting *COF-366-Co* was tested in the

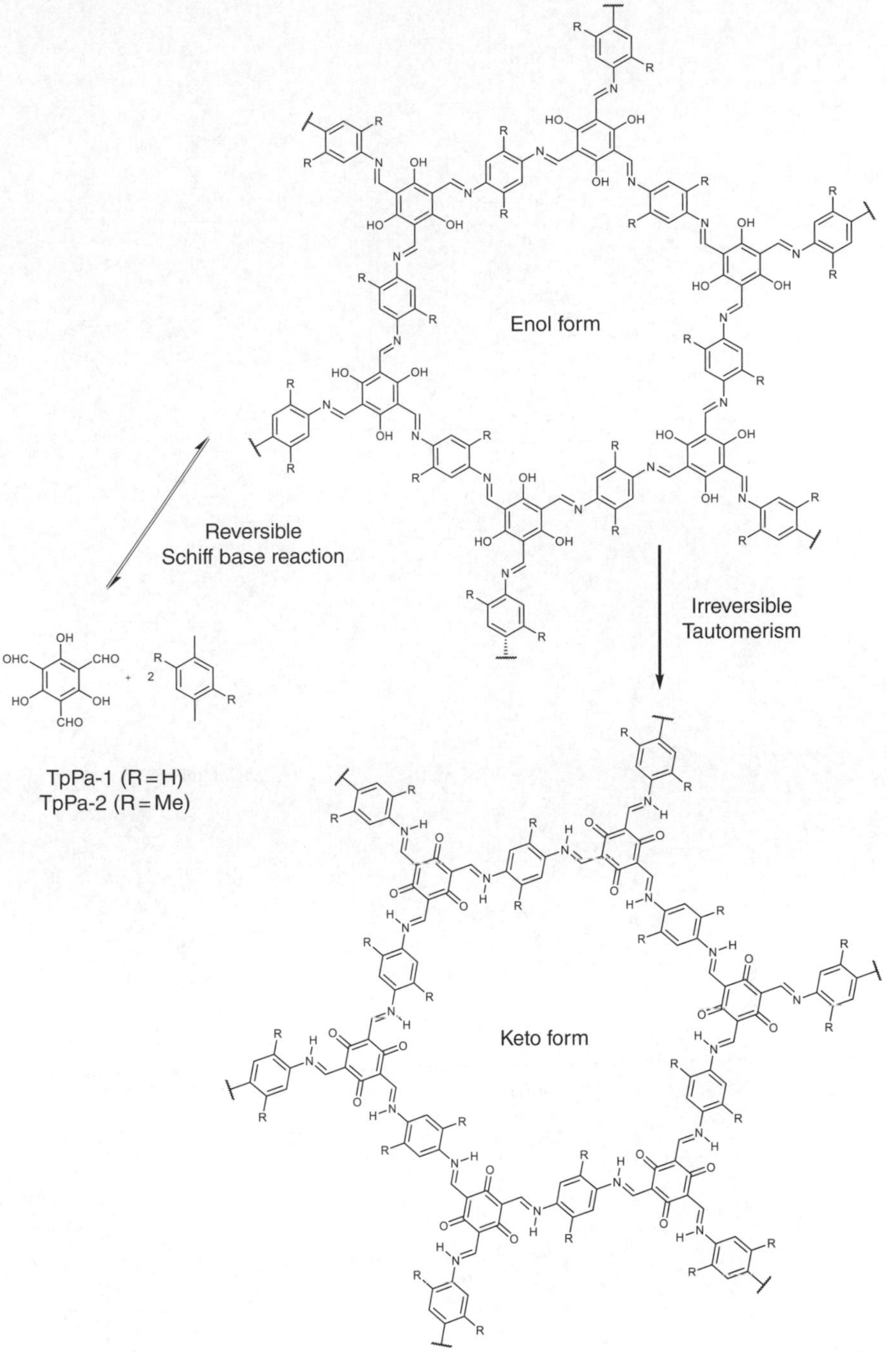

Figure 9.23
Synthesis of highly chemical stable COFs using a combination of reversible and irreversible reactions.
Source: Reproduced with permission of ACS [30].

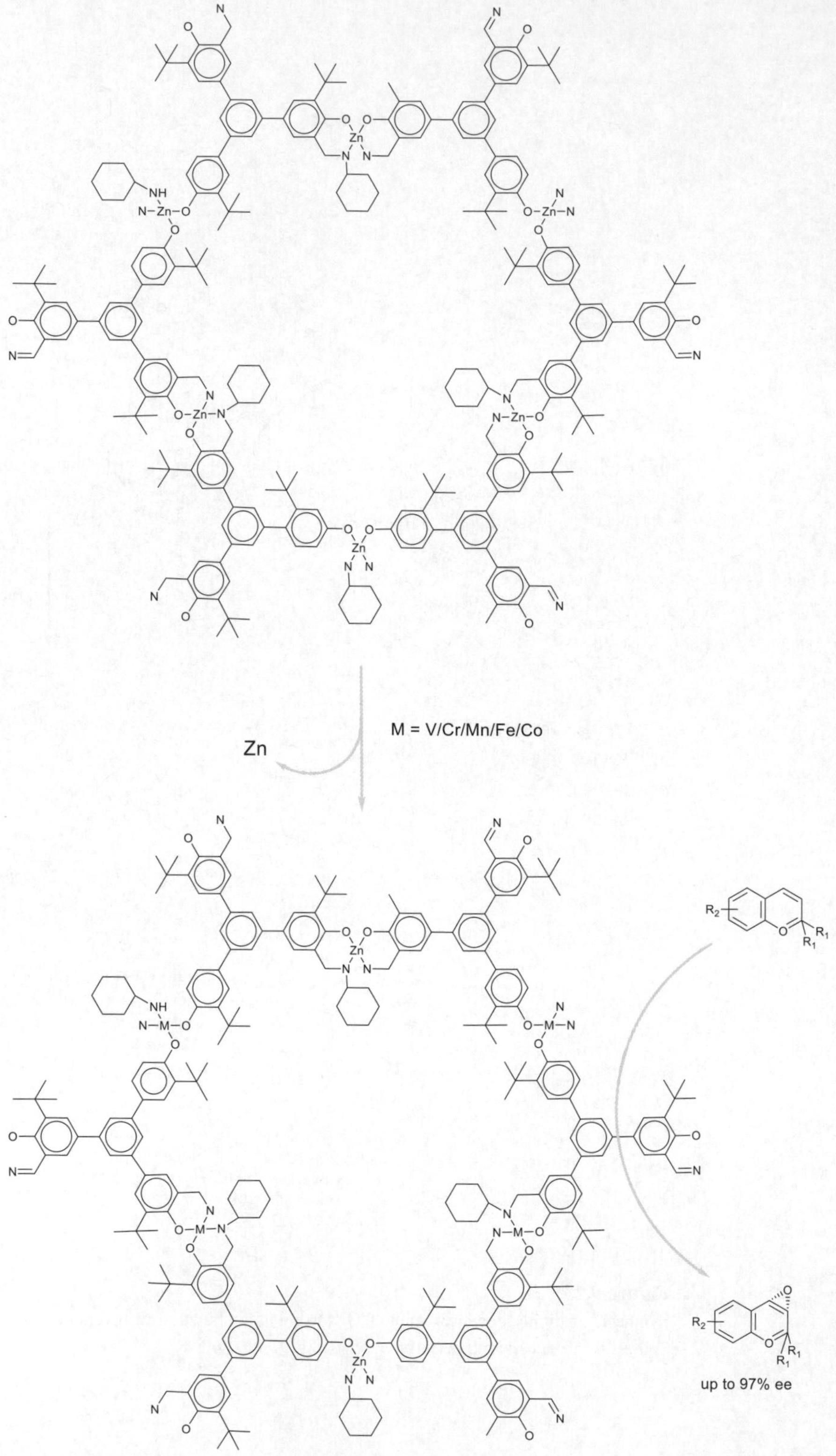

Figure 9.24
Representation of the chiral bimetallic imine COF used in the one pot synthesis of amino alcohols. Source: Reproduced with permission of ACS [31].

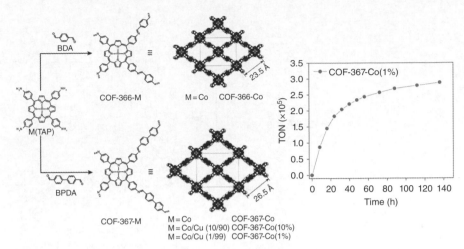

Figure 9.25

Construction of the Co-porphyrin-based imine COFs for the electrochemical conversion of CO_2. Source: Reproduced with permission of AAA [32].

electrocatalytic conversion of CO_2 to CO. The COF material was deposited onto a conductive carbon fabric. After saturation of the neutral aqueous solution with CO_2, the *COF-366-Co* displayed a distinct reduction potential producing CO as the major product. To further enhance the catalytic performance, the authors synthesized an isoreticular variant of COF-366 using the extended aldehyde building unit, biphenyl-4,4'-dicarboxaldyde, which resulted into *COF-367-Co*. The larger pore size results in more accessible Co-porphyrin sites and thus an enhanced activity due to the enhanced CO_2 adsorption within the framework. The TON value for the CO_2 conversion was 48.000 taking only the accessible Co-porphyrin sites into account. The researchers also prepared a bimetallic variant, the *COF-367-CoCu*. TON and TOF values of 296 000 and 9400 h^{-1} could be obtained that correspond to a 26-fold increase in activity in comparison to the molecular cobalt complex.

9.5.4.3 Click Chemistry for CO_2 Adsorption

A systematic study on the pore surface engineering of imine COFs toward use in CO_2 adsorption was performed by Jiang and co-workers [33]. Four imine linked porphyrin-based COFs were synthesized having various ethynyl contents by mixing different molar ratios of 5,10,15,20-tetrakis(*p*-tetraphenylamino)porphyrin, 2,5-bis(2-propynyloxy)terephthalaldehyde and 2,5-dihydroxyterephthalaldehyde (see Figure 9.26). In the following step, click reactions were carried out between the ethynyl groups of the COF material and azide compounds having various functional groups including ethyl, acetate, hydroxyl, carboxylic acid, and amino groups. Twenty COFs were synthesized having various functional groups on their pore wall ranging from hydrophobic to hydrophilic and from basic to acidic. This allowed the authors to systematically tune the surface area and pore volume of the resulting COFs but also to change the pore size from mesopores to supermicropores. The

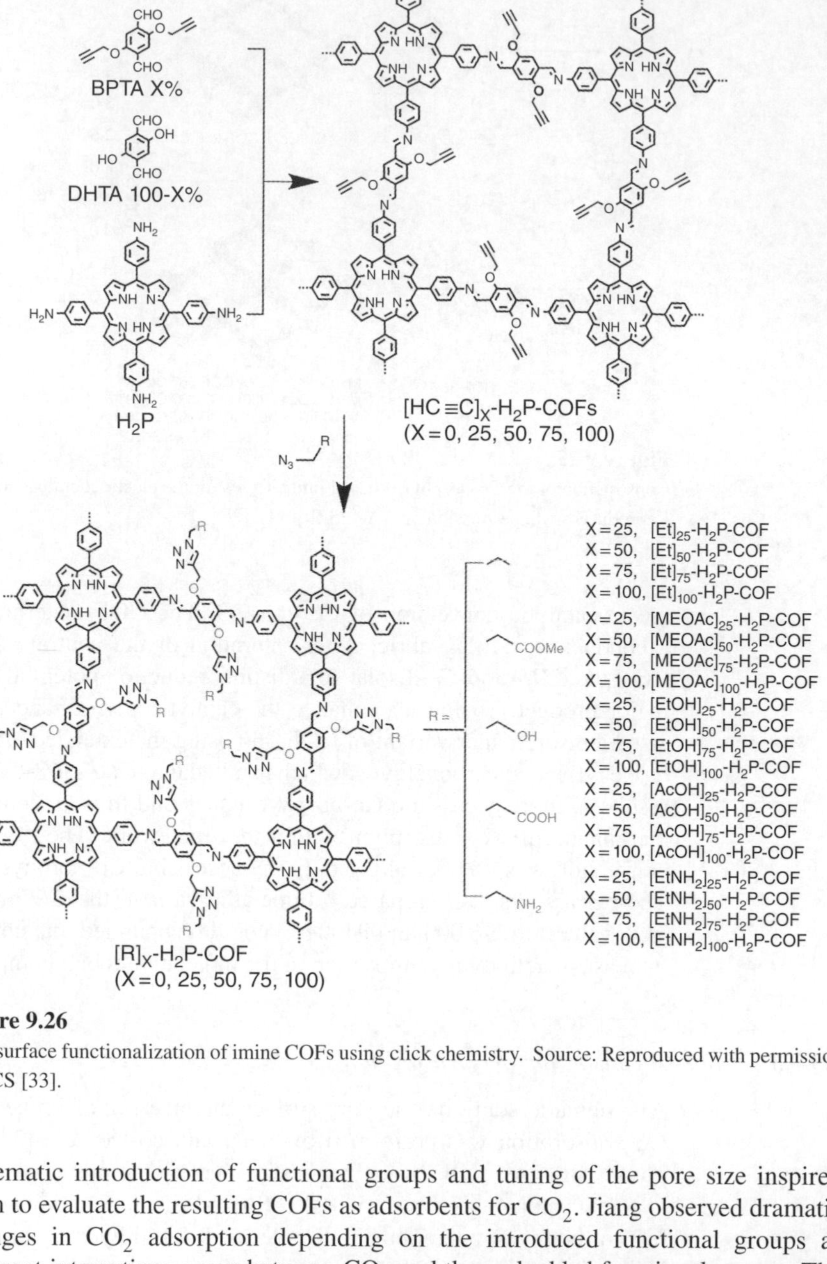

Figure 9.26

Pore surface functionalization of imine COFs using click chemistry. Source: Reproduced with permission of ACS [33].

systematic introduction of functional groups and tuning of the pore size inspired them to evaluate the resulting COFs as adsorbents for CO_2. Jiang observed dramatic changes in CO_2 adsorption depending on the introduced functional groups as different interactions occur between CO_2 and the embedded functional groups. The ethynyl and ethyl-based COFs only showed a weak CO_2 adsorption capacity but the ester functionalized COF materials showed a much higher affinity for CO_2 because of dipole interactions. Carboxylic acid and hydroxyl pore functionalized imine COFs exhibit even higher CO_2 affinity as not only dipole but also hydrogen-bonding interactions can occur. The highest CO_2 adsorption capacities were obtained for the amino functionalized COFs as also acid–base pairs with CO_2 can be formed.

Donglin Jiang is born in 1966. He obtained his Master's degree at the University of Zhejiang in 1992. In 1998 he received his Ph.D. at the University of Tokyo under the supervision of professor Takuzo Aida. In 2005 he became Associate Professor at the Institute of Molecular Science & Sokendai and from 2016 to now, he is a professor in the field of environment and energy at the Japan Advanced Institute of Science and Technology (JAIST). Professor Jiang explored the application of COFs in the field of supercapacitors and lithium-ion batteries besides their use as proton conducting materials toward their use in fuel cells, organocatalysts, and the capture of CO_2. In 2006 he obtained the Young Scientist Award, which is one of the most prestigious prizes for academic achievements in Japan.
Photograph: https://wol-prod-cdn.literatumonline.com/cms/attachment/ad58eb1b-ccbf-4c26-b2eb-5b71adceec29/mauthor.jpg

9.5.4.4   Nanocarrier for the Transportation of the Drug Quercetin

Another interesting application of imine-based COFs is their use as nanocarriers for drugs. Bettina Lotsch's group published a nice example in 2016. They prepared a water stable imine COF having free electron pairs on the imine nitrogens available for the reversible noncovalent interactions for guest molecules [34]. This imine COF was synthesized by using a mixture of triazine triphenyl aldehyde and triazine triphenyl amine in mesitylene/dioxane in the presence of acetic acid. The resulting crystalline imine COF exhibited a very high BET surface area of $2197\,m^2\,g^{-1}$, which is among the highest values reported for 2D COFs. The material was completely water stable. The interaction behavior of this imine COF with the anticancer drug Quercetin was studied by soaking the COF for 16 hours in a drug loaded THF solution. Solid state NMR measurements demonstrated that one-third of the imine nitrogens exhibited H-bonding corresponding to a highly saturated pore structure of one *Quercetin* molecule per unit cell. The COF loaded Quercetin as well as the pristine COF and

Quercetin were introduced into human-breast carcinoma MDA-MB-231 cells. The authors noted that the COF as such possesses no cytotoxicity in comparison to the control experiment. A significant decrease in the number of cancer cells was observed after only 1 day, whereas the direct Quercetin injection showed an induction period of 2 days.

Bettina V. Lotsch is born in 1977 and completed her Master's degree in chemistry at the University of Munich in 2002. In 2006, she obtained her doctoral degree at the same University. She then performed a post-doctoral stay in the group of G.A. Ozin at the University of Toronto in Canada. In 2009, she became Assistant Professor in Munich up until to the beginning of 2017. Meanwhile, in 2011, she became independent group leader at the Max Planck Institute for Solid State Research. In 2017, Lotsch was appointed Director of the same institute in Stuttgart (Germany). In the same year, she was appointed as Honorary Professor at Munich University.

Photograph: http://www.fkf.mpg.de/171992/Curriculum_Vitae

Exercises

1 Describe the advantages of COFs in comparison to MOFs. What are the main differences between both materials?
2 Describe the different COF linkages. Which linkages are the most stable?
3 It is well-known that boron-based COFs show a limited stability in the presence of water. How can this stability be improved?
4 Which method can be applied to scale up the synthesis of CTFs and imine COFs?
5 For the given CTFs, what are the starting substrates used to synthesize the corresponding CTFs?

6 For these CTFs, calculate the theoretical C, H, and N weight percentages and the molar C/N and C/H ratios.

Material	C (%)	H (%)	N (%)	C/N	C/H
CTF-1	75	3.15	21.86		
HCTF	67.91	5.7	26.4		

7 In an experiment, HCTF was made through ionothermal synthesis at 400 °C and 500 °C using $ZnCl_2$ as the catalyst. The substrate: $ZnCl_2$ ratio was 1 : 10. After synthesis, the material was cleaned with hydrochloric acid, distilled water, and tetrahydrofuran. It was then dried under vacuum at 120 °C for 24 hours. Elemental analysis on the materials yielded the following results.

Material	C (%)	H (%)	N (%)	Unknown
HCTF-400-1 : 10	70.9	1.9	8.9	18.3
HCTF-500-1 : 10	73.7	1.5	8.4	16.4

Calculate the carbonization percentage. Describe the possible chemical changes that could occur to cause such carbonization. What is the relationship between the chemical changes and synthesis temperature? Explain the use of the results of elemental analysis. What could be the unknown?

Answers to the Problems

1 COFs possess in general a lower density then MOFs as they are constructed of solely light elements. Moreover, in general COFs possess a higher thermal and chemical stability than MOFs allowing an easier pre- or postsynthetic modification without introducing a significant change in the porosity and loss in crystallinity.

2 B—O-based linkages which include the boroxine-based linkages, boronate ester linkages, borosilicate linkages, and spiroborate-linked ionic COFs. C—N-based linkages including imine-based COFs, hydrazone linkages, squaraine COFs, aromatic C=N-based COFs including the triazine and phenazine frameworks, and imide and amide-based linkages. The C—C-based linkages and borazine (B—N) linkages. The CN-based linkages are the most stable.

3 The water stability of boron-based COFs can be enhanced by introducing hydrophobic alkyl groups in the pores in the COFs; for example, methyl, ethyl, and propyl. Another attractive alternative for the instable boroxine and boronate ester COFs is to use spiroborate-based COFs, which are ionic derivatives of boronic acids exhibiting a good stability in water, methanol, and in basic conditions.

4 The mechanochemical synthesis and continuous flow synthesis.

5 Dicyanobenzene; trans-3-hexenedinitrile.

6

Material	C (%)	H (%)	N (%)	C/N	C/H
CTF-1	75	3.15	21.86	4	2
HCTF	67.91	5.0037	26.4	3	0.99

7

Material	C (%)	H (%)	N (%)	Unknown	C/N	C/H
HCTF-400-1 : 10	70.9	1.9	8.9	18.3	9.3	3.1
HCTF-500-1 : 10	73.7	1.5	8.4	16.4	10.2	4.1
HCTF (Theoretical)	67.91	5.7	26.4	–	3	0.99

At high synthesis temperatures, partial carbonization can occur due to side reactions. Due to excess of $ZnCl_2$ (10 times) and high temperatures, irreversible reactions (Diels–Alder-type reactions) are possible, which decrease the crystallinity of the material and causes partial carbonization.

HCTF-400-1 : 10:

On comparing the theoretical and experimental C/N and C/H ratios, it can be said that there is a 3.1 times higher carbon content due to carbonization.

HCTF-500-1 : 10:

On comparing the theoretical and experimental C/N and C/H ratios, it can be seen that the C/N ratio is 3.4 times higher than the theoretical C/N and the C/H ratio is 4.1 times higher. In addition to an increase in the carbon content, this also suggests that at 500 °C more —CN cleavage occurs than at 400 °C as the N content is lower and C content is higher. Similarly, at 500 °C, a higher number of C—H bonds break easily than they do at 400 °C.

The unknown could be $ZnCl_2$ residue, which cannot be completely removed with the cleaning steps.

References

1. Cote, A.P., Benin, A.I., Ockwig, N.W. et al. (2005). *Science* 310: 1166–1170.
2. Zeng, Y.F., Zou, R.Y., Luo, Z. et al. (2015). *J. Am. Chem. Soc.* 137: 1020–1023.
3. Diercks, C.S. and Yaghi, O.M. (2017). *Science* 355: 923–925.
4. Liu, Y.Z., Ma, Y.H., Zhao, Y.B. et al. (2016). *Science* 351: 365–369.
5. Ma, Y.X., Li, Z.J., Wei, L. et al. (2017). *J. Am. Chem. Soc.* 139: 4995–4998.
6. Spitler, E.L. and Dichtel, W.R. (2010). *Nat. Chem.* 2: 672–677.
7. Campbell, N.L., Clowes, R., Ritchie, L.K., and Cooper, A.I. (2009). *Chem. Mater.* 21: 204–206.
8. Lanni, L.M., Tilford, R.W., Bharathy, M., and Lavigne, J.J. (2011). *J. Am. Chem. Soc.* 133: 13975–13983.
9. Du, Y., Yang, H.S., Whiteley, J.M. et al. (2016). *Angew. Chem. Int. Edit* 55: 1737–1741.
10. Abrahams, B.F., Price, D.J., and Robson, R. (2006). *Angew. Chem. Int. Edit* 45: 806–810.
11. Doonan, C.J., Tranchemontagne, D.J., Glover, T.G. et al. (2010). *Nat. Chem.* 2: 235–238.
12. Singh, H., Tomer, V.K., Jena, N. et al. (2017). *J. Mater. Chem. A* 5: 21820–21827.
13. Dogru, M., Handloser, M., Auras, F. et al. (2013). *Angew. Chem. Int. Edit* 52: 2920–2924.
14. Kuhn, P., Forget, A., Su, D.S. et al. (2008). *J. Am. Chem. Soc.* 130: 13333–13337.
15. (a) Hug, S., Tauchert, M.E., Li, S. et al. (2012). *J. Mater. Chem.* 22: 13956–13964. (b) Liu, L., Xia, Y.J., and Zhang, J. (2014). *RSC Adv.* 4: 59102–59105.
16. (a) Bojdys, M.J., Jeromenok, J., Thomas, A., and Antonietti, M. (2010). *Adv. Mater.* 22: 2202; (b) Katekomol, P., Roeser, J., Bojdys, M. et al. (2013). *Chem. Mater.* 25: 1542–1548.
17. Ren, S.J., Bojdys, M.J., Dawson, R. et al. (2012). *Adv. Mater.* 24: 2357–2361.
18. Troschke, E., Gratz, S., Lubken, T., and Borchardt, L. (2017). *Angew. Chem. Int. Edit* 56: 6859–6863.
19. Palkovits, R., Antonietti, M., Kuhn, P. et al. (2009). *Angew. Chem. Int. Edit* 48: 6909–6912.
20. Bukhtiyarova, M., Lunkenbein, T., Kahler, K., and Schlogl, R. (2017). *Catal. Lett.* 147: 416–427.
21. Periana, R.A., Taube, D.J., Gamble, S. et al. (1998). *Science* 280: 560–564.
22. Gunasekar, G.H., Park, K., Ganesan, V. et al. (2017). *Chem. Mater.* 29: 6740–6748.
23. (a) Hull, J.F., Himeda, Y., Wang, W.H. et al. (2012). *Nat. Chem.* 4: 383–388; (b) Himeda, Y., Onozawa-Komatsuzaki, N., Sugihara, H. et al. (2004). *Organometallics* 23: 1480–1483.
24. Bavykina, A.V., Goesten, M.G., Kapteijn, F. et al. (2015). *Chemsuschem* 8: 809–812.
25. Uribe-Romo, F.J., Hunt, J.R., Furukawa, H. et al. (2009). *J. Am. Chem. Soc.* 131, 4570.
26. (a) Liu, J.L., Chen, H., Zheng, S.R., and Xu, Z.Y. (2013). *J. Chem. Eng. Data* 58: 3557–3562; (b) Wang, B.Y., Lee, L.S., Wei, C.H. et al. (2016). *Environ. Pollut.* 216: 884–892.
27. (a) Lin, S., Lu, D.N., and Liu, Z. (2012). *Chem. Eng. J.* 211: 46–52; (b) Zhu, H.J., Jia, Y.F., Wu, X., and Wang, H. (2009). *J. Hazard Mater.* 172: 1591–1596; (c) Wang, C., Luo, H.J., Zhang, Z.L. et al. (2014). *J. Hazard Mater.* 268: 124–131.
28. Peng, Y.W., Wong, W.K., Hu, Z.G. et al. (2016). *Chem. Mater.* 28: 5095–5101.
29. Das, G., Shinde, D.B., Kandambeth, S. et al. (2014). *Chem. Commun.* 50: 12615–12618.
30. Kandambeth, S., Mallick, A., Lukose, B. et al. (2012). *J. Am. Chem. Soc.* 134: 19524–19527.
31. Han, X., Xia, Q.C., Huang, J.J. et al. (2017). *J. Am. Chem. Soc.* 139: 8693–8697.
32. Lin, S., Diercks, C.S., Zhang, Y.B. et al. (2015). *Science* 349: 1208–1213.
33. Huang, N., Krishna, R., and Jiang, D.L. (2015). *J. Am. Chem. Soc.* 137: 7079–7082.
34. Vyas, V.S., Vishwakarma, M., Moudrakovski, I. et al. (2016). *Adv. Mater.* 28: 8749–8754.

Index

Introduction to Porous Materials, First Edition.
Pascal Van Der Voort, Karen Leus and Els De Canck.
© 2019 John Wiley & Sons Ltd. Published 2019 by John Wiley & Sons Ltd.